Space-Time, Relativity, and Cosmology

Space-Time, Relativity, and Cosmology provides a historical introduction to modern relativistic cosmology and traces its historical roots and evolution from antiquity to Einstein. The topics are presented in a non-mathematical manner, with the emphasis on the ideas that underlie each theory rather than their detailed quantitative consequences. The tests and experimental evidence supporting the theories are explained together with their predictions and their confirmation.

The discussion of the Special Theory begins by stating the Principle of Relativity and its roots in the ideas of Galileo and Newton, followed by deriving its main consequences, including the relativity of simultaneous events, time dilation and length contraction and the equivalence of mass and energy. The General Theory of Relativity and its consequences when applied to the large-scale structure of the universe are discussed, including its tests and striking predictions. The discussion of modern relativistic cosmology includes the Cosmological Principle, possible geometries of space-time, and the consequences of Hubble's observations leading to the Big Bang hypothesis. The last section of this chapter presents a brief overview of some of the most exciting research topics in the area of relativistic cosmology, concluding with a description of the deficiencies of the Big Bang and a possible resolution.

This textbook is intended for undergraduate students undertaking a science course in non-science majors. It is also accessible to advanced high school students, as well as the non-scientist layman who is concerned with science issues.

JOSE WUDKA is Professor of Physics at the University of California, Riverside, where he has been a faculty member since 1990. Since gaining his PhD from the Massachusetts Institute of Technology in theoretical particle physics in 1986, he has held positions at the University of Michigan and the University of California, Davis. During 1992–3, he was the Superconducting Supercollider Laboratory Fellow at Riverside. He regularly attends conferences in the area of particle physics phenomenology and is a regular contributor to journals including *Physical Review Letters*, *Physical Review D*, and *Nuclear Physics B*.

Space-Time, Relativity, and Cosmology

Jose Wudka
University of California, Riverside

CAMBRIDGE
UNIVERSITY PRESS

CAMBRIDGE UNIVERSITY PRESS
Cambridge, New York, Melbourne, Madrid, Cape Town, Singapore, São Paulo

Cambridge University Press
The Edinburgh Building, Cambridge CB2 2RU, UK

Published in the United States of America by Cambridge University Press, New York

www.cambridge.org
Information on this title: www.cambridge.org/9780521822800

First published 2006

Printed in the United Kingdom at the University Press, Cambridge

A catalog record for this publication is available from the British Library

ISBN-13 978-0-521-82280-0 hardback
ISBN-10 0-521-82280-7 hardback

To Mariana

Contents

Acknowledgments

This book grew out of a course that was created by Frederick Cummings and Peter Kaus in 1982, and which I "inherited" after their retirement in the early 1990s. The course format makes it easy to include new developments in the subject matter, and uses these changes to provide a beautiful example of the evolution of scientific theories. This attractive combination has proved a successful lure for students, who will fill a large lecture hall every time the course is offered. I have strived to maintain a similar approach in the present publication, so any success it might have should be credited to the original designers.

I am very grateful to both Cummings and Kaus for having introduced me to the challenging and rewarding field of teaching non-science majors, as well as for the many discussions about the foundations of relativity, the development of scientific theories and physics in general.

This book is dedicated to the memory of my dear friend and colleague, Lynne Deutsch, who opened new eyes to the sky.

1
The scientific method

Motivation

Most of you, the readers, have probably tuned in to the news at one point or another to hear the anchor person begin a story with the phrase "Today scientists reported," probably followed by a description of something completely outside of your everyday experience: colliding microscopic particles, genetic engineering, or statements about the behavior of giant galaxies or the Big Bang. How do scientists arrive at their conclusions on such topics? And perhaps more importantly, why should *you* believe anything they say?

The answers are important, for the topics covered in this book are indeed such things as the creation of the Universe, the violent death of massive stars, and so forth, which you will (probably) never personally experience. So an understanding of how science is done and why we trust the results is relevant to what will follow. Equally important is realizing that scientific results always represent a qualified "truth" (as opposed to an absolute truth). New discoveries that change how science views the universe are always just around the corner. The "scientific method" provides the recipe for generating a consistent set of unprejudiced ideas about how the Universe works; it is also the method by which the current ideas are replaced by new ones, based on observations of the real world.

In addition, the scientific method has a practical aspect that cannot, and should not, be neglected. Many aspects of our society have become increasingly intertwined with a manifold of technological improvements that lengthen and enrich life,[1] and we have come to expect (and demand) that every new such improvement will perform safely and reliably. To insure the safety and efficacy of these products most countries have instituted mechanisms to *test* whether these expectations are fulfilled and to insure that failing products are not distributed among the public. We expect these tests to be unprejudiced, repeatable and consistent, and so the standard approach in devising them must follow the scientific method.

Despite these protective mechanisms the market pullulates with wonderful life-enriching, disease-curing, products whose powers are said to derive from some ancient magical or mystic principle; or else whose healing powers are supported by a befuddling mishmash of scientific "facts" judiciously taken out of context from appropriately chosen publications. Having had no unprejudiced, repeatable and independent verification of their properties, all such products (and their advocates) should be considered suspect. In the best cases these items are harmless, but sometimes they are dangerous,[2] and often they are fraudulent.

The intellectual and practical applications of the scientific method make it important for the public to have a basic understanding of the manner in which scientific theories are obtained, tested and accepted. With this type of knowledge the public can protect itself from non-scientific quackery, and impress upon the government the need to create and *enforce* policy against it. In many cases the lives of people depend on the reliability and thoroughness of this scrutiny.

It is not my intention to imply that all human intellectual pursuits should be centered on a scientific approach. But when gathering quantitative information about Nature, systematizing the data, and deriving from it a deeper understanding of the world around us, the scientific approach described below is the best method available.

Brief history

In his work "The History of Animals" the Greek philosopher Aristotle claimed that men and women had a different number of teeth, and that their internal organs were also different. These assertions were supported by long arguments. No verification was sought and, indeed, Aristotle believed that none was required, since the arguments were based on the author's philosophy, which was consistent and perfectly satisfactory. The possibility that these statements might actually be *wrong* never occurred to Aristotle, nor did the thought that careful observation and experimentation might be needed to decide the issue.

Early Greek philosophers (e.g. Thales of Miletus) noted that, in many instances, Nature exhibits a regular behavior. Most noticeable among these are, perhaps, the change of day into night and the progression of the seasons. This eventually gave rise to the idea of natural phenomena being determined by a set of rules, or by logical conclusions derived from these rules. The realization that Nature is not driven by the whims of the gods marks the beginning of science.

Accepting the existence of rules that describe the world around us, it is natural to ask how can we best discover these rules. One possible method is to argue that since the world around us can be explained by drawing conclusions from a set of first principles, these first principles ought to be part of a coherent and logically constituted philosophy. So all we have to do is guess the right philosophy, from it the behavior of the whole of Nature can be deduced.

The notion that the properties of the world around us can be obtained from a series of ideas, a philosophical edifice, is called the *deductive* method of reasoning.

Deductive reasoning

Argument is based on a rule, law, principle, or generalization. In other words, "I'm right because I said so."

Aristotle conducted some of the earliest studies of the world around us, interpreting his observations using this deductive approach. He did this by choosing a set of first principles, which he considered eminently clear, obvious and natural. Unfortunately they were also inappropriate for the task, as illustrated by the above example. Still, such was Aristotle's preponderance, that his deductive method was blindly followed for more than 13 centuries after his death, even today it is used (perhaps unwittingly) by many people. Deductive reasoning might be a good rule to follow when dictating human activities (*a la* Sherlock Holmes), but it is a very unreliable way of obtaining information about Nature.

A good example of the sort of logical traps hiding in deductive reasoning can be seen in what is known as a *syllogism*. A syllogism is a logically incorrect generalization. For example, one might state "All cats have fur," and then state "A dog has fur; therefore a dog is a cat." Though it may seem silly to us now, this type of argument resulting from the deductive approach formed the basis of science for centuries.

By the end of the Middle Ages the deficiencies inherent to the deductive method became increasingly apparent. It was during this period that the foundations for the modern approach to science were laid. While writing a series of treatises on papal power and civil sovereignty, William of Ockham published his "Razor": a prescription for cutting away any unnecessary ideas present in a description of Nature.[3] Later, the Renaissance witnessed the final demise of the deductive approach in scientific inquiries; in its stead an *inductive* approach was adopted, based on experimentation and careful analysis of the data.

Inductive reasoning

Arguments are based on experience or observation. In other words, "I'm right and I can do an experiment to prove it."

The shift in science from deductive to inductive reasoning was prompted by the various writings of Francis Bacon, and perhaps more forcefully by the *results* obtained by Galileo and Newton. The same basic approach used then (with minor alterations) is still followed today in most research. The reliability of scientific results we have come to expect is due to the inductive approach. Through it, the US Food and Drug Administration can determine whether a given medicine is safe, and under which conditions it is dangerous. By following the scientific method, we obtained knowledge about the resistance of materials which is used by architects and civil engineers when declaring a building safe or unsafe. Rules obtained from the method ensured that the Galileo spacecraft, setting out from Earth at the appropriate speed and direction, would rendezvous with the planet Jupiter on December 7, 1995, after a journey of over six years and 3 billion kilometers.

The basic assumptions behind the modern scientific method are the existence of a reality outside our own bodies, and that said reality can be understood by the human mind. It is, of course, conceivable that we are all characters in somebody's dream, but this is something we cannot prove or disprove (were it true one cannot but commend the dreamer on the richness and vividness of his/her imagination ... and hope the alarm-clock does not go off). The reality of the universe around us is something that ultimately must be taken on faith. Still one can provide plausibility arguments such as the following one by W. Churchill[4] that considers the possibility that our Sun is the result of our fertile imaginations:

> ... happily there is a method, apart altogether from our physical senses, of testing the reality of the Sun. It is by mathematics. By means of prolonged processes of mathematics entirely separate from their senses, astronomers are able to calculate when an eclipse will occur. They predict by pure reason that a black spot will pass across the sun on a certain day. You go and look, and your sense of sight immediately tells you that their calculations are vindicated ... We have got independent testimony to the reality of the Sun. When my metaphysical friends tell me that the data on which the astronomers made their calculations were necessarily obtained originally through the evidence of their senses, I say 'No'. They might, in theory at any rate, be obtained by automatic calculating machines set in motion by the light falling upon them without admixture of the human senses at any stage ... I am also at this point accustomed to reaffirm with emphasis my conviction that the Sun is real, and also that it is hot – in fact as hot as Hell, and that if the metaphysicians doubt it they should go there and see.

In this book I will follow the development of modern cosmology starting with ancient creation myths, and ending with the currently accepted ideas based on Einstein's theory of relativity and the currently accepted theories of elementary particle physics. Throughout the narrative I will point out the gradual shift from the deductive to the inductive approach, discussing the reasons behind this shift and its consequences.

Science's approach to knowledge

The scientific method can be described as a recipe for providing a level of understanding of a part of Nature. The recipe itself is simple, but, as is often the case, the devil is in the details: the way in which the recipe is applied and the context within which this application occurs. I will start by listing my version of the steps, and then I will discuss each separately.

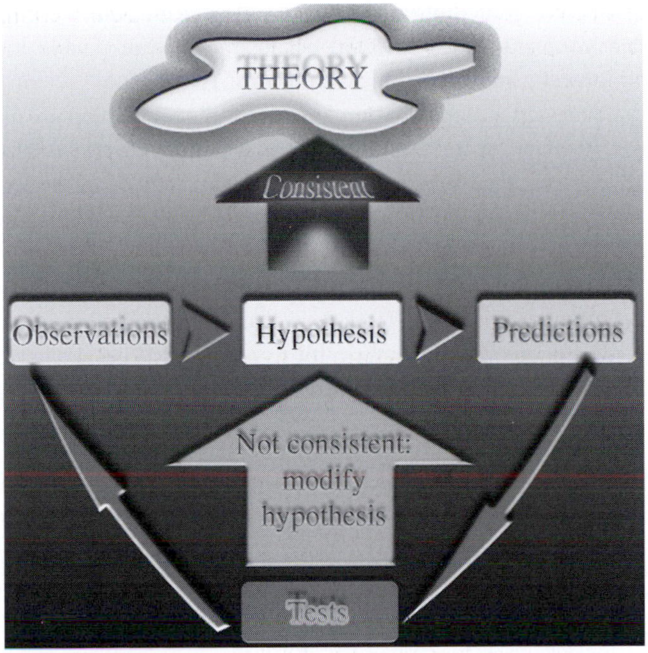

Figure 1.1 Diagram illustrating the scientific method.

The steps in the method (Figure 1.1) are the following:

1. Observe some aspect of the Universe.
2. Invent a tentative description, called a *hypothesis*, which is consistent with what you have observed. This might range from a fine tuning of existing ideas to a complete revamping of accepted knowledge (sometimes called a "paradigm shift").
3. Use the hypothesis to make predictions.[5]
4. Test those predictions by experiments or further observations, and modify the hypothesis in the light of these results.
5. Repeat steps 3 and 4 until there are no discrepancies between theory and experiment and/or observation. Once this type of consistency is achieved the hypothesis is *validated* and accepted as a new theory.

This method has several very attractive features:

- It is unprejudiced in the sense that the data from experiments and/or observations is the sole arbiter of whether a hypothesis or theory survives.
- It does not require superhuman powers. Any person with the patience and money to perform the experiments can test the reliability of a theory.

There is a common misconception that the scientific method is the "method of science," that is, the *way* science is done, as opposed to *why* we believe scientific results. Many of you may recall your elementary school science fair projects, where you were told to follow the "scientific method," defined as the process by which hypotheses are verified. Perhaps in the course of your project, you discovered something unexpected and exciting, but were disappointed to find that your grade was lowered because you were unable to prove your initial hypothesis, and so failed to faithfully follow the "scientific method." Such incidents are all too common in science education, and are all the more unfortunate because, in truth, the actual practice of science is anything but methodical, being rather fairly chaotic. The "scientific method" is the recipe from which we can make sense of this chaos, choosing hypotheses that will be kept as "theories," and discarding those that are inconsistent with Nature.

This recipe, however, also raises many questions. For example, how are hypotheses concocted? How do we select among competing hypotheses? How do we obtain predictions from the hypothesis? What if a hypothesis explains the initial data but has no predictions? I will deal with these issues below.

Paradigms

When studying a part of Nature the observer does so with a certain set of loose preconceived ideas, a certain amount of background knowledge that he/she uses to make sense of the observations. This is often called a *paradigm*. This background knowledge can range from the very basic (e.g. far-away things look smaller) to the complex (e.g. the propagation of light in matter is determined by the interaction of light with subatomic particles). Using this background knowledge, the observer attempts an explanation of the phenomena and produces a hypothesis. It is important to note that this hypothesis, though created within the context of a certain body of knowledge, need not *agree* with that knowledge. Thus Einstein studied the properties of light using Newton's ideas of motion, but the hypotheses that flourished into the Special Theory of Relativity disagreed with Newton's basic assumptions about space and time, which were almost universally accepted as providing the basic description of the workings of Nature; Harvey obtained his revolutionary description of blood circulation only after careful anatomical observations motivated by Galen's ideas, etc.

Hypotheses and theories

As with all creative processes it is difficult, if not impossible, to explain or even describe how hypotheses are generated. Researchers have historically come up with their hypotheses in every conceivable way: inspiration, laboriously investigating data, even dreams.[6] The *origin* of the hypothesis, however, is immaterial (except perhaps as a field of study in psychology), what matters is whether the hypothesis describes the data, and can be used to make new accurate predictions that are subsequently verified. If these conditions are satisfied the hypothesis is accepted, otherwise it is discarded.

A little thought shows that no hypothesis can be proved *absolutely* true, for that would entail testing it in *all* possible ways under *all* possible circumstances.

In contrast, a hypothesis *can* be proved false for it makes predictions that may or may not be verified by experiment. To be scientifically useful, hypotheses must be *falsifiable*. Thus *all* scientific theories are constantly in peril of being proven wrong by new data or observations: experiments are the sword of Damocles for theories. This is a positive aspect of all scientific investigations for it provides a natural mechanism for testing and improving scientific knowledge.

An example of a falsifiable hypothesis is Newton's universal law of gravitation that predicts the planetary orbits. These predictions are verified, within experimental accuracy, for all members of the Solar System *except* Mercury. The orbit of this planet exhibits a slight deviation from the Newtonian predictions: Newton's hypothesis has been falsified. The General Theory of Relativity also provides predictions for all the planetary orbits, and these agree with observations, *including* the case of Mercury. Because of these (and many other) observations the General Theory of Relativity is now accepted as the best description of gravitation. Yet this does not mean that future experiments need to agree with its predictions. Should a discrepancy be confirmed, this theory will have been falsified and the search for a more accurate description of gravitational phenomena will begin.

An example of a non-falsifiable hypothesis is the one that claims the Moon is densely populated with little green men. These beings are then assumed to be perfectly in tune with human intentions so that whenever we attempt a way of finding them they would know of this in advance and thwart us. Should anyone on Earth look at the Moon, the green men have foreknowledge of this and they *all* hide. When the Apollo spacecrafts landed on the Moon they all moved to the dark side (of the Moon) beforehand, and obliterated all traces of their existence. If we were to observe the whole surface of the Moon simultaneously then, by the time our equipment is set, they would have dug tunnels deep into the Moon (again perfectly covering their tracks) and would stay there until we stopped observing. If we were to obliterate the Moon they would again know of our intentions ahead of schedule and leave, timing their trip so that we could not see them, etc. It is clear that by construction the existence of these little green men cannot be disproved, their existence has to be taken on faith and might be a matter for mystical or philosophical musing, but it has no place in a scientific discussion.

The experimental verification of hypotheses is paramount, and it is in this realm that many controversies arise. It is possible for an experiment to provide an apparent confirmation of a prediction, but this verification is later found to be the result of poor experimental design, a misinterpretation of the data, the result of extraneous effects that were not properly taken into account, etc. In order to minimize these potential problems scientists test predictions using many different experimental designs. The confirmation of a prediction by a given experiment will usually set off a flurry of new experiments also aimed at verifying this prediction, but using many different approaches. In addition the details of every such experiment are published, and the experimenters open themselves to public scrutiny, thus insuring that the experiments are reproducible and unprejudiced.

On some rare occasions an experiment will verify a prediction that is *not* verified by other contemporary experiments. Yet, in the long run, this single result turns out to be correct. This is the case where the experimental techniques of the first researcher are much superior to any contemporaries'. Even though the positive claims are verified by future careful and more sophisticated experiments it is undeniable that the original positive results are criticized and even ridiculed by contemporaries. The pain associated with the creation or confirmation of scientific theories is almost always relegated to footnotes, if mentioned at all.[7]

The predictions made by the theory should apply to all phenomena that the theory claims to describe. If the majority of such predictions, but not all, are confirmed, the hypothesis should be modified in order to properly describe the discrepancies, or to explain why they do not fall within its scope. Theories cannot pick and choose the phenomena they aim to describe by selecting successes and rejecting failures.

Box 1.1

Continental drift

The idea of continental drift, proposed by the German geophysicist Alfred Wegener (1880–1930), was originally motivated by the observation that fossils of identical plants and animals are found on opposite sides of the Atlantic. The standard explanation at that time (1911) postulated that land bridges, now sunken, had once connected far-flung continents, but this could not explain certain geological features such as the close fit between the coastlines of Africa and South America, the match between the Appalachian mountains of eastern North America with the Scottish Highlands, and the fact that the distinctive rock strata of the Karroo system of South Africa were identical to those of the Santa Catarina system in Brazil. Using these arguments and a variety of fossil records, Wegener boldly proposed that all these observations could be explained by assuming that about 300 million years ago all the continents themselves had once been massed together in a single super-continent he called *Pangea* (from the Greek for "all the Earth") which has since been fractured into the current geography. Wegener was not the first to suggest this, but he was the first to present extensive evidence for it.

Though some scientists supported this hypothesis, the reaction from the geophysical community was almost uniformly hostile, and often exceptionally harsh and scathing. Part of the problem was that Wegener provided no convincing mechanism for how the continents might move: he envisioned the continents plowing through the Earth's crust driven by tidal and centrifugal forces, but his opponents noted that these forces are too weak for the task, and that the shape of

Box 1.1 (cont.)

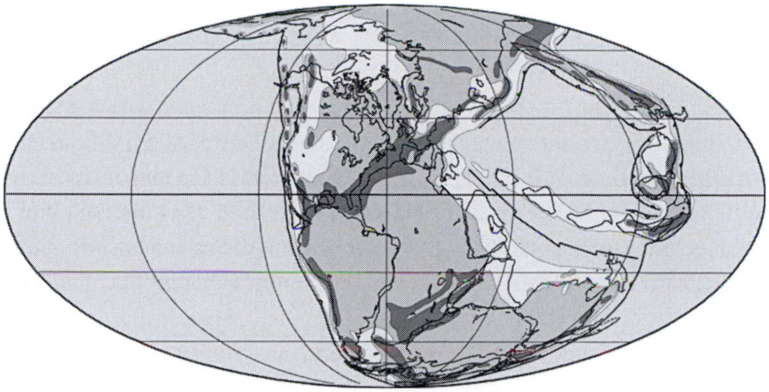

Figure 1.2 A modern reconstruction of the original super-continent Pangea. (courtesy of C. R. Scotese, PALEOMAP Project (www.scotese.com))

the continents would be distorted beyond recognition should they plow through the hard crust. In addition Wegener's original data contained errors that led him to make impossible claims (e.g. that North America and Europe were moving apart at over 250 cm per year – about 100 times faster than the actual number).

Despite this unpromising beginning, Wegener's basic idea has been fully vindicated by modern investigations. These experiments show that the Earth's crust is composed of a series of mobile plates, called *tectonic plates*, floating on a bed of molten rock that behaves as a viscous fluid (a result of the enormous pressures it is subjected to). The various continents lie atop the tectonic plates and so move about with them. When plates separate the underlying molten rock oozes out (as happens in the middle of the Atlantic Ocean); and when plates collide they generate regions of intense earthquake activity (as in the west coast of North America), and often give rise to stupendous mountain ridges, such as the Himalayas.

Trivial as the above may sound, such practices are sometimes adopted as policy in pseudo-scientific research. For example, the official policy of the *Journal of Parapsychology* is to reject publications containing negative results since these are considered as failures and hence of little use; the editors of this journal assume a priori the existence of parapsychological phenomena and deny they should be subject to testing.

Hypotheses and theories within the exact sciences (and in some cases within the social sciences) are often cast in mathematical language. The derivation of the predictions comes then from manipulation of the mathematical expressions within the theory as determined by logic. Such manipulations correctly describe Nature in tantalizing fashion (in E. Wigner's words, "the unreasonable effectiveness of mathematics") and there is no obvious reason why this should be so, but it is. One advantage is that the predictions and assumptions are very precise; a disadvantage is that the mastering of the techniques requires time and effort.[8]

Ockham's Razor

A description of a phenomenon often requires a set of initial conditions in order to provide predictions. For example, in order to predict the motion of all bodies in our Solar System, we need to know the positions and velocities of all of them at one time (any one instant will do). With this information the future behavior of the planets can be predicted by solving Newton's (or Einstein's) equations. One might, of course, ask what circumstances dictated those initial positions and velocities, and even formulate a hypothesis for that. But one should not use the accuracy of Newton's law of gravitation to justify the second hypothesis. For example, suppose we claim that the planets have their present orbits because a race of advanced aliens put them at the appropriate places and gave them the required initial pushes. So far this is a valid hypothesis. But this does not warrant the claim that our alien hypothesis is *supported* by the great accuracy with which we can predict the planetary motion. We can predict planetary motion thanks to Newton's genius in devising his law of gravity, this has *nothing* to do with how the planets got where they are. The aliens (or lack thereof) are irrelevant in predicting the planets' motion.

Top-loading a theory with a new hypothesis and claiming the successes of the theory justify the hypothesis is common, but it is clearly wrong. This was repeatedly emphasized and justified by the medieval philosopher/theologian William of Ockham, who summarized this reasoning with the statement:

> *Pluralitas non est ponenda sine neccesitate.*
> (Entities should not be multiplied unnecessarily.)

The above is referred to as Ockham's "Razor," because it "cuts" between those components necessary for a theory, and those which do not add anything except complication. Ockham's Razor does *not* mean "always keep hypotheses simple," rather "eliminate the dead wood." Einstein's theory of gravitation is considerably more complicated than Newton's, yet it provides a much better description of Nature. Neither of them attests to the presence of aliens in our universe.

The validity of scientific theories

Scientific theories have, by their very nature, a limited applicability. As I mentioned above, even if we find a theory that is absolutely true in some metaphysical sense, we would be unable to demonstrate this admirable quality: we cannot possibly verify the infinity of the predictions such a theory would provide.

When investigating a new set of phenomena, researchers come armed with the available successful theories. Using these theories they try to describe the new experiments and observations. If they are successful the range of applicability of the theory is extended and researchers move to other areas of interest. If, on the other hand, the theory is *not* successful in explaining the new results we must search for an alternative. The new theory aims at explaining both the new results as well as old ones for which the old theory was perfectly adequate.

When accurate measurements of Mercury's orbit showed a small deviation from Newton's predictions, several hypotheses were proposed to explain these observations. These included the possibility of the Sun being not quite spherical, being surrounded by a tenuous gas cloud, etc. All such attempts were unsuccessful. Finally, in 1916 Einstein showed that the General Theory of Relativity accurately predicts planetary motion, *including* that of Mercury. The range of applicability of Newton's theory was shown to be inadequate for sufficiently large gravitational forces and high experimental precision.

The new experiments then mark the limit of applicability of the old theory. One does not say that the old theory is "wrong" for it was assumed from the start that it had limited applicability.[9] One merely has found the boundary theory's domain. Theories are perishable commodities: new experiments are continually being done, testing theories in both familiar and new environments. Eventually, through an increase in experimental accuracy or by foraying into new territories, discrepancies are observed. The limits of the theory are then found, and a flurry of activity ensues, attempting to replace the old theory with a newer one explaining both old *and* new results.

I mentioned repeatedly the import of experimental accuracy in testing the predictions of theories and hypotheses. Precision is an important point often overlooked when discussing verification of a theory: no experiment has infinite accuracy. Even the numerical predictions from a theory have a certain error since the calculations require some experimental input. For example, Newton's law of gravity states that the force is proportional to the masses of the objects, but what *are* the values of these masses? Well, these are determined by weighing the objects and the result has a certain error since the balance (*any* balance) is not infinitely precise. If the reporting of an experiment is to be done honestly, it should include a measurement of the error. Let's examine the issue of error further, because it is important in interpreting scientific results.

Errors and averages: testing hypotheses

It is the heart-felt desire of every parent that his/her children should be all above average. Yet, were this to be granted, this (apparently) laudable attitude has unintended consequences, for in order for some children to be *above* average, some others must lie *below* average. Thus every parent is not only wishing the best for his/her children, but also hoping to condemn other children to a life of substandard capabilities. If the wish of *every* parent were granted, all our children would be merely average.

Keeping this in mind, suppose we would like to find out how well we are teaching arithmetic in elementary schools. One way is to give students a test containing a variety of problems in each of the areas covered in the course, and then grade each problem keeping careful notes as to the performance of each student in each area. One can go further, finding out the average grade in each section and how the grades are distributed about this average. If the overwhelming majority of students get the same grades (within a fraction of a percent, say), and if this result is reproduced in test after test, we can use this measure to gauge the effectiveness of the instruction methods. Moreover, if the test results show that the great majority of students solved all problems, except those involving fractions, we can infer that our way of teaching fractions is ineffective.

Suppose now that we instead decide to grade students on a pass/fail method for the full test. We can then get a broad feeling as to the effectiveness of the instructional methods and materials, but we are unable to conclude whether failures are due to an overall deficiency, or whether there is a particular area of weakness.

Finally suppose that the test results range over all possible values with some students failing in one area and others in another, without any clear pattern, then very little can be inferred. Perhaps the teaching methods should be completely revamped, perhaps the tests are ill designed (e.g. if the instructions are so poorly worded that the students do not understand what they are supposed to do).

A similar situation occurs when analyzing experimental data of all types. Well-designed experiments can accurately test a prediction only up to the experimental precision available. Two competing hypotheses making predictions that differ by amounts below this precision simply cannot be differentiated by this experiment;[10] for example, competing hypotheses of cellular structure cannot be tested using low-magnification microscopes that cannot clearly distinguish the features inside cells. Ill-designed experiments usually test nothing and provide little information. In contrast, well-designed experiments often also suggest the modifications in the hypothesis that are needed in case discrepancies are found.

The statement "Newton's theory of gravitation accurately predicts the data" should be qualified by saying "within the errors." The above-mentioned discrepancy in the motion of Mercury represents a deviation of less than 1%, so that the observations had to reach that degree of accuracy before any problem with that theory could be determined.

There are many sources of errors. There is, of course, the possibility of flaws in the experimental design. These can induce deviations from the predictions that have nothing to do with the validity of the theory. This type of error is, of course, preventable, and experimentalists continually strive to eliminate them or, if this is impossible, to estimate their effects. This can be done by designing the experiment to both test the hypothesis *and* also measure some other well-known quantities, and then use the results of these last measurements to estimate the accuracy of the hypothesis test. For example, the same experiments that measure the charge of certain subatomic particles called quarks also measure the charge of the electron (known to an accuracy of 40 parts per billion); the errors in the quark charge can be gauged by comparing the measured electron charge to the accurate value known from other experiments.

In addition there are errors due to unpredictable events. For example, the line voltage might fluctuate (due to a slight increase on the city's electric grid) and this would reflect in a slight deviation in the data. These errors are not preventable but, since they are random, they can be estimated using statistical analysis.

In order to minimize such problems, experiments are frequently repeated with different designs, yet testing the same hypothesis/theory. If a contradiction is found (such as when some experiments confirm and others refute a hypothesis), a careful redesign is done again and again until a clear unambiguous result is obtained. In addition, this process exposes the reasons for the original discrepancies. Experimental results measuring a certain quantity provide a value together with an estimated error, which can then be compared with the predictions. The hypothesis/theory is either validated or not. Validation does *not* mean that the theory being supported is the absolute truth. It merely means that its predictions agree with the observations *within the experimental accuracy*. Deeper theories may be devised in the future, and these may disagree with the old one, yet agree with the experimental results as they fall inside the error interval for the measurement. In this case more precise experiments are needed (and devised whenever practical) to determine the validity of any new hypothesis.[11]

Experimental designs also have additional problems. It is possible for the researcher to be unconsciously swayed towards one or another result, stopping an experiment only if the favored hypothesis appears confirmed. It is also possible for the experiment to measure extraneous effects that have no relation to the main object of the experiment. In order to avoid these pitfalls researchers use, whenever possible, what are called *double-blind* experiments.

A double-blind experiment is really two almost identical experiments. The idea is best described by an example. Suppose a careful analysis of the structure of DNA suggests that exposure to X-rays produces cancer. One might want to investigate this using, say, cockroaches. So we place the victims inside an isolated cell and give them food and water and then irradiate them; after two days 90 percent of the roaches exhibit tumors . . . but do we *know* this is due to the X-rays? Perhaps the illness is the result of some other effect such as the food we gave them containing carcinogenic substances, or the cardboard of the box in which they were kept having some chemical that produced the tumors, etc.

In order to determine whether the cancer was caused by the X-rays we then select another group of roaches, as similar to the ones in the first group as possible, and we put them in the same type of box, and feed them the same stuff, and just as often, etc. The only difference is that this group gets no X-rays. This second group is called the *control group*. Finally, when looking for tumors we arrange things so that the person probing the roach does not know from which group it came (so they would not be examined with particular care if irradiated but only summarily for the controls). Only after all the roaches are examined do the researchers determine whether there is any correlation between exposure to X-rays and the incidence of cancer in roaches. This type of experiment, where the persons taking the data are unaware of whether the subject is from the control group or not, is called a double-blind experiment. In a world of impersonal scientists guided by pure logic double-blind experiments might not be needed (though a control group would), but our reality dictates otherwise.

There are some predictions that cannot be tested using double-blind experiments. Perhaps the most dramatic of these are the ones stemming from the Big-Bang hypothesis that provides a quantitative description of the average properties of the Universe (such as the density of matter) and the manner in which they evolve. These predictions are tested against a variety of observations, but it is impossible to change the Universe and see whether the modified Universe will behave as predicted by this hypothesis. For example, the Big-Bang hypothesis states that if there is a sufficiently large amount of matter in the Universe the current expansion will eventually come to an end, and the Universe will contract, ending in a state of infinite density (the so-called "Big Crunch"). It would be nice to test this, but to do that we would have to create (at least) two universes in the laboratory (one to serve as the control), then add to one different amounts of matter and determine whether its evolution agrees with the predictions. In such cases double-blind experiments simply cannot be done, and we have to console ourselves with accepting a hypothesis when its predictions are validated by the observations. In the end, researchers can only do their best with the budget they have.

The evolution of scientific theories

Scientific theories, as many creations of the human mind, are far from static. They are born, tested, and modified and eventually replaced by newer, better descriptions of Nature.[12] Failure to understand this can lead to severe misinterpretations of the results. For example, in the 1930s a man called Wilbur Glen Voliva published a special issue of the periodical of the Catholic Apostolic Church in Zion (Illinois), which, among other things, claimed that the Earth is flat[13] and is also at the center of the Solar System. To support his claims, Voliva noted that the proponents of the heliocentric model do not agree among themselves: Copernicus claimed that the Sun was stationary, while the eighteenth-century astronomer Sir William Herschel claimed that the Solar System as a whole was moving. Voliva then stated: "It is asserted by these advocates of the Copernican system of astronomy that it is an *exact science* – and yet these two great men, Copernicus and Herschel, contradict each other, Copernicus saying that Herschel is a liar and Herschel saying that Copernicus is a liar – and Voliva agrees that they are both right."

The discrepancy between Copernicus and Herschel's hypotheses concerning the motion of the Solar System as a whole is used to discredit the description of the motion of the planets *inside* the Solar System. The two are, from the point of view of Copernicus, quite unrelated. Ockham's razor clearly shows the error in Voliva's argument. *However* the motion of the Solar System around the Galaxy *is* related to planetary motion from the point of view of Newton's theory of gravity, so much so that a stationary Sun would be extremely difficult to account for. During the three centuries separating Copernicus and Herschel (so that the former could not have personally called the latter a liar) the heliocentric model of the Solar System, which originally provided a description of the motion of the five planets closest to the Sun, provided the motivation for the development of the universal theory of gravitation. This theory predicts that the same force responsible for planetary motion should force our Sun to orbit the center of the Milky Way, and this has now been measured and is seen to proceed according to the predictions of the theory. Thus the original hypothesis of Copernicus has been replaced by the more general theory of gravity; only in this sense is Copernicus "wrong."

There are many ways in which scientific theories are modified or discarded. They can fizzle out, or they can be debunked in light of a dramatic experiment. In either case the theory is replaced due to its inability to explain the data. As I have argued, disagreement with more precise experimental results defines the limits of applicability of the old theory: it delimits the conditions under which we can hold it as a good description of Nature. If the predictions of a new hypothesis prove capable of explaining the new results *and* the old ones as well, this new hypothesis is accepted as a new description of the phenomena at hand, having an extended predictive power compared to the old theory. We then say that the new theory has *subsumed* the old one.

This "changing of the guard" is most dramatic when it involves a change of paradigm: the new description requires a radical change in our perception of Nature. There have been many such cases: when Pasteur demonstrated that illness can be caused by microscopic organisms; when Watson and Crick described the helical structure of DNA; when Einstein provided a description of space and time as relative and dynamic objects, etc. Such paradigm changes are accepted only after a very careful investigation of the new hypothesis and after it is confirmed by numerous experiments; as pointed out by Carl Sagan, "extraordinary claims require extraordinary evidence." Note that these paradigm shifts are based on the experimental evidence, not on the philosophical arguments presented (though these *can* be of importance, they are *not* decisive), so it was with this in mind that J. Moore noted: "Darwin is famous not for what he said, but for what he said that he could *prove*."

The examples of paradigm changes illustrate another feature in the evolution of scientific theories. Newer theories are not necessarily simpler than the older ones (for what is simpler than to say, "It is like that because the gods want it that way"?). The newer theories are adopted because they provide a unified consistent (and falsifiable) description of a wider set of phenomena than the old ones. For example, a virtue of the theories of relativity is that they provide a unified description of mechanics, gravity, and electromagnetism, and they are used in the quantum description of phenomena down to a scale of 10^{-18} m, even though mathematically they are considerably more complex than Newtonian mechanics.

Science and the public

It is easy to find reports of scientific discoveries in the press. These reports contain a brief description of the result often accompanied by a short history of the problem and a few illustrative diagrams. By its nature, this type of reporting concentrates on the most dramatic claims, some of them quite controversial. If the reporting is done in a levelheaded way, it will also include a discussion of the opposing views and several cautionary sentences. Unfortunately this can make the articles very dull, people stop reading them, and the reporter gets in trouble with his editor; dramatic claims with even more dramatic extrapolations, on the other hand, sell newspapers. So in this type of reporting, journalists often throw caution to the wind and mix proven results with the speculations (and even the hopes and dreams) of the researchers, without any clear distinction. The impression is that *all* of this is currently (or soon will be) backed by actual data.

An example is provided by the newspaper articles on a Martian meteorite which exhibited formations similar to the ones seen in certain kinds of very ancient Earthly fossils. The similarity was repeatedly used to conclude that Mars had supported life, but it was seldom mentioned (if at all) that, despite this being a tantalizing possibility, there *are* non-biological processes that can produce similar formations. Yet, most newspaper reports presented this possibility of extraterrestrial life as the most likely explanation; most failed to emphasize that rigorous verification was (and is) lacking and that the origin of the meteorite's features was a subject of heated controversy. Had they been reported, such discussions among researchers would have been a very useful portrait of the scientific process; it would have made clear that this was a situation in flux, where various hypotheses were being tested and accepted or discarded *according to the data*. As it was, the impression given was of a melee where the most interesting ideas are discarded in favor of a boring status quo.

This type of anecdote illustrates the advantages and problems of having the scientific process exposed in all its phases. On the one hand the general public gets a glimpse of the rigorous process used to arrive at reliable theories. It demystifies this process and illustrates its strengths and weaknesses. The disadvantage is that the process is easy prey for sensationalistic and exaggerated reporting. This has a counterproductive effect, giving the impression of a fickle community that will adopt outrageous claims only to dispose of them without much rhyme or reason. It is, of course, the duty and responsibility of the scientific community to ensure that the first approach is publicized, and the second condemned. Sadly scientists often skirt this responsibility,[14] and this has serious consequences: the public can become mistrustful of scientists and the theories they support, which is partly to blame for the abysmal support for science education.[15] It also provides fertile ground for the growth of pseudo-scientific and mystical ideas based on a mixture of rhetoric and mysticism that provide "alternative" explanations to a variety of phenomena.

Box 1.2

Cold fusion

One example that illustrates the close scrutiny given to new scientific ideas or results is provided by the case of "cold fusion." In 1989 two University of Utah scientists, S. Pons and M. Fleischmann, claimed to have discovered a way of creating nuclear fusion in a test tube, without the need of the large and complicated machines that make this process an impractical energy source. If viable (and correct), this would open the possibility of solving any future energy crisis, so the economic implications were enormous. Perhaps because of this the initial claim was presented as a certain result: a fact; absolute truth. The political and publicity machineries then started to digest it: public monies were allocated, congressional (state and federal) committee meetings were devoted to it, press conferences were scheduled, etc.

Simultaneously other scientists, believing these results were suspect, were assiduously trying to reproduce the claims coming from Utah. The politics surrounding the issue were so complicated that some important data from the original experiment were kept secret. For example, initially the experimental setup was not fully disclosed.

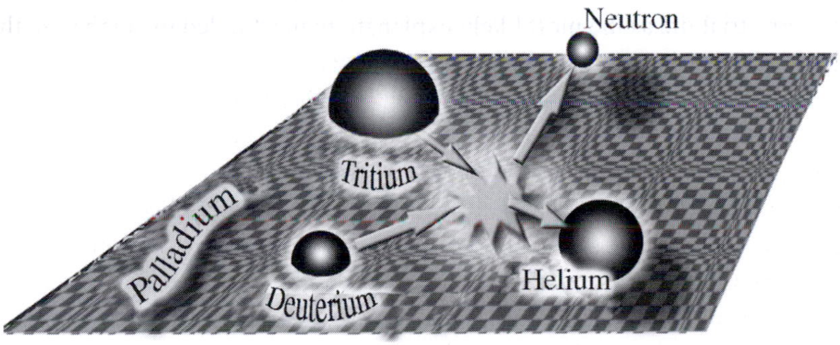

Figure 1.3 The reaction hypothesized to be responsible for the cold-fusion phenomenon.

Box 1.2 (cont.)

In the end a variety of experiments showed unquestionably that the original data had nothing to do with fusion, and after a few more fireworks the whole issue died out. If this had been treated as a run-of-the-mill claim, and had been tested outside the very intense public scrutiny it received, the fallacy of the initial results would have been found much sooner and would have resulted in a few embarrassing moments for Pons and Fleischmann. As it was, in the resulting fallout their careers were destroyed.

The original cold-fusion hypothesis assumed that if sufficient amounts of tritium and deuterium are placed on a palladium surface, this metal would act as a catalyst and induce a nuclear reaction through which helium and neutrons are released as well as a considerable amount of energy. This hypothesis was finally disproved when no trace of helium could be found.

There is sometimes a certain amount of drama involved in the testing and subsequent acceptance or rejection of new scientific ideas, a process that is often far from friendly. A claim that is suspected to be erroneous will be subject to very rigorous testing and, if it is indeed found wanting, it will often arouse the most predatory instincts among researchers who publicly denounce the errors contained in the now defunct hypothesis in a rather unflattering (sometimes even insulting) way. After this the interest dies away.

There are, however, instances where political or other non-scientific constraints limit or even forbid independent scrutiny of a new idea, and this can lead to serious, negative effects. A clear example is provided by the scientific studies sponsored by the tobacco companies that investigated the possible harmful effects of cigarette smoking. These studies found no harmful effects; they also kept the data and the conditions under which the experiments were carried out away from independent investigators. The enormous political and economic interests maintained this state of affairs for many years (in serious detriment of the health of the public); only through one of the most intense public-health campaigns (and the intervention of "whistle-blowers") was the fallacy of these claims exposed.

The development of scientific ideas cannot be separated from the social environment in which these ideas germinate, as illustrated by the above stories. Scientists are people of their time and are affected by societal pressures. Despite this, the results obtained through the process described above do provide information about Nature. It is true that the *topics* that are most assiduously studied can be (and often are) determined by the interests of the time, and that the fascinating question of one century becomes the quaint footnote of the next one. Nonetheless the *results* obtained are valid (within a certain range of applicability) and provide unprejudiced information about the world around us . . . even for the quaint areas of research.

Some authors, however, use this to claim that *all* of science (its methods, topics, and results) is just a social phenomenon. If this argument were carried to its logical conclusion, one would have to believe, for example, that the current theory of gravitation would not be valid in another social environment. This runs counter to the underlying philosophy of scientific investigations. The credo is that there is a world around us that we can probe and investigate, and whose regularities we may try to comprehend; that these properties of Nature exist independently of us, and would continue to determine the motion of the Universe quite undisturbed if we were to disappear from the face of the Earth.

Scientific knowledge aims at being objective, at least to the greatest possible extent. But we humans require other types of knowledge, other experiences that *are* subjective and allow us to understand the way in which we interact with one another, the way in which society changes and reacts, etc. There should be no moral ranking of these different types of knowledge, nor should the methods of obtaining one type be used outside of their realms. If one blindly applies the scientific method to human problems one can easily end up with morally unacceptable conclusions. Similarly, the deep wanderings of philosophy are not to be used in testing scientific theories lest we end up describing our whims and weaknesses and ascribing them to Nature. Scientific knowledge complements these other ways of looking at the world around us, it does not, cannot, and should not replace them.

Before closing this section I would like to comment on another fallacy associated with scientific research. There are many portrayals of scientists in the media as sober, impersonal, almost inhuman beings that can wrestle secrets from Nature. Such inhuman characterizations are erroneous. There is no distinguishing moral "property" exclusive of scientists. These are people with the same expectations, hopes and disappointments as any other person. The same desires of the "average" person for fame and fortune, for recognition, or for peace and tranquility also find a place in the heart of a scientist. What is remarkable is that the *results* obtained by scientists are so separate from these human peculiarities. This is what makes scientific knowledge so reliable: it is not the people, but the method they follow.

The other side

It is a common occurrence to hear news of some new wonderful and tantalizing discovery whose mere presence appears to debunk some of the most cherished scientific principles. Thus we have prescriptions for travelling faster than light, for running automobiles on a single 9-volt battery (complemented by transforming a certain amount of matter into energy . . . using a regular electric motor). There are claims of "psychic healing" which use novel medical techniques where operations are performed without surgical instruments, and where the wound heals instantaneously and without any scarring, etc.

Such claims, when they reach the public, inevitably receive a certain amount of attention and are frequently fraudulent. There is nothing wrong with a magician who performs an "operation" where blood, tissues and organs flow aplenty, though the patient evidences no cut or discomfort afterwards. What is wrong is to make people believe that this is an *actual* operation that will have the same effects as the regular ones, and then charge the same amount as for a hospital procedure. One can callously say that a person so foolish as to believe such claims deserves to be made a dupe, but this fails to understand that the people that fall for such scams are often vulnerable, being strained physically and sometimes financially. Such "pseudoscience" should be denounced and stamped out whenever possible.

The ideas and claims that populate the vast country of pseudoscience inevitably violate the constraints imposed by the scientific method: the results cannot be reproduced, they require certain secret expertise that the author cannot divulge, the experimental techniques are not disclosed, etc. Though these desirable properties are often absent, (and perhaps in order to make up for this fatal deficiency) the authors often provide alternative bonuses; for example, the new "theory" is supposed to encompass widely different fields. In these claims no leap of faith is too perilous, no logical chasm so deep that they cannot be spanned by some statement. Authors of such ideas are undeterred by exposure or fraud, they soon assimilate the criticism or find reasons to ignore it. Thus, for example, the *Journal of Parapsychology* no longer accepts any papers where no parapsychological effects are reported. All negative results are rejected. Generally speaking, *anything* goes.

The scientific community takes a very dim view of the often fraudulent and sometimes outright dangerous claims made by pseudoscientists. This then creates a "conspiracy theory" frame of mind among many pseudoscientists, they feel that the "establishment" is conspiring against them, attempting with vast (if nebulous) resources to suppress their novel results, presumably because of the vested interests of all scientists in preserving the status quo; it then follows that anyone criticizing them is a member of this vast conspiracy.

There *is* a large inertia among the scientific community when faced with a dramatic change in our understanding of Nature, such reluctance should be met with testable consequences and incontestable evidence, not with rhetorical outbursts. The hallmark of the scientific method is that even seemingly outrageous claims are ultimately accepted *if supported by the data*. There are a few instances where claims that have the full ring of being pseudoscientific stories turn out to be perfectly respectable theories; the best example being the theory of continental drift. But the shift from one group to another occurs only after predictions are tested, and the theory of continental drift did provide definite predictions that were confirmed by observations.

I am not so dull as to believe that all people interested in pseudoscience are true believers of what they read. There is a certain amount of entertainment value in these endeavors. The trick is not to forget that this is *only* entertainment, until it provides predictions that can be tested and are verified.

Notes

1. I will not dwell on the question of whether humans are happier in this type of society, nor discuss the shameful inequalities in quality of life between developed and underdeveloped countries and regions.
2. For example, the "dietary supplement" ephedra is linked to heart attacks, seizures and death, but manufacturers were forced to clearly state this on the label only after deaths and ensuing lawsuits forced them to; no thorough testing of the dangers of this drug were carried out before making it available to the public. Ginseng should not be taken by people with high blood pressure, but the manufacturers of this herbal supplement are not required to state this clearly on the product's label.
3. Ockham's Razor will be discussed fully below.
4. W. S. Churchill, *My Early Life – A Roving Commission* (London: The Hamlyn Publishing Group, Odhams Press, 1947).
5. It is interesting to note that the process of making predictions from a hypothesis is actually *deductive* reasoning! However, these predictions are considered to have little weight until verified by the inductive process of testing them through experiment.
6. In 1865, F. A. Kekule proposed that the benzene molecule had the structure of a ring. This hypothesis was motivated by a dream he had, where he saw a snake swallowing its own tail. Later, of course, he provided clear experimental verification of this idea.
7. The Austrian physicist Ludwig Boltzmann (1844–1906) created the Theory of Statistical Mechanics, which describes the way large conglomerates of atoms behave collectively as a macroscopic body. His ideas were ridiculed by some of the leading figures in the German scientific establishment especially by the chemist W. Ostwald (who did not believe in atoms!). Boltzmann, depressed by these attacks and by his own poor health, committed suicide in 1906. His ideas were experimentally verified shortly after his death.
8. This is different from saying that the mastering of mathematics is open only to the "chosen ones." *Anyone* can approach the subject and check the results claimed by the proponents of a hypothesis, though this verification might require significant effort. The scientific method makes the process open, it does not claim to make it easy.
9. Even if the adjective "wrong" is often used colloquially, it merely means, "it disagrees with the data, given some experimental precision." It is not meant as a moral judgement.
10. Though they could be distinguished if they both make predictions that differ from the observations by amounts *larger* than the experimental precision.
11. In the case where two theories make all of the same predictions (within experimental accuracy), we would use Ockham's Razor to choose the simpler theory as preferred.
12. In some cases theories do die, sometimes ignominiously; such is the case of the N-ray theory (I. M. Klotz, The N-ray affair. *Scientific American* **242** (1980)).
13. This church evolved into the currently active Flat Earth Society (see, for example, http://www.hist.unt.edu/09w-ar7n.htm).
14. Though the press is often also remiss in publishing corrections to misguided or inflated stories concerning scientific issues, and even if the corrections are published, their implications and significance are seldom explained.
15. I hasten to note that we believe that while misinformation or distortion of scientific ideas *supports* these societal trends, they do not *cause* them.

2
From antiquity to Aristotle

Creation myths: the dawn of cosmology

The first cosmologies spurted from, or were a part of, a creation myth, in which one or more deities made the Universe out of sheer will, or out of their bodily fluids, or out of the carcass of some god they defeated. Such descriptions can hardly be called hypotheses since all is subject to the whim of powerful gods and when a prediction *is* provided it can always be violated if a god is out of sorts that day. Thus an eclipse might signal one god's displeasure while a bountiful harvest another's response to an agreeable sacrifice. By their very nature these descriptions of Nature and of the origin of the world are not falsifiable. This is hardly surprising; in the dawn of humankind people were very vulnerable to the vicissitudes of living in constant contact and at the mercy of the elements, which are as fickle as the will of the gods.

The more we construct the cocoon of civilization around us the more leisure we have to observe Nature's regularities, and infer from these the mechanisms that generate changes in the world around us. From such knowledge, testable and falsifiable cosmologies were born. The road was far from smooth, with many wrong turns and branching paths that led nowhere. In this chapter we relate some of the first such forays.

The early Greeks devised the first scientific model of the Universe more than 2000 years ago. The pinnacle of these investigations was reached in Ptolemy's *Almagest*, a work that survived for ten centuries as the best description of our Solar System. These achievements were born of the observations of Egyptian astronomers and, more importantly, the careful and extensive data compiled by Babylonian astronomers who kept careful records of their observations. Thus it is these earliest of civilizations that I turn to first.

One can trace a line connecting the early Greek cosmologies, to Copernicus, to Galileo and Newton, and then to Einstein; and because of this I will devote the bulk of this chapter to this part of the story. The richness of human imagination was, of course, not restricted to the Greek culture. Indian, Chinese, pre-Columbian American (to mention but a few) cultures also had their own cosmologies and creation myths. Many of these cultures observed the regularities in the progression of heavenly bodies, developed the mathematical tools necessary to describe them, and then used these to construct accurate calendars. And yet these early ideas did not lead to more and more predictive scientific hypotheses, but to mystical and/or religious pursuits, and so I will recount them only briefly.

Though creation myths are not scientific hypotheses (as defined in the previous chapter), they do have the virtue of explaining that most central of all questions, why is there a Universe at all, and do so in a charming and engaging manner. In order not to maintain the main theme of the book without completely ignoring these creation stories, I have opted to include a select number of them in an appendix at the end of this chapter.

Babylonia

In the third millenium BC there arose in a fertile plain between the Tigris and Euphrates rivers one of the earliest (human) civilizations. First constituted by a series of city-states, it then evolved into a coherent monarchy centered on Babylon and endured for over 23 centuries, until conquered by the Persians (530 BC). Despite the many destructive upheavals, which seem to be the curse of the region, the inhabitants enjoyed, during times of peace, a sophisticated system of government and of commerce, and developed the practical mathematics that facilitate these endeavors. This knowledge survived in thousands of clay tablets marked with cuneiform (wedge-shaped) writing from which we get a glimpse of this very early civilization. Among these tablets we find one of the earliest (surviving) law canons, the code of Hammurabi, on which many subsequent codes of conduct, including the Biblical, are based. We also find one of the first poems, the epic of Gilgamesh. In addition, the tablets provide a record of everyday life: business deals, complaints before the law, land measurements, property titles, school homework, etc. These menial records also tell of the tools used in commerce, in particular we learn that the Babylonians divided the week into 7 days, the day into 24 hours, and each hour into 60 minutes, a custom that has survived 40 centuries.

Babylonian astronomers kept detailed records over many centuries[1] of the motion of heavenly bodies. In order to facilitate the observation and recording of such phenomena they grouped stars into constellations and divided the night sky into regions, each associated with a sign, and thus gave rise to the signs of the zodiac. These observations lead them to notice that the motion of the Moon and the planets exhibited some regularities which they were able to describe using arithmetic formulae, which they then used to *predict* astronomical events with remarkable accuracy – as was the case for lunar eclipses (in contrast no such attempts seem to have been made by Egyptian astronomers). These efforts were motivated in part by practical necessities such as predicting the sowing and harvesting times; in addition they needed to specify the appropriate times for religious observances.

Connected with these observations came the realization of the influence of the Moon and Sun on Earth and its creatures (the seasons, the tides, the – coincidental – correlation between the lunar phases and menstrual cycles, etc.) and so grew the notion that heavenly bodies influenced and even determined the ups and downs of human affairs. In this way the use of astronomical observations for prophesying was born; it is from the Babylonians that we inherited astrology.

Egypt

The valley of the Nile, with its remarkable fertility and the protection offered by the surrounding desert, provided a safe and nurturing environment for the development of civilization, from which we have unearthed signs 50 centuries old. In the following paragraphs we will leave aside the fascinating history of Egypt and most of its achievements and will only mention briefly their mathematical and astronomical achievements.

Egyptian mathematics was developed as an answer to a variety of practical problems, such as business transactions and land measurements (the flooding of the Nile regularly erased property markers, so that property boundaries had to be re-determined year after year). Because of these underpinnings it is likely that for these ancient mathematicians a number always signified a specific collection of objects, without any abstract meaning. We know that they invented accurate methods for arithmetic computation because examples have survived in several papyri, but the calculations were often tedious, for example to obtain $41 \cdot 59 = 2419$, nine operations had to be performed (all additions and subtractions). Still they were able to calculate areas and estimate the number π (Figure 2.1).

The ancient Egyptians, as their Babylonian contemporaries, found it necessary to invent an accurate calendar for religious and agricultural reasons; the regular flooding by the Nile provided a very strong motivator, for people needed advanced warning in order to prepare for the event. Early on, Egyptian star-gazers noted that such floods occurred every 365 days, at the time when the star we now call Sirius reappeared in the night sky; this suggested a calendar of 365 days, which they adopted. Still this is not quite accurate, for the year is closer to 365¼ days (that is why we have leap years), and so the Egyptian calendar fell behind one day every 4 years. After 730 years the calendar was off by 6 months and only after 1,460 years did this calendar again agree with the physical year. Egyptian astronomers were aware of this problem, but were unwilling or unable to provide a systematic fix. Presumably they shifted various observances as the need arose.

The Egyptians used a method for keeping time using water clocks or *clepsydrae*, and every effort was made to maintain their accuracy. In fact, the oldest known water clock dates from the reign of Thutmose III (about 1550 BC) and is now in the Berlin Museum. But Egyptian astronomy progressed no further. Apparently there were no attempts to systematize or model the workings of

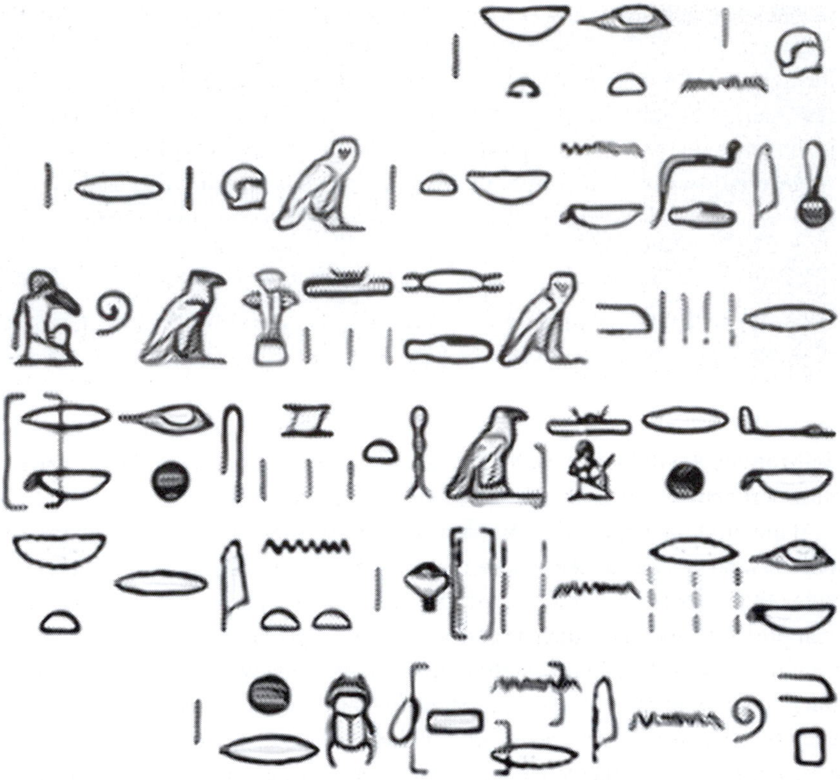

Figure 2.1 The Hieratic version of the Moscow papyrus. Example of calculating (the surface area of) a basket (hemisphere). You are given a hemisphere with a mouth of 4 + 1/2. What is its surface? Take 1/9 off of 9, the basket is half an egg (hemisphere). You get 1. Calculate the remainder (when subtracted from 9), which is 8. Calculate 1/9 of 8. You get 2/3 + 1/6 + 1/18. Find the remainder of this 8. After subtracting 2/3 + 1/6 + 1/18, you get 7 + 1/9. Multiply 7 + 1/9 by 4 + 1/2. You get 32. Behold this is its surface! You have found it correctly. (from: *The crest of the Peacock: Non-European Roots of Mathematics*, George Gheverghese Joseph)

the heavenly bodies, and their view of the constitution of the heavens was simplicity itself: they believed the sky was a roof placed over the world supported by columns placed at the four cardinal points. The Earth was a flat rectangle, longer from north to south, whose surface bulged slightly and with Egypt and the Nile as its center, naturally). On the south there was a river in the sky supported by mountains, and on this river the Sun god made his daily trip (this river was wide enough to allow the Sun to vary its path as it is seen to do). The sky spanned the top of the box like an iron ceiling (flat or vaulted depending on the narrator). From this ceiling hung lamps, held by strong ropes, distributed randomly: these were the stars, whose motions were apparently ignored. Four strong columns, connected by a continuous mountain chain, kept the sky aloft (though their sturdiness was a source of sporadic worry). Such a world was supposed to be the creation of the gods, but there is no single version of this myth; an example is provided in the appendix to this chapter.

India

It would be ludicrous to even pretend to summarize the scientific achievements of the Indian civilization in such a small space as this; so, as for the other

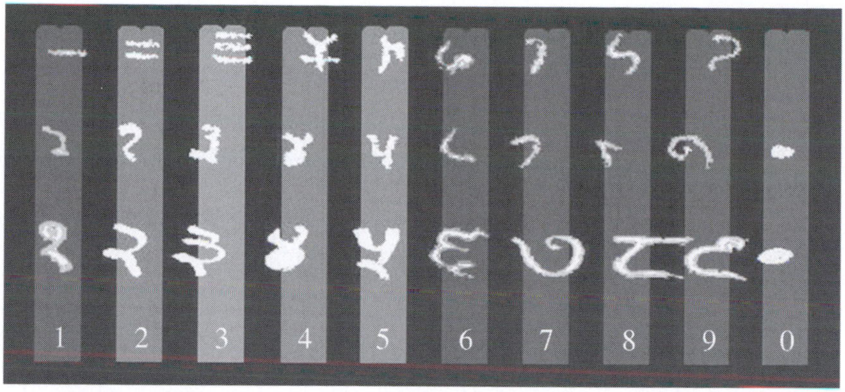

Figure 2.2 The evolution of Hindu numerals. Top, Brahmi (300 BC); middle, Gwalior (AD 876); bottom, Devanagari (eleventh century AD). (from: *Mathematics*, David Bergami and the Editors of LIFE (Time Inc., 1963, New York); image by M. Saris)

civilizations, I will limit myself to an enumeration of some of the developments most pertinent to our ideas of space, time, and cosmology.

At the time when the civilizations of Mesopotamia and Egypt prospered, early Indian cultures, such as the one in Mohenjo-daro, flourished. These peoples, as most of their contemporaries, were satisfied with the mathematics needed to deal with practical matters – such as measuring the height of altars. Yet, as in other parts of the world, religious and practical needs soon forced the creation of a much more sophisticated science. By AD 499 Aryabhata, the most famous of the early Hindu astronomers and mathematicians, was able to discuss in verse (!) such tender topics as quadratic equations, trigonometric functions and the value of π. At that time Indian mathematicians were already conversant in algebra (which apparently was developed almost independently in India and Greece), in the decimal system, and the positional use of the zero. These ideas were later exported to China, the Islamic world (during the Muslim domination) and from there eventually to Europe. It is also to Indian mathematicians that we owe the so-called "Arabic" numerals, which were devised before 256 BC (Figure 2.2). Some areas of Indian science and mathematics, developed as a result of Greek influences – Alexander conquered the north of India in 329 BC; but many of the developments mentioned above were quite original.

Indian astronomers were able to calculate the diameter of the Moon, to predict Lunar and Solar eclipses as well as the motion of the major stars. In a remarkable anticipation of the Copernican system Aryabhata stated that "The sphere of stars is stationary, and the Earth, by its revolution, produces the daily rising and setting of planets and stars." His successors, however, rejected these ideas. It is also worth mentioning that there were advocates of an atomistic hypothesis contemporary with Democritus (though the dates are highly uncertain), and with essentially the same ideas: the whole of what we see is made of small atoms separated by a void. As is the case in many such prescient predictions, they were presented without any experimental backing; this and their apparent opposition to "common sense" observations made most people disregard them.

Mesoamerica before Columbus

Of the rich variety of cultures that flourished in Mesoamerica, the Mayan carries the distinction of having had the most developed mathematical system, which they used very successfully in developing a very accurate calendar. Still, despite these achievements one must keep in mind that all the predictions made by these ancient star-gazers were based on numerology, without the benefit of a clear picture of how the Solar System, let alone the Universe, works.

A rough picture of the world among these cultures consisted of a flat Earth shaped as a square, each corner associated with a cardinal point, and with a specific color and deity. These gods also served as columns for supporting the sky, which was represented as a double-headed serpent whose skin features included the stars and planets; the sky was believed to contain 13 layers, each one also with its own ruling god or goddess. This whole Earth–sky contraption was sometimes pictured as resting on a pool of water surrounded by water lilies.

Below ground was the underworld, a nine-layered horror where most souls dwelled. As for the sky, each infernal level had its own godly ruler. Usually a dark and dismal place, it was enlivened by the Sun and Moon that traversed it in order to appear in the east at the appropriate time every day. In some cases the underworld was represented as a nine-layered inverted pyramid, whose base matched that of its 13-layered heavenly counterpart, with our world being the place where these two pyramids touched.

Among the variety of these Mesoamerican cultures the Mayan had, by far, the most sophisticated calendar with the mathematics needed to develop and interpret it. The Mayan mathematical alphabet consisted of only three symbols: , •, and – representing the numbers 0, 1, and 5 respectively (sometimes an

Box 2.1

Thales of Miletus (624–546 BC)

Was born and died in Miletus, Turkey; traveled extensively throughout the Orient. He was the first known Greek philosopher, scientist and mathematician, and the first natural philosopher in the Milesian School, believed to have been the teacher of Anaximander. Though none of his writing has survived, he is credited with five theorems of elementary geometry that he applied in practical matters: (1) a circle is bisected by any diameter; (2) the base angles of an isosceles triangle are equal; (3) the angles between two intersecting straight lines are equal; (4) two triangles are congruent if they have two angles and one side equal; (5) an angle in a semi-circle is a right angle. Though many legends have been collected around Thale's name, still it is generally accepted that he predicted a Solar eclipse in 585 BC.

additional symbol for the number 20 was added). The value of any symbol in a numerical expressions depended on its position,

represented 115 (the ordering was from top to bottom, not left to right). With this tool the calendar devised became quite accurate; it consisted of a 365-day year without explicit corrections for leap years (though it is believed some ad-hoc fixes were implemented to deal with this problem).

The nightly motion of the stars and the variations of the stellar field with the seasons were considered of great importance by most Mesoamerican cultures, but especially among the Aztecs. They believed that the heavenly regions were the kingdoms of the gods Quetzalcoatl, the Sun god (in one of its forms), and of Tezcatlipoca, the personification of evil. Regular stellar motions were paramount for the continuation of this world, and any disruption was the harbinger of disaster. Most worrisome was the possibility that the stars we know as the Pleiades would stop in their tracks, in which case our world would come to an untimely end. To prevent such a disaster and to propitiate a smooth stellar path, the ceremony of the "new fire," with its fresh human sacrifice, had to be performed at the appropriate time in the prescribed manner.

In all Mesoamerican cultures the task of obtaining the pertinent observations, cataloging them and then extracting their regularities fell on the priestly caste, for which purpose they had but a few primitive astronomical instruments. Despite this, in some cases they reached exquisite accuracy in their computations. Mayan astronomers were able to measure the lunation period to 0.002 percent accuracy, and the period of Venus to 0.02 percent. In addition there are indications in the Dresden codex that they followed the motions of Jupiter and were able to predict its future positions. This ability to predict the behavior of the heavens gave enormous prestige and power to the priestly class, and cemented their preponderance in all aspects of political and social life.

China

The mathematical and astronomical achievements of ancient China are summarized in the *Zhou Bi*, a collection of ancient texts on these subjects compiled during the Western Han dynasty around 100 BC. In it there is a description of a cumbersome but accurate calendar for a year 365¼ days long, and a measurement of the lunation period (accurate to a part in a thousand). The mathematical tools presented include Pythagoras' theorem, various ways of measuring areas and an approximation of the number π by 3 (this was significantly improved later on). Early Chinese mathematics presents us with the first known mention of a negative quantity.

Technologically this society produced a variety of important discoveries such as gunpowder and paper; the compass is also attributed to Chinese ingenuity. According to various historians, in the first millennium BC the Duke of Chou

Box 2.2

Pythagoras of Samos (560–480 BC)

Born Samos, Greece, died in Italy. Pythagoras was a Greek philosopher responsible for important developments in mathematics, astronomy, and the theory of music (though of his actual work nothing is directly known). He founded a philosophical and religious school in Croton that had many followers. This school practiced secrecy and communalism making it hard to distinguish between the work of Pythagoras and that of his followers.

presented certain foreign ambassadors with chariots that were to be used on their way back home, and, we are told, were equipped with certain "south-pointing needles" for the ambassadors not to lose their way. This is the first mention of what must have been the use of magnetic rocks or loadstones for guidance purposes. This knowledge was then apparently lost in the vast history of the region, to be rediscovered around the first century AD by the astronomer Chang Heng; but not until the twelfth century is the compass needle mentioned as a useful tool for mariners.

In ancient times there was but a vague view on the shape and form of the Universe, merely associating a square with the Earth and a sphere with the sky. This simple view evolved over the two centuries from 100 BC to AD 100 into a Universe composed of a stationary Earth resting on water, and a variety of heavenly bodies that circled it regularly (similar descriptions were adopted throughout Eastern Asia). Independently of the picture of the world, China has had a very long history of astronomical observations reaching back to the thirteenth century BC. They noted Solar eclipses as well as supernova events (exploding stars). The most impressive of these events was the observation in AD 1054 of such a supernova event that lasted for two years, after that the star dimmed and disappeared from view. The astronomical observations were sufficiently precise for later astronomers to determine that the location of that exploding star is now occupied by the Crab Nebula; it is now known that this nebula is a violently expanding cloud of gas and debris and, extrapolating backwards, that this expansion began after the cataclysmic destruction of a star roughly ten centuries ago.

The rise of individualism

Myths are the ancient seeds of modern cosmology, not because some of the specific ideas there contained have survived through the millennia, but because they are the first indications of a desire to understand and describe in a rational way the world around us; the slow but steady development of modern cosmology

Box 2.3

Anaxagoras of Clazomenae (499–428 BC)

Born in Ionia of Greek origin, lived most of his life in Athens. He was imprisoned for claiming that the Sun was not a god and that the Moon shines by reflecting sunlight. While in prison he tried to solve the problem of squaring the circle, that is, constructing with ruler and compasses a square with area equal to that of a given circle (this is the first record of this problem being studied). He was saved from prison by Pericles but had to leave Athens.

stemmed from the mythological tradition. The first steps in this direction were carried out by generations of pioneering Babylonian and Egyptian (and to a certain extent, Indian) astronomers and mathematicians, whose names, however, are now lost to us.[2] But this is not the case for more modern philosophers and scientists, and so the achievements of Greek science will be associated with specific individuals; first in a rather vague way, later in precise writings that have survived the 20 centuries that separate us from them.

The Greeks were apparently the first people to look upon the heavens as a set of phenomena amenable to human comprehension, separate from the fickle whims of the gods. The gods might have remained as primary causes of the phenomena, but the regularities of the phenomena themselves were assumed to proceed according to impersonal laws. Knowing these, the behavior of objects could be predicted. They were able to extract a great amount of information using nothing but very elementary observations, existing data, reasoning based on logic (a field they also invented), and the imagination to create new hypotheses.[3] One of the most dramatic of these results is attributed to Thales of Miletus, who used the careful data on Solar eclipses gathered by Babylonian astronomers to *predict* an eclipse that was observed in Asia Minor in 585 BC. This feat was the more remarkable given that he had no concrete description of what caused the eclipse, let alone one for the motion of planets and satellites around the Sun. Several decades later, Anaxagoras noted that fallen meteorites are made of molten iron, inferring that these bodies are made of the *same* materials as those found on Earth. He then came to the startling conclusion that *all* heavenly bodies have this property, and that the materials we see around us are also the building blocks of the whole Universe. Anaxagoras supposed that the reason for this commonality is that the heavenly bodies came from Earth, and were spun off due to the Earth's rapid rotation. He also conjectured that lunar eclipses occur when the Earth happens to be between the Moon and the Sun, and concluded that the Moon shone only by reflecting sunlight.

The fact that some of the proposed ideas were wrong does no discredit to their originators; they were still *bonafide* hypotheses subject to verification and amenable to falsification and provided the first steps towards a deep understanding of Nature. These achievements, however, were not the result of an altruistic thirst for knowledge, but of an evolutionary process motivated (as for Egyptian and Babylonian science) by very practical desires, such as the need to improve and simplify commerce and land ownership. The early Greeks were tireless sailors, who traded in most of the Mediterranean ports and often beyond those; in these travels they came into contact with various aspects of the civilizations they visited, and in this way they learned their astronomy and obtained whichever artifacts had been devised for better and safer navigation.

Before proceeding with the momentous development achieved by the Greeks, it is worth noting that many results are labeled as "attributed" to the philosopher in question, with the possibility remaining that it was actually the result of one of his students. This is due in part to the great span that separates us, during which many manuscripts have been lost, so that some achievements reach us only thorough oral tradition. In addition, these ancient researchers loathed to go through the menial task of writing their results in a coherent way, and left such activities to their followers.

Early Greek cosmology

Almost from its inception, Greek science was based on the deductive method. This facilitated a description of the world based on a unifying philosophical set of principles, but, as I mentioned previously, this very generality is prone to very serious errors. Despite this, deduction was almost universally accepted as the standard approach to science until the late Renaissance. Mythology certainly had a very strong influence on the first descriptions of the world devised by the early Greeks. The basic picture was that of a disk with solid ground in the middle, surrounded by an ocean; below this disk was Tartarus (where the spirits of all dead humans resided forever), while above the disk was a region of air and, beyond that, the ether where stars dwelt. Thales subscribed to this view, adding his belief that the whole contraption rested on a cosmic ocean, and that the water from this ocean was the primordial substance, giving rise to all other substances through different "condensations." This picture of the world was not universally accepted, for example, Parmenides believed in a spherical Earth. Anaximander believed the Earth to be a cylinder floating in space surrounded by a series of spheres made of mist, beyond which lay a great fire; the Sun, Moon and stars were glimpses of this fire through the mist. In a more poetical vein, Empedocles believed the cosmos to be egg-shaped and governed by alternating reigns of love and hate. Such early Greek cosmological theories did explain all the data available at the time, though they made no predictions. Yet, even with these deficiencies, this period is notable for the efforts made to understand the workings of Nature using a rational basis. This idea was later adopted by Plato and is the basis of all modern science.

Figure 2.3 Pythagoras' theorem. The areas of the squares attached to the smaller sides of the triangle equal the area of the largest square.

Further developments in cosmology were very strongly affected by increasingly more careful observations. It was known since Babylonian times that the great majority of objects in the sky are seen to revolve around the Earth with their relative positions fixed, as if they were nailed to an enormous sphere that rotated about our planet. These bodies then came to be known as the "fixed stars," and the sphere to which they were supposed to be attached, the "sphere of fixed stars." Yet, not all celestial bodies were in this way constrained: a few of them wandered about the firmament in regular but mysterious patterns, and these came to be known as planets (from the Greek, *planasthai*, "to wander"). Since the motion of stars was regular and predictable, all efforts were directed at understanding the rules dictating the planetary meanderings about the heavens. At first a god was provided for each of them, who would then decide which route the planet would take (and, apparently, stick to this decision for all time). Later a more rational explanation for their behavior was sought (and found).

So it was that about five centuries BC there lived in Samos, Greece, a philosopher, mathematician and astronomer named Pythagoras, whose school of philosophy was to have a most profound influence on the further developments in cosmology. Pythagoras is best known for the theorem in mathematics that bears his name, and relates the sides of a right-angled triangle to its hypotenuse (Figure 2.3). Though the Babylonians used this result 1000 years earlier,[4] Pythagoras is credited with being the first to prove it (allegedly, for no manuscript remains). Of the many other discoveries attributed to Pythagoras, the observation that there is a relationship between music and mathematics is of particular relevance here. Legend has it that while taking a casual walk,

Pythagoras happened upon a blacksmith at work; tarrying a while he remarked that two anvils when struck at the same time sometimes produce agreeable sounds that matched those heard in harmonic chords. Upon investigation he noted that this happened when the anvils were made of the same material, and if the ratios of their weights were certain rational fractions (for example, if one was twice as heavy as the other the sounds were an octave apart). He then showed that similar results can be obtained for vibrating strings (if equally tensed, they produce harmonious tones when the ratios of their lengths are whole numbers). This discovery of a relationship between a physical effect (the harmonizing of sounds) and mathematics (the ratios of weights) impressed Pythagoras enormously and lead him to the daring and momentous conclusion that Nature and mathematics are intimately related. It is from this realization that we inherited the very fruitful belief that mathematics is the language of Nature, a concept that has proved central in the development of both science and mathematics.

Later Pythagoras and his followers expanded this idea into the statement that Nature and number are not only related, but in fact one and the same. From this they claimed (but failed to prove) that the Universe could be understood in terms of whole numbers, with special importance given to a few of them: a point was associated with one, a line with two, a surface with three, and a solid with four. Their sum, ten, was sacred and omnipotent.[5]

The Pythagoreans inherited (apparently from Anaximander) the idea that certain geometrical figures, and more specifically, circles and spheres, must play a central role in the description of the heavens. This presumably came from a belief that perfection, as defined in terms of circles and spheres, was an essential component of the motion and shape of planets, moons and stars. In one way or another, this idea survived 21 centuries, until Kepler's description of planetary orbits and Galileo's observations of the Moon's surface in the 1600s. Thus the Pythagoreans burdened all heavenly bodies with the quality of perfection,

Box 2.4

Democritus of Abdera (*c.* 460–370 BC)

Best known for his atomic theory, he was also an excellent geometer. Very little is known of his life except that Leucippus was his teacher. He's believed to have traveled widely, perhaps spent a considerable time in Egypt, and he certainly visited Persia. Democritus wrote many mathematical works but none of his original manuscripts has survived. He claimed that the Universe was a purely mechanical system obeying fixed laws, that space is infinite and eternal, and that it contains an infinite number of atoms. He also explained the origin of the Universe through atoms moving randomly and colliding to form larger bodies and worlds. Democritus' philosophy contains an early form of the conservation of energy.

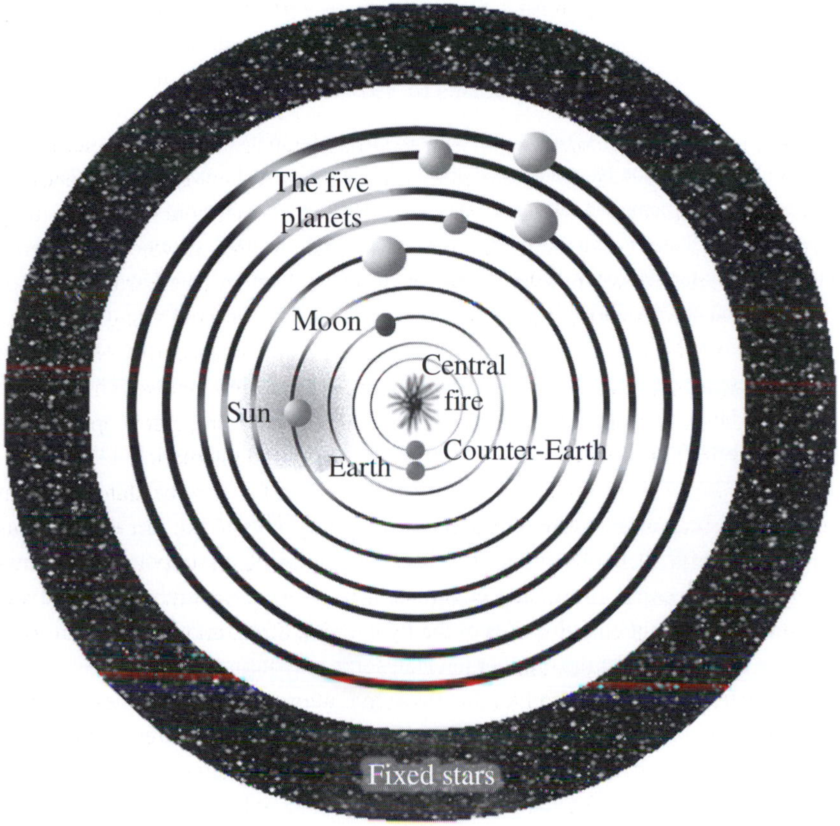

Figure 2.4 The Pythagorean universe.

requiring them to be perfect spheres moving on perfect circles, and, setting aside the fixed stars, numbering precisely ten, which was also deemed a perfect quantity.[6] They knew, however, of only eight such celestial bodies: the Sun, Moon and Earth, and five planets (Mercury, Venus, Mars, Jupiter and Saturn). Then, in modern language, the Pythagoreans used their hypothesis to *predict* the existence of two other celestial bodies: a "central fire" and a sixth planet. This fire was supposed to reside at the center of the Universe, around which all other bodies revolved – it was different from the Sun, and in fact was believed to have engendered all that we see. The sixth planet, the "*counter-Earth*," was supposed to move in perfect harmony with our planet, in such a way that it perpetually blocked the central fire from our sight, while at the same time remaining invisible to us (Figure 2.4).

Whirling about the central fire at enormous speeds, the planets and the sphere of fixed stars were supposed to produce sounds, "the harmony of the spheres." We could not easily discern this ethereal music, since we were used to it from childhood (a sort of perpetual harmonious cosmic hum), though we would certainly notice if anything went wrong! The Pythagoreans also included music into the unity of Nature so that numbers, cosmos and music were believed to be one and the same object.

Box 2.5

Empedocles (*c.* 492–432 BC)

Born in Acragas (now Agrigento; a center of culture up to its destruction by the Carthaginians in 406 BC) into a rich and aristocratic family, Empedocles traveled extensively participating fully in the passion for learning prevalent at the time. Aristotle attributed to him the invention of rhetoric, and Galen that of medicine in Italy. Best known for his theory that all objects are made of a combination of four elements, he also had other remarkable ideas including a rudimentary theory of evolution based on the survival of the fittest and a form of the law of conservation of energy. There are many legends regarding his life, stemming mostly from his own poems, where he endows himself with god-like powers (it is not known whether this was an issue of style or belief). It is from these works we get the most direct impression of the man.

Over the centuries astronomers have diligently compared the Pythagorean hypothesis with their observations, without obtaining any supporting evidence for either the central fire or the counter-Earth, so we can safely disregard this hypothesis. This not only disposes of the Pythagorean description of the Universe, but also of the idea that the number ten is in some way intimately connected to our Solar System. Of course, the fact that these hypotheses have little scientific merit does not imply that they are of no historical importance. The following discussion will, I hope, make it amply clear that the Pythagorean ideas concerning the perfect quality and motion of the heavenly bodies were destined to have a profound and lasting effect on the development of astronomy and cosmology.

The composition of things

Greek philosophers proposed all manner of hypotheses concerning the origin and composition of matter. For the most part these were, like their early cosmologies, descriptive, with little or no predictive power; the specific mechanism by which the basic building blocks of Nature engendered the prodigious variety of objects was left unspecified. Some proposals, like the atomistic idea, were tantalizingly close to some modern views, while others had distinct religious overtones.

One school of thought held that there was a single primordial substance that engendered all we see through various processes; Thales believed it to be water, Anaximenes to be air, and Heraclitos and also the Pythagoreans to be fire. In contrast, a different school, represented by Anaximander, believed in the existence of an infinite number of basic "building blocks." A third school, whose exponents were Xenophanes, Parmenides, and Zeno, proposed a mystical/religious view of the origin of matter, claiming the existence of an abstract and absolute principle that was responsible for the origin of the Universe; such a principle could be reached by reason alone, it permeated the Universe and was one with it.

But the most popular view, commonly believed to have been first proposed by Empedocles, was that all substances were made of a combination of four elements: fire, air, water and earth; a belief that survived 2000 years. Despite its popularity and longevity, this notion was not universally accepted. Most notably, and during Empedocles' lifetime, Democritus and Leucippos proposed a radically different possibility: that all types of matter could be understood as different combinations and arrangements of discrete indestructible particles made of a primordial substance and in the shape of geometrical figures, the now famous atoms (from the Greek *a*, "not" + *thomos*, "cut"). The atoms constituting any one body were closely tied together, but did not necessarily touch; in between there was a void.[7] Aristotle carefully examined these contrasting views, and conclusively favored the four-element hypothesis; he noted that there was no evidence of the graininess of matter implied by the atomistic view and, in addition, he saw fundamental problems in the existence of a vacuum in between atoms.[8]

Heliocentrism in antiquity

The rotation of the celestial sphere adorned with stars, the wandering motions of planets, and especially the daily trips of the Sun and Moon, strongly suggest that all these bodies do precisely what is observed: they move around the Earth. Now we know this is but an illusion caused by the rotation of our planet around its axis as it travels around the Sun, a description commonly ascribed to Copernicus that has been verified to exquisite accuracy. It is then quite remarkable that 17 centuries before Copernicus, and based on basic geometry (and a certain amount of common sense), Aristarchus of Samos proposed what is essentially the Copernican system.

Box 2.6

Aristarchus of Samos (310–230 BC)

Born and died in Greece. A mathematician and astronomer celebrated as the exponent of a heliocentric Universe, and for his pioneering attempt to determine the sizes and distances of the Sun and Moon. He was a student of Strato of Lampsacus, head of Aristotle's Lyceum, coming between Euclid and Archimedes. Little evidence exists concerning the origin of his belief in a heliocentric system; the theory was not accepted by the Greeks and is known only because of a summary statement in Archimedes' *The Sandreckoner* and a reference by Plutarch. His only surviving work, *On the Sizes and Distances of the Sun and Moon*, provides a remarkable geometric argument, whereby he determined that the Sun was about 18.9 times as distant from the Earth as the Moon, and 18.7 times the Moon's size. Both these estimates were an order of magnitude too small (due to instrumental inaccuracies, not faulty reasoning). He also accurately determined the length of the Solar year.

Using a clever argument Aristarchus estimated the size of the Sun and concluded it must have a radius about 6.75 times larger than Earth's (the actual number is about 108.9, the discrepancy is due to inaccuracies in his observations *not* due to incorrect reasoning). He then argued that it was inconceivable that such a behemoth would slavishly circle a puny object like our planet. Having assumed this, he concluded that the Earth must rotate on its axis in order to explain the (apparent) motion of the stars. Unfortunately Aristotle soundly rejected these ideas, and provided such weighty arguments against them that they were all but forgotten until the Copernican revolution of the sixteenth century.

Aristotle's arguments against the heliocentric model are not trivial. For example, if the Earth rotates, how is it that a stone thrown upwards falls on the same place? For, during the stone's trip in the air, the Earth should move from under it (were it rotating). When riding a cart pulled by a fast horse, the driver feels a wind in his face. How is it then that the Earth's rotation does not generate such a wind? The proponents of the heliocentric model were unable to answer these objections and thus the hypothesis was rejected. So the matter lay until Copernicus and then Galileo answered these criticisms.

The shape and size of things

Among the many beautiful and elegant mathematical achievements of the Greek civilization, geometry has proved to be one of the most enduring and useful. Since early in the fourth century BC we have records of applications of geometry to a variety of situations: Thales used the properties of similar triangles to determine the heights of objects by measuring their shadows, Aristarchus determined the relative sizes of the Earth, Moon and Sun, and used these results to support his heliocentric theory. But perhaps the most accurate of such calculations was done by Erathostenes of Cyrene, who around 200 BC determined the size of the Earth to within a few percent. The argument is so simple and elegant that it is worth a slight detour.

Eratosthenes noted that at noon on every equinox (when day and night have the same length on the equator) the Sun's light reached the bottom of a deep well in Syene (now Aswan, Egypt), but that this did not happen on any other time of the year. He reasoned that this happened because on this day the Sun was straight overhead the well. But then he noted that on that same day and at that same time, in Alexandria, a city directly north of Syene, a tall column *did* cast a shadow, which then implied that the Sun was *not* straight overhead there. This raised a puzzling question: how could the Sun be straight over one city and not over another? The answer is that this happens because the Earth is curved approximately like a sphere.

By measuring the angle at which the Sun's rays rained down on Alexandria, Eratosthenes determined the fraction of the Earth's circumference that separated these cities (Figure 2.5). He then only needed to measure the distance from Syene to Alexandria in order to determine the size of the Earth. Now we know

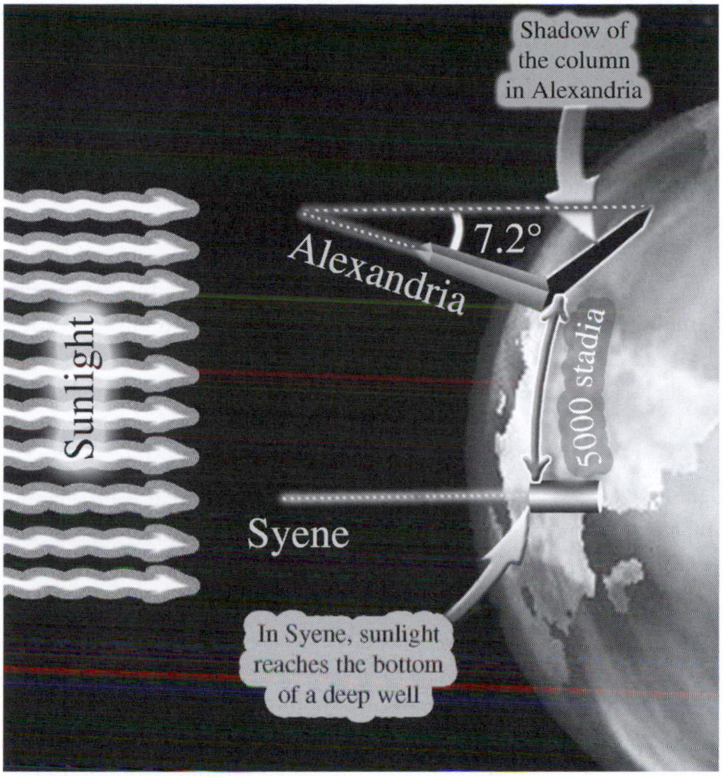

Figure 2.5 Eratosthenes' procedure for measuring the Earth was based on the fact that columns at different locations will cast different shadows due to the curvature of the Earth. He measured the angle of the shadow cast by a column at Alexandria on the equinox (when a similar column casts no shadow on Syene, which is on the equator); *assuming* that the Earth is a sphere, the two cities are 7.2° apart. Of the 360° needed to travel around the world, the two cities are separated by 360/7.2 of that distance, which he knew equals about 500 miles; he concluded that the Earth's circumference must be roughly 25,000 miles.

Box 2.7

Eratosthenes of Cyrene (276–197 BC)

Of Greek origin, lived most of his life in Alexandria. He was born in Cyrene, which is now in Libya. He worked on geometry and prime numbers and is best remembered for his prime number "sieve," which, in modified form, is still an important tool in number theory research. Eratosthenes measured the tilt of the Earth's axis with great accuracy, and compiled a star catalog containing 675 stars. He also suggested that a leap day be added every fourth year and tried to construct an accurately-dated history. In his old age he became blind and is said to have committed suicide by starvation.

that this distance is about 500 miles, but 23 centuries ago measuring this number was a non-trivial enterprise. Eratosthenes accomplished it by convincing King Ptolemy to let him borrow the royal "walkers," men trained to take steps of the same length even after walking many a weary mile. In this way he learned that 5,000 stadia separated Alexandria from Syene (1 stadium = 157.5 meters), and concluded that the Earth's circumference measures about 254,000 stadia, or 25,000 miles; correct to within two percent!

It is important to remember that the realization that the Earth was round was not lost to the following centuries; neither Columbus nor any of his (cultivated) contemporaries had any fear of falling off the edge of the world when traveling westward trying to reach the Indies. The controversy surrounding Columbus' trip was due to a disagreement on the *size* of the Earth. Columbus had, in fact, seriously underestimated the radius of the Earth and so believed that the tiny ships he would command had a fair chance of getting to their destination at the other side of the globe. He was, of course, unaware of the interloping piece of land we now call America; had this continent not existed, Columbus and his crew would probably have perished miserably in the middle of the ocean.

From Aristotle to Ptolemy

There are few instances of philosophers that have had such a deep influence as Aristotle, or of cosmologists whose theories have endured as long as Ptolemy's. Aristotle's influence is hard to underestimate, for two millennia philosophers and scientists followed the path laid by him, ignoring, mocking and often condemning those who strayed. His ideas defined Greek philosophy from the third century BC on, they prevailed through the Roman empire, were translated into Arabic in the ninth century, and compelled Moslem thinkers to seek a reconciliation between his thought and that of Islamic doctrine. In the twelfth century thinkers such as Maimonides and Ibn Daud strived to harmonize Aristotelian philosophy with Judaism; and in the thirteenth century numerous Christian philosophers of the stature of Thomas Aquinas attempted the same synthesis with the Catholic theology. Surely Aristotle would be amazed and pleased to see so many world-shaking faiths pay homage to his thought![9]

The span of Aristotle's work is enormous, ranging form logic to biology. Many of his ideas have endured the test of the centuries, but his cosmology, based on a geocentric system, is not one of them; in the words of W. Durant:

> His curious mind is interested, to begin with, in the process and techniques of reasoning; and so acutely does he analyze these that his Organon, or "Instrument" – the name given after his death to his logical treatises – became the textbook of logic for two thousand years. He longs to think clearly, though he seldom, in extant works, succeeds; he spends half his time defining his terms, and then he feels that he has solved the problem.

It must be noted, however, that at least where science is concerned, Aristotle's arguments are, for the most part, logical and based on reason, accepting or rejecting hypotheses based on the predictions they produced. For example, he argued forcefully that the Earth is spherical, noting that only a spherical Earth can account for the shadow seen on the Moon during lunar eclipses. He also

Box 2.8

Aristotle (384–322 BC)

Born Stagirus, Greece, died Chalcis, Greece. In 367 he entered Plato's Academy in Athens and soon became a teacher there. After Plato's death in 347 BC, Aristotle joined the court of Hermias of Atarneus; in 343 BC he became tutor to the young Alexander the Great at the court of Philip II of Macedonia. In 335 BC Aristotle founded his own school, the Lyceum in Athens. The Academy had become narrow in its interests after Plato's death, but the Lyceum under Aristotle pursued a broader range of subjects with prominence to the detailed study of Nature. After the death of Alexander the Great in 323 BC, the anti-Macedonian feeling in Athens made Aristotle retire to Chalcis where he died the following year.

Aristotle was not primarily a mathematician but made important contributions by systematizing deductive logic. He wrote on physical subjects; and parts of his *Analytica Posteriora* show an unusual grasp of mathematics. He also had a strong interest in anatomy and the structure of living things in general, which helped him to develop a remarkable talent for observation.

considered, and rejected, Aristarchus' heliocentric system because it could not answer a series of objections that seemed to follow from the assumption that the Earth rotates. In modern language, Aristotle rejected a hypothesis whenever its predictions were not confirmed by observations. Of course, such predictions were obtained using Aristotle's paradigm of the world around him, and were therefore sometimes incorrect; in the case of the heliocentric hypothesis they were made based on erroneous ideas about the properties of motion. So Aristotle came down squarely in favor of the geocentric model of the world, and humanity followed him for 20 centuries.

Ptolemy enlarged Aristotle's cosmological ideas creating a very involved model of the Solar System in which the whole of the Universe rotated around the Earth. This insistence on a geocentric universe came not (at least, not originally) out of blind faith or sheer pig-headedness, but was supported by the fact that this was a model that described all observations with great accuracy, despite it being so awkward and complicated. It was the increase in observational accuracy that eventually rendered this model untenable, and favored the heliocentric description of the Solar System.

The motion according to Aristotle

One of the fundamental propositions of Aristotelian philosophy is that there is no effect without a cause. Applied to moving bodies, this proposition dictates that there is no motion without a force. Specifically, Aristotle proposed that the speed should be directly proportional to the force applied and inversely proportional to any resistance to the motion:

$$\textit{force} = (\textit{resistance}) \times (\textit{speed})$$

The original reasoning equaled resistance with the object's mass, but it was also associated with a property of the medium in which the object moves. This is because a body will take a shorter time to traverse a thinner medium than a thicker one of the same length: things will go faster through air than through water, as verified by direct experience. However, Aristotle did not provide a *quantitative* relationship between the properties of the medium and the "resistance," nor did he prescribe any way of measuring or quantifying forces. Hence the equation presented above could not be used in numerical predictions.

The notion that a force is needed for an object to maintain any fixed speed is, apparently, quite reasonable. Take, for example, the case of an ox pulling a cart: the cart only moves if the ox pulls, and when the ox stops pulling, the cart stops. To all appearances this stoppage is not caused by anything acting on the cart, it merely slows down and eventually stops of its own accord. So this suggests that when forces are absent the speed will vanish, in agreement with Aristotle's rule of motion. And yet, despite this apparent verification, most of Aristotle's ideas concerning motion are quite wrong, in particular, the above relationship between force and speed is incorrect. A correct understanding of motion had to wait almost 2000 years and was the result of the experiments and insights of Galileo and the genius of Newton.

Aristotle also noted that not all motion is of the ox-pulling-the-cart type. For example, if the cart is dropped (with or without the ox) from a great height, it will move downwards with ever increasing speed and will apparently not stop unless impeded by something (such as the Earth's surface). This motion again appears to occur by itself, without anything acting on the cart.

Such considerations led Aristotle to conclude that the second type of motion must in some way be natural to the cart since – apparently – it happens by itself, without any coaxing. On the other hand, the first type of motion is the result of the ox's efforts, and is unnatural or forced in the sense that the cart will not pursue it if left to its own devices. Generalizing, he proposed that *all* objects have *intrinsically* a natural state of motion, which they will perform unless acted upon by some external agency; this natural motion is *not* the result of some external effect acting on the body, but a property of the object itself. Thus a lump of earth naturally moves downward while a flame naturally moves upward.

Aristotle then combined this idea with his belief in Empedocles' hypothesis that all objects are made from various combinations of four elements, and concluded that each of these elements has a natural motion embedded in its nature: earth towards the center of the world, water floating on the surface of the Earth, air hovering over the water, and fire towards the heavens. Different proportions of these elements in a given object determine then its natural motion:

Box 2.9

Aristotle's law of motion (from *Physics*, book VII, chapter 5)

*"Then, **F** the movement have moved **M** a distance **D** in a time **T**, then in the same time the same force **F** will move ½**M** twice the distance **D**, and in ½**T** it will move ½**M** the whole distance for **D**: thus the rules of proportion will be observed."*

In modern terms we have:

- ***F***: a force acting on an object
- ***M***: the mass of the object
- ***D***: the distance traveled by the object
- ***T***: the time needed to travel this distance

Then the statements translate into:

- The distance is determined by the force ***F***, the mass ***M*** and the time ***T***.
- Given a force ***F*** that moves a mass ***M*** a distance ***D*** in a time ***T***, it will also move half the mass by twice the distance in the same time.
- Given a force ***F*** which moves a mass ***M*** a distance ***D*** in a time ***T***, it will also move half the mass the same distance in half the time.

These three rules imply that the product of the mass and the speed depends only on the force. In particular, the larger the mass the smaller the speed (for a fixed force), so that mass measures the "resistance" to motion.

It follows from this hypothesis that a body of constant mass under the action of a constant force will have a constant speed. This is *wrong* (the speed will increase with time), though the technology available at the time was probably incapable of demonstrating this fact (had Aristotle thought of doing the experiment).

earth dominates in the composition of a rock and so it will naturally fall; fire dominates in smoke and so it rises, etc.

But not all objects fit in this simple scheme: Aristotle had only to look at the Sun and Moon to realize that these bodies do not move upwards or downwards, nor do they float or hover, but move in *circles* (or something like it) around the Earth. Then, since he fully accepted the idea that *all* bodies must move according to their composition, and since no earthly body moves naturally in circles, he concluded that celestial bodies are made of a substance not found on Earth: the fifth element. Thus the final version of the table of elements according to Aristotle looked like Table 2.1. This fifth element, argued Aristotle, must have markedly different properties from the other four. All objects on Earth suffer change and decay, while heavenly bodies apparently do not (they actually do, but the timescales are usually much too large for these changes to become apparent during one man's life). Therefore Aristotle associated the fifth element

Table 2.1. *Aristotlion table of elements*

Objects on Earth	Earth
	Water
	Air
	Fire
Heavenly objects	The fifth element

with a degree of perfection unheard of on Earth: objects made solely out of it will remain constant and immutable for all time, moving in circles around the Earth, for that is their nature.

Having adopted these ideas as reasonable hypotheses, Aristotle proceeded to draw conclusions (or predictions) from them; here I will mention only two. The first is the impossibility of a vacuum: for, he argued, in a vacuum there is no resistance to any motion, and so a minuscule force will produce an arbitrarily large velocity; but having infinite velocities resulting from arbitrarily small effects is clearly ridiculous, so a vacuum cannot exist. This then led him to reject the atomist hypothesis of Democritus and Leucippus, who imagined geometrical atoms separated by empty space.

A second conclusion concerns the motion of falling bodies. He observed that a falling rock gains speed; this he attributed to a gain in weight, for then the "earthy" content of the rock would increase and its tendency to move towards the center of the Earth would grow, leading to a larger velocity (how this was accomplished was not elucidated). But, if weight determines the speed of fall, then a heavier object will fall faster than a lighter one when dropped from the same height, and the time difference will be in proportion to the two weights; a ten-pound weight would reach the Earth by the time a one-pound weight had fallen one-tenth as far. This conclusion was again consistent with his hypothesis on the composition of things: the heavier object is more "earthy" and, having a smaller proportion of the other elements, there will be less in its nature to detain its motion downwards.

Yet the idea that only a force produces unnatural motion soon ran into trouble. The simplest example is the motion of an arrow: how come the arrow does not simply stop almost immediately after leaving the bow? After all, there is no force acting on it, so its horizontal velocity should quickly vanish (its *vertical* motion should not, for it's natural to it). In order to explain this Aristotle claimed that what happens is that in order to fly, the arrow's tip must continually push air particles away, these particles then travel to the back of the arrow and push it along in their effort to return to their original positions. Eventually this process runs itself down, and the arrow stops and falls; this cessation of horizontal motion was believed by Aristotle to happen rather suddenly (Figure 2.6).

Figure 2.6 Illustration of Aristotle's explanation of an arrow's motion. The particles displaced by the tip return to push the arrow forward. (fairy image courtesy of Mats Öhrman)

Leaving aside the contrived nature of this idea, one can ask why is it that arrows have the pleasure of being impelled by the air they displace, while our old friend the ox-and-cart experience no such effect? Despite such problems, Aristotle's ideas on motion survived the Roman Empire and the Middle Ages until Galileo demolished them with a few simple (but very clear) arguments. Up to that time every treatise on the subject of motion took the above ideas for granted (though there were a few notable dissidents).

Aristotelian cosmology

Aristotle's cosmological work *On The Heavens* is the most influential treatise of its kind in the history of humanity. It was accepted for more than 18 centuries from its inception (around 350 BC) until the works of Copernicus in the early 1500s. In this work Aristotle discussed the general nature of the Cosmos and certain properties of individual bodies; he provided a unified and logical view of the Universe consistent with his philosophy, and motivated by the observations available at the time, as colored by his ideas on motion. This theory of the Universe was able to make predictions that were verified by subsequent observations; it was flexible, in the sense that as new and more precise data on the motion of celestial bodies became available, it could be modified to accommodate it without compromising its basic underpinnings. It was also wrong.

Aristotle's basic idea was that all objects move in reference to the Earth, either up or down or around it: the Earth is the center of the Universe. He came to this geocentric description guided by several precepts of his philosophy, precepts that he considered obvious and inescapable. The two main ones were, first, his hypotheses that the free motion of objects is determined by their composition (their "nature"), and, second, that the motion of heavenly bodies is intimately related to mathematics and, more specifically, to geometry. Observations (even at that time) indicated that heavenly bodies move in circle-like trajectories around the Earth, then, Aristotle argued, they must be made only of the fifth element and, therefore, must also be perfect. Because of this perfection their trajectories must also be perfect and, since the circle is a perfect geometric figure (which is another preconception drawn from his philosophy), heavenly bodies must move in circles around the Earth. In a modern parlance this would be called a prediction to be verified by observation, at the time it was accepted since it was supported by weighty arguments and did not contradict the data.

Since the stars and planets were made of this exalted fifth substance and hence "must" move in circles, it was natural, according to Aristotle, for these objects to be spheres also, another "perfect" geometric object. The Cosmos "must" then consist of a central Earth (which he accepted as spherical) surrounded by the Moon, Sun, and stars all spherical and moving in circles around it; this conglomerate he called "the world." Note the strange idea that all celestial bodies are perfect, yet they must circle the imperfect Earth that is also spherical, despite its deficient nature.

The closest heavenly body to the Earth was supposed to be the Moon, followed by the Sun, Venus, Mercury, Mars, Jupiter, Saturn, and, finally, the stars moving in circles by virtue of their being affixed to their sphere. Aristotle in fact imagined that *all* the other celestial bodies were also fixed to spheres, so that the Aristotelian Universe consisted of a series of ethereal spheres, each carrying one or more heavenly bodies, moving perpetually and majestically about the Earth. Much later these spheres took a more physical character and astronomers began talking about "crystal" spheres; and this became a problem when, eventually, the data required such spheres to intersect: why didn't they shatter?

On the specific description of the heavens, Aristotle created a very involved system containing 55 spheres (!) which, despite its complexity, had the virtue of explaining and *predicting* most of the observed motions of the stars and planets. This model had then all the characteristics of a scientific theory: starting from the hypothesis that heavenly bodies move in spheres around the Earth, Aristotle painstakingly determined their size and rate of rotation using the available observations, until all data could be accurately explained. He then used this theory to make predictions (such as where Mars will be a year from now), which were subsequently confirmed.

The Earth represented the most imperfect part of the Universe: were a modern Odysseus to travel from the Earth to the distant fixed stars, he would witness a

Box 2.10

Ptolemy (AD 100–170)

Born in Ptolemais Hermii, Egypt, died Alexandria, Egypt where he probably spent most of his life. Ptolemy was one of the most influential Greek astronomers and geographers of his time. He made a large number of astronomical observations from Alexandria during the years AD 127–41, and he used this data to construct a geometric model of the Universe based on the Earth-centered system advocated by Aristotle. Using this theory he was able to accurately predict the positions of all significant planets and stars using combinations of circles known as epicycles. This theory became the accepted description of the Universe and survived almost unchallenged for 1400 years. These results were presented in his treatise known as the *Almagest*. In another book, the *Analemma*, he discussed the projection of points on the celestial sphere. In his third surviving work, the *Planisphaerium*, he was concerned with stereographic projection. Ptolemy also devised a calendar that was followed for many centuries despite its inaccuracies (it required corrections of about one month every six years) which generated complications, especially for religious and agricultural pursuits.

gradual transformation into perfection as he reached the Moon. Thereafter he would see only perfect objects composed of the fifth element. In the region beyond the sphere of the fixed stars, the Universe continued into the spiritual realm where material things cannot be. If heavenly bodies were to be perfect they must be eternal, since either their being composed at one point in time or facing decay in the future would contradict this property. Hence Aristotle believed that the world had existed and would continue to exist forever. No dramatic moment of inception by godly effort or flight of fancy generated the world, it simply was, had been, and would be.

This eternal heavenly motion must, argued Aristotle, have a primary cause. There should be an agency that caused the motion of planets and stars, determined their speed, and perhaps also their position. This would be the primordial and ultimate cause of motion, the mover that generated all change in the Universe. By virtue of causing motion in the perfect heavens, this prime mover must itself have been perfect, so all its properties, including its location, should be fixed and unchanging; despite its causing everything else to move, the prime mover did not itself move. Thus Aristotle postulated the existence of a static or unmoved prime mover, lying beyond the sphere of the stars, and whose action caused the sphere of the stars to turn, which motion thereafter "trickled" down to the lower spheres, and generated the majestic dance of the planets and stars around the Earth.

Aristotle also addressed the question whether this world is unique or not, that is, whether there is another privileged spot like Earth in the Universe with its own retinue of celestial objects humbly circling around it. He answered this in the negative: the Earth must be the *unique* center of the Universe. The reason, he argued, was that if it were not so, an earthly object (like a rock) would not know which way to move, should it go down toward our Earth, or should it follow a path towards the other privileged object somewhere in the heavens? But it is a fact that objects do not show such hesitations or confusions, but invariably fall toward *our* Earth, which must then be the unique center of the Universe. One of the great astronomical discoveries of Galileo came from his observation of the Jovian system of five satellites slavishly moving around Jupiter (*not* around Earth). These observations demolished one of the most resilient of the pillars of Aristotelian cosmology: the Earth cannot be the center of the Universe. Aristotle was wrong not in the logic, but in the initial assumptions: things do *not* have a natural motion. Once this was realized the whole edifice proposed by Aristotle and perfected by Hippocrates and Ptolemy collapsed.

The Aristotelian universe provided a coherent consistent paradigm for the understanding of the Universe. When modified by subsequent astronomers, it managed to explain accurately all observations with the precision available up to the seventeenth century. The conclusions derived from this model were used by clergy and layman to regulate the yearly activities; its success was beyond question. And it was this success that created such uproar when this theory was rediscovered by the religious societies of Islam (in the ninth century), and Christian Europe (twelfth century), for there are direct contradictions between the philosophical underpinnings of Aristotle's theory and the various religious descriptions of the Universe and its creation (if nothing else Aristotle demanded an eternal Universe). Yet Aristotle's theory, being so useful and accurate, could not be simply dismissed; some of the best minds of the Middle Ages devoted an enormous amount of effort attempting to reconcile Aristotle with the Biblical scriptures.

Ptolemy

The Aristotelian description of the world was accepted almost universally soon after its inception. Yet, over time, and as more precise observations were made, it became obvious that the theory needed to be modified in order to accommodate the new data that concerned the planets, and in particular Mars. From very early times it was known that planets generally move from east to west, though at rates different from that of the fixed stars; still one could track, night after night, the position of any planet with reference to the fixed stars. Most planets were then seen to gain or lag behind the stars in every observation, but not Mars, which lagged behind the motion of the fixed stars most of the year, but over a period of about three months it reversed direction and started gaining on them. After this period it resumed its previous sedate motion. This behavior is known as the *retrograde* motion of Mars (Figure 2.7).

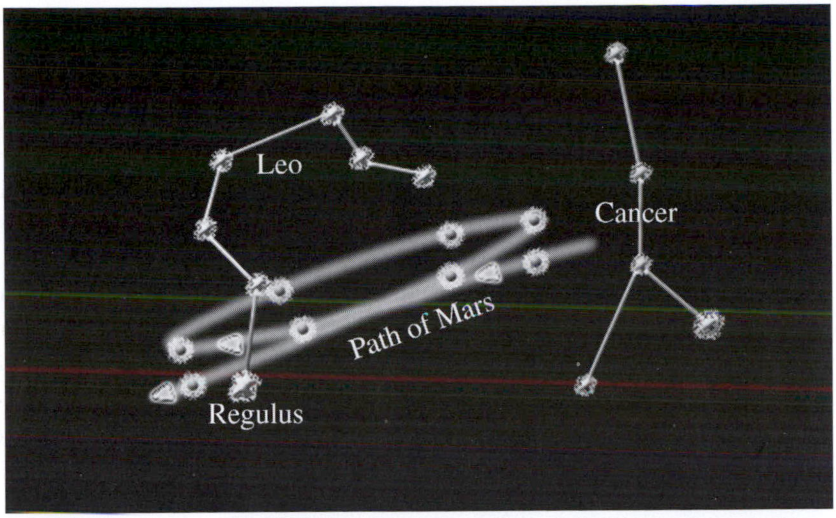

Figure 2.7 Ptolemy assumed that Mars moves in an epicycle; it would then reverse its motion periodically, as observed.

Aristotle's theory, as appealing and satisfying as it was, could not explain this cosmic indecision. One could not, for example, state that Mars simply stopped in its tracks and decided to go the other way, for that would imply a lack of perfection in this planet, which is inconsistent with the philosophy behind the theory and its contents. Thus it became necessary to devise a modification of the Aristotelian model that preserved the ideas associated with the perfection of the celestial bodies, and at the same time explained the motion of Mars.

The necessary modification is attributed to Hipparchus and was popularized and perfected by Ptolemy. The basic idea is to note that, accepting that the planets are to move in circles, one might imagine that the centers of such circles are not fixed, but instead themselves move around circles, and that the center of this second circle is the Earth (Figure 2.8); these piggy-back circles are known as *epicycles*.

The device of having planets move on epicycles retains the idea that planets "must" move in circles, and that ultimately the motion is around the Earth, *and* it explains the data. Yet, with yet more precise data, even the original epicycle model was unable to explain the observations and so, sometimes a third circle was required (the motion of the planet was then described as a circle moving on a circle that moves on a circle around the Earth). For other cases the necessary modification was for the second circle not to have the Earth at precisely its center; for other planets the motion around the circles was not uniform, sometimes speeding up and sometimes slowing down.

In the end the model was exceedingly complex with planets following complicated paths determined by strings of circles with offset centers and varying speeds. In addition, none of the quantities that determined these circles, the radius, displacement of its center, or rate of speeding up and slowing down were predicted by the theory, they had to be obtained from the observation. The theory

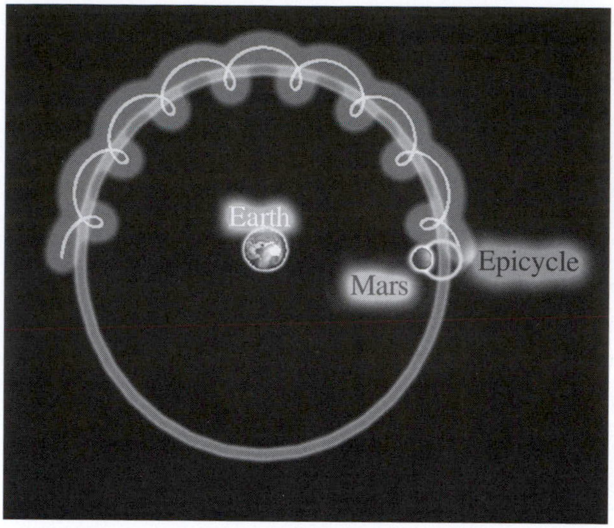

Figure 2.8 Mars' retrograde motion with respect to the fixed stars.

of epicycles also compromised its underlying philosophy: the planets did not move precisely around the Earth and, in addition, their speeds varied; such non-uniformities were, strictly speaking, not in accordance with Aristotle's ideas. The final version of this convoluted model was perfected by Ptolemy and published in his treatise the *Almagest* ("The Great System"). It soon became the standard astronomical reference for the next 14 centuries, accurately predicting the motion of the planets, the Moon and the Sun, and of the stars, and this virtue explains and justifies its longevity.

Greek science

There is much more to Greek science than the developments of the first cosmologies. Galen and Hippocrates provided the basis of medicine, Aristotle and others the basis of Biology; and the social sciences are also well represented. But especially relevant to the developments of the ideas about space and time were the discoveries made in geometry and algebra that we associate with Euclid and Archimedes.

Euclid

Anyone trying to evaluate the dizzying variety of work we have inherited from the Greeks often faces the daunting task of attempting to understand ideas that come to us only in partial form, or through some brief reference in more recent publications, so that one must infer (and sometimes guess) their full significance. Of the exceptions where complete works[10] are available, one of the most notable is Euclid's *magnum opus*, *The Elements*.

Using Euclid's results we can measure land, calculate distances, and even understand the workings of our Solar System (in deriving the consequences of his "system of the world," Newton used only geometrical arguments). *The Elements* also provides one of the clearest examples of logical deductive reasoning and has served as a model for the construction of mathematical theories for more than two millennia; it starts with a set of five postulates that are taken as natural and self-evident, and using these Euclid derives a large portion of the mathematics we inherited from the Greeks.

For many centuries the five postulates proposed by Euclid were believed to be the smallest self-consistent set one could devise in order to construct a complete description of geometry. A smaller number of postulates would not allow us to prove certain results that, based on elementary observations, appeared to be true. A larger number of postulates were shown to be redundant (some could be

Box 2.11

Euclid of Alexandria (*c.* 323–285 BC)

He was the most notable mathematician of antiquity, best known for his geometry treatise *The Elements*, whose influence persists today. His date and place of birth are uncertain, and almost nothing is known of his personal life (to the point that it is not known whether he was a historical character, the leader of a team, or even a *nom de plume* of a team of Alexandrian mathematicians). *The Elements* is written without any preface, so no glimpse of the author's character is available, though Pappus refers to him as a man of good disposition and the highest scholarly stature. There are comments indicating that Euclid studied at Plato's academy, where he learned Eudoxus' and Theatetus' geometry. It is likely that none of the proofs in *The Elements* are Euclid's, but his are the organization and exposition of the material. The treatise starts with a set of definitions followed by five postulates, from which a wealth of theorems and propositions is derived (including the main results of Eudoxus and Pythagoras, theorems in plane and space geometry, and in number theory and irrational numbers) in a clear, rigorous, and elegant manner (though with the aid of some tacit assumptions); this format serves as the model for mathematical texts to this day. The last two postulates are of particular interest in that they hold only in flat homogeneous spaces; geometries where these postulates do not hold were developed in the nineteenth century. More than 1000 editions of *The Elements* have been published since it was first printed in 1482.

Euclid also wrote: *Data*, dealing with the properties of figures; *On Divisions*, that discusses figure sections; *Optics*, which is the first Greek work on perspective; and *Phaenomena*, an introduction to astronomy with some results on stellar positions and movements. Of his books the following have been lost: *Surface Loci*, *Porisms*, *Conics*, *Book of Fallacies* and *Elements of Music*.

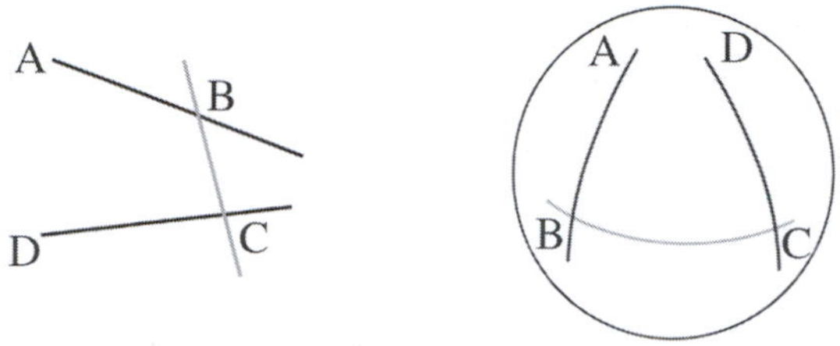

Figure 2.9 Euclid's postulate on parallel lines; for a plane surface (left), and a curved one (right).

proved by using a smaller set). Thus Euclid's or *Euclidean* geometry was considered for centuries to be the *only* possible one.

So matters remained until the middle of the nineteenth century, when Lovachevsky, Riemann and others developed alternative geometries that used only four of Euclid's postulates. The resulting geometry was not incomplete, as was believed up to that time, but described instead the properties of lines and angles on curved surfaces. The one postulate that was not included reads

> That, if a straight line falling on two straight lines makes the interior angles on the same side less than two right angles, the two straight lines, if produced indefinitely, meet on that side on which are the angles less than the two right angles.

Translated into English this means that (refer to the left diagram in Figure 2.9) if the angles **ABC** and **BCD** do not add to *precisely* 180° the two black lines will cross. If the angles do add up to exactly 180°, then the lines **AB** and **DC** are parallel.

Now constructing a geometry without this postulate would imply, for example, that the black lines will meet even if the angles add up to 180°, which might seem very peculiar (nay, impossible), until one imagines drawing lines on curved surfaces. Consider, for example, a sphere (refer now to the diagram on the right in Figure 2.9) where we draw two meridians starting from the equator; referring to the figure on the right, the angles **ABC** and **BCD** are both of 90° (one goes straight north along a meridian, traveling perpendicular to the equator), and yet the lines meet at the North Pole!

We conclude that Euclidean geometry describes correctly flat surfaces (and by extension volumes). If the space is curved, a generalization is required, which is provided by curved or *Riemannian* geometry. I go into this for two reasons. First, it illustrates how a change of paradigm (in this case allowing the underlying surface to be curved) can lead to novel conclusions; and second, because, as it turns out, our Universe is described by the geometry of Riemann and not that of Euclid: space is curved.

Archimedes

Greek science also had a very practical aspect, with Archimedes as the most notable figure. His discoveries include such useful machines as the lever and the

Box 2.12

Archimedes (*c.* 287–212 BC)

Born in Syracuse (Sicily), son of Phidias, an astronomer, and related to King Hieron II of Syracuse; killed during the capture of Syracuse by the Romans. Information about Archimedes comes from the prefaces to his books and from the works of Plutarch and Livy. He probably studied under a pupil of Euclid in Alexandria and kept in touch with Alexandrian mathematicians (who sometimes plagiarized his results). He was a prolific inventor, credited with devising Archimedes' screw, the lever, and several war machines; he also invented a notation for large numbers (Arabic or Hindu numerals were centuries away). Though practical creations gave him high renown, he believed that pure mathematics was his only worthy pursuit. Geometrical problems regularly commanded all his attention (even during the somewhat infrequent times he was persuaded to take a bath). His surviving works are: *On Plane Equilibriums*, dealing with mechanics; *On Floating Bodies*, which among various issues concerning fluids contains his famous principle that gives the weight of a body immersed in a liquid (and which in a famous anecdote he used to determine the quality of a golden crown); five books (*Quadrature of the Parabola*, *On the Sphere and Cylinder*, *On Spirals*, *On Conoids and Spheroids*, *Measurement of a Circle*) covering various topics of geometry, containing also a method of integration that he used to find areas and volumes of regular bodies, and a good approximation to the value of π and various square roots; and *The Sandreckoner*, containing a novel numbering system capable of expressing numbers as large as 8×10^{16}, and a recount of Aristarchus' heliocentric hypothesis. In 1906, a tenth manuscript, *The Method*, was discovered. There exist references to other works now presumably lost.

screw, now so widely used that they are taken for granted. Archimedes was not only brilliant in devising these machines, but also in using them in practical situations; the most dramatic example was his defense of Syracuse.

In 212 Rome assailed Syracuse by land and sea, and though Archimedes was a man of 75, he superintended the defense on both fronts. Behind the walls that protected the harbor he set up catapults able to hurl heavy stones to a considerable distance, and the rain of projectiles was so devastating that the Roman troops retreated and then were forced to advance only by night. When ships approached the shore, bowmen harassed the sailors (shooting at them through the holes that Archimedes' aides had pierced in the walls of the city). Moreover, the inventor had arranged within the walls great cranes turned by cranks and pulleys, which, when the Roman vessels came within reach, were made to drop upon the ships heavy weights of stone or lead that sank many of them. Other cranes, armed with gigantic hooks, grasped vessels, lifted them in the air and dashed them against the rocks, or plunged them end-foremost into the sea. The

Roman fleet withdrew and all hopes were turned to an attack by land. But Archimedes again bombarded the troops with large stones thrown by catapults, to such an effect that the Romans fled, and, saying that they were opposed by gods, refused to advance again.

So the Romans resigned themselves to taking Syracuse through a blockade, and after a siege of eight months the city finally surrendered. In the slaughter and pillage that followed the Roman commander gave orders that Archimedes should not be injured, but this was not to be. During the sack a Roman soldier came upon an aged Syracusan absorbed in some figures he had traced in the sand, the Roman commanded him to present himself at once to the commander, but Archimedes refused to go until he had worked out his problem; he "earnestly besought the soldier," says Plutarch, "to wait a little while, that he might not leave what he was at work upon inconclusive and imperfect, but the soldier, nothing moved by his entreaty, instantly killed him." It is through the carelessness of such fools that the best we have to offer is sometimes lost.

Appendix: a creation myth pot pourri

A creation myth describes just that: the process of creation. It introduces the forces behind the act of creation, and sometimes also the reason for the creative act. These are hardly scientific theories: they predict nothing more than the fact that there is a universe around us, for any observation can be explained by simply assuming it is the result of the will of one or more of the gods. Yet these myths are the ancient seeds of modern cosmology, not because some of the specific ideas there contained have survived through the millennia, but because they are the first indications of a desire to understand in a rational way the world around us.

Babylonia

The Babylonian creation myth states that in the beginning there was only chaos. Eventually two gods decided to join powers to create what would evolve into our world. "In the time when nothing which was called heaven existed above, and when nothing[11] below had yet received the name of earth, *Apsu*, the Ocean, who first was their father, and Chaos – *Tiâmat*, who gave birth to them all, mingled their waters into one, reeds which were not united, rushed which bore no fruit" (*sic*). This initial mass was not a very fructiferous primordial soup, it tarried in generating life, and when it eventually did so, it was of a feeble nature, with but a few gods appearing over a long time. Yet, as this generation of life progressed, the vitality of the created increased, and their attributes became more and more clearly defined. These gods eventually paired into male and female couples, and produced litters of little gods and goddesses, including, for example, the nine planets. In time this pantheon elected *Anu* as their leader.

But there came a time when Tiâmat felt that all these newcomer upstarts were encroaching on her domain, and proceeded to create an army of offspring to do battle against the other gods. These children of Tiâmat had nightmarish shapes with a haphazard mix of heads, and human and animal members. Tiâmat armed them with terrible weapons, and sent them off to war against the other gods, under the command of her husband *Kingu*.

The other gods were all terrified of the invaders (cousins as they were) and knew not how to react; only *Marduk* believed himself strong enough to oppose the attackers. The other gods were only too glad to have a champion; they declared him king, bestowed on him all their powers, and they sent him off with the command to "... cut short the life of Tiâmat, and let the wind carry her blood to the hidden extremities of the universe."

And so, Marduk made a bow and a spear; he placed a thunderbolt before him and filled his body with a devouring flame; he trapped the four winds, and using them created the hurricane, the evil wind, the storm and the tempest; so armed he galumphed to war. He went through the ranks of monsters surrounding Tiâmat and soon engaged her in battle. When she opened her mouth to swallow him he thrust a bag full with hurricane winds into her, so that her maw split and she swelled. Taking advantage of her indisposition, Marduk pierced her with his lance and killed her. He had the winds carry her blood to hidden places in the north and then he split her carcass, making the lower half the earth and the upper the heavens.

And in this way Marduk created the Universe as people have known it since. The Earth consists of a big circular plane surrounded by a river, which no human may cross. Beyond looms an impassable mountain barrier; and the whole thing rests on a cosmic ocean. The mountains support the vault of heaven, made of a very strong metal. There is a tunnel in the northern mountains that opens to the outer space and which also connects two doors, one in the east and one in the west. The Sun comes out through the eastern door, travels under the metallic heavens and then exits through the western door; he spends the nights in the tunnel. Marduk regulates the ebb and flow of the Universe through the motion of the Sun.

Having settled the shape and motions of the universe, Marduk made living creatures of many kinds, but they died, unable to withstand the light from the Sun. So Marduk begged his father *Ea* to behead him, and to then mix his blood with clay and fashion new beasts and men strong enough to live. And this Ea did do. Living creatures were made from the very essence of divinity (an irony given the cavalier attitude of these peoples towards human and animal suffering); it was because of such gifts that it was our duty to adore the gods, especially Marduk (or his memory thereof).

Egypt

One of the most colorful versions of the Egyptian creation myths tells of how, in the beginning of the world, *Nuit*, the goddess of the night, was in a tight embrace with her husband *Sibû*, the earth god. Then one day, when Nuit and Sibû were engrossed in this agreeable occupation, the god *Shû*, without an obvious reason, grabbed her and elevated her to the sky (to become the sky) despite the protests and painful squirmings of Sibû. But Shû felt no sympathy for him, and instead froze Sibû even as he was thrashing about, and he became the Earth. And so Sibû remains to this day, his twisted pose generating the irregularities we see on the planet's surface. Over time he has been clothed in verdure, and generations of animals prospered on his back, but his pain persists. Nuit is still supported by her arms and legs, which became the columns holding the sky. The result was the universe as described above.

An expanded version of this myth imagines that in the beginning the god *Tumu* suddenly cried "Come to me!" across the cosmic ocean. A giant lotus flower then appeared with the god *Ra* inside. Ra then separated Nuit and Sibû, and the story proceeds as above. It is noteworthy that creation did not come through muscular effort, but through Tumu's voiced command. This later evolved into the belief that the creator made the world with a single word, then with a single sound (yet the creation through pure thought was not considered).

After creation the gods taught the arts and sciences to the Egyptians. In particular *Thot* taught the Egyptians how to observe the heavens and the manner in which the planets and the Sun move, as well as the names of the (36) constellations (though he apparently neglected to tell them about eclipses, which are never referred to).

India

The Indian creation myth is one of the most profound and intriguing. In it the Universe undergoes a cyclic process of creation and destruction over a period of 4.32 billion years, a "day of Brahma." At the beginning of each of these periods Brahma wakes up and finds the Universe void, and, seeing nothing, decides to create the world; stars, planets and living beings thus appear. Brahma, satisfied at having accomplished this, goes to sleep again. With Brahma asleep the Universe evolves through natural means over one thousand "great ages" (each divided into four "ages") over which humankind gradually deteriorates. At the end of the last age, the god Shiva performs a terrible dance with which he destroys the whole Universe. Then, in the resulting void, Brahma wakes up, and seeing nothing, decides to create the Universe again.

There is no final purpose towards which the Universe moves, there is no progress, only endless repetition. We do not know how the Universe began, perhaps Brahma laid it as an egg and hatched it; perhaps it is but an error or a joke of the Maker. No other creation myth contains such an endless cycle of creation and annihilation processes, nor deals with such enormous time periods (for example, in Navajo mythology, living things change worlds in the space of months, let alone billions of years). It is a curious coincidence that, under certain initial conditions (apparently not realized in our Universe), the Einstein equations have a solution describing a universe that goes through an endless cycle of expansions and contractions.

Once created, the behavior of the Universe is more mundane. In the *Surya Siddanta* it is stated that the stars revolve around the cosmic mountain *Meru*, at whose summit dwell the gods. The Earth is a sphere divided into four continents. The planets move by the action of a cosmic wind and, in fact, the Vedic conception of nature attributes all motion to such a wind. It was noted that the planets do not move in perfect circles and this was attributed to "weather forms" whose hands were tied to the planets by "cords of wind."

Greece

The Greek creation myth version outruns its competitors in dramatic value, and it also illustrates the very human characteristics of the Greek gods. This is a story of jealousy, incest, murder, and mutilation.

In the beginning, so the story goes, seven beings appeared at the creation of the Universe. Then, either through intermarriage or parthenogenesis, these initial seven gave birth to a dizzying variety of offspring. Of particular relevance to this story is *Gea*'s creation of *Uranus*, the heavens, who became one of his mother's mates, covering her on all sides. Uranus and Gea had 18 children: the three *Cyclops*, the three *Hecatoncheires*, and twelve *Titans*. Uranus, however, was a terrible father and husband. He hated the Hecatoncheires and imprisoned them by pushing them into hidden places of Gea's womb, where they were guarded by a hideous nymph called *Kampe*, probably a daughter of Gea (by a different father). This imprisonment displeased Gea, but of all her children only the youngest Titan, *Cronus*, would help her punish Uranus. And so it happened that as Unranus lay with Gea, Cronus grabbed his father and castrated him, throwing the severed genitals into the ocean. From his spilt blood came the *Giants*, the *Ash Tree Nymphs*, and the *Erinyes*, and from the sea foam where his genitals fell came *Aphrodite* (nothing is wasted in such stories). The fate of Uranus is not clear, in one version he simply dies; in another he withdraws from our world; in a third one he exiles himself to Italy. Before disappearing, however, he prophesized that Cronus' rule would come to an end much as Uranus' did: deposed by one of his children.

In the meantime Cronus became the next ruler. As his first action he imprisoned the Cyclops and the Hecatoncheires in Tartarus, for he trusted none but his closest siblings, the Titans. He married his sister *Rhea* with whom he had many offspring (the other Titans were equally productive), and so he ruled happily for many ages. Yet he did not forget the prophecy made by his father, and attempted to foil his destiny: in order to avoid being overthrown by his sons, he ate them as they were born (he apparently did not *chew* them, and they were still alive inside him). Rhea was angry at this treatment of her children and, like her mother-in-law, plotted against her husband. So, when it came time to give birth to her sixth child, Rhea hid herself, and after the birth she secretly left the child to be raised by nymphs. To conceal her act she wrapped a stone in swaddling clothes and passed it off as the newborn, and Cronus swallowed it.

This child was *Zeus*, who grew into a handsome youth while hiding in Crete. When he became a young god *Metis* suggested to him a plan to defeat Cronus: Zeus was to disguise himself as a cupbearer and give Cronus a potion that would force him to vomit up the other children, they would become Zeus' allies, and together they could fight and hopefully overthrow Cronus and the other Titans. The first part of the plan worked like a charm: the other five children were vomited up, they were thankful to Zeus, and made him their leader.

But Cronus was yet to be defeated. He and the Titans, except *Prometheus*, *Epimetheus*, and *Oceanus*, fought to retain their power. *Atlas* became their leader in battle and for some time it looked as though they would defeat the young gods. However, Zeus was cunning; he went down to Tartarus, killed Kampe and freed his uncles and aunts the Cyclops and the Hecatoncheires, who, together with Prometheus, joined Zeus' cause. The Cyclops provided their liberator with lighting bolts. With such weapons, and with guile and courage, Zeus drew the Titans forth into what turned out to be a very successful ambush. For, at the appointed place and with perfect timing, the Hecatoncheires rained down on the Titans hundreds of boulders with such a fury that they thought the mountains were falling on them. They broke ranks and ran, giving Zeus the victory. Zeus exiled the Titans who had fought against him into Tartarus. Except for Atlas, who was singled out for the special punishment of holding the world on his shoulders until the end of time (except for a short vacation at Heracles' expense).

However, even after this momentous victory Zeus was not safe. Gea was now angry that her children had been imprisoned, and gave birth to a last child, *Typhoeus* (Typhon, for short), who was to avenge Cronus' downfall. Typhon was a winged monstrosity with matted hair and beard, pointed ears and fire-spitting eyes, and so large that his head was said to brush the stars. He had human form down to the thighs, but boasted two hissing serpents instead of legs. Some say he had 200 hands, each with 50 serpents for fingers; and also 100 heads, one human and the rest in the shape of bulls, boars, the ubiquitous serpent, lions, etc. This was such a fearsome creature that all the gods fled before him, all that is, except Zeus. The newly crowned king of the gods valiantly faced the monster, and flinging his lighting bolts was able to kill it, or at least subdue it (opinions vary). Dead or alive he was buried under Mount Etna in Sicily, yet before suffering this ignominy he had time to leave the world such charming offspring as *Kerberus*, the three-headed dog that guards the gates of Hell, and *Hydra*, the nine-headed serpent.

Much later the Giants presented a final challenge to Zeus' rule. They went so far as to attempt to invade Mount Olympus, piling mountain upon mountain in an effort to reach the top. But by then the gods had grown strong and, with the help of Heracles, the Giants were subdued or killed. In the ensuing peace the gods multiplied and, in the ripeness of time, Zeus and his retinue asked Prometheus to create man, and Epimetheus, his brother, the other mortal creatures. Epimetheus proceeded to bestow upon his creations all manner of qualities, making them strong, resourceful and self-sufficient. Prometheus then sought an equivalent gift for man, and chose fire. In order to bestow it he sneaked up to the sky, and stole a flame from the Sun's chariot. Zeus was not amused and, using his power, contrived to retrieve the gift; but Prometheus rose and stole the flame again and secretly gave it to man. Enraged Zeus had his old ally chained to a rock in the Caucasus Mountains where vultures were to devour his livers for all time without his ever dying. Such were the pains endured by mythological parents on behalf of their mortal children.

Epimetheus had a different fate. As a reward for his creating all the animals Zeus gave him a wife name *Pandora* ("all the gifts"), a charming woman whose dowry was but one box, given to her by the gods. This box came with the command never to be opened, which she of course proceeded to do, only to see all the blights of humankind pour out and scatter to the four winds. She tried to close the box, and eventually succeeded, but by then she only managed to retain hope inside. And so we received all the ills that darken our life, and the one tool to fight them.

Before leaving this brief recount of Greek mythology let us mention three (pathenogenous) grandaughters of Chaos, the three *Fates*:[12] *Clotho*, "who spins," *Lachesis*, "who measures," and *Atropos*, "who could not be avoided." The first would spin the threads of life, one for each person; the second would measure their length, determining the lifespan of each and all; the last, and most terrible, would cut it at the allotted moment, ending life and releasing the soul. The rulings of the Fates could not be changed, not even the gods themselves (even Zeus) could sway them; for the first time some occurrences were not subject to the whims of the gods.

Aztec

In Aztec mythology the world has passed through four ages, each with its representative main deity, and each ending in destruction. We now live under the fifth Sun.

The first age was the age of *Tezcatlipoca*, patron of evil and sorcerers, god of the night, omnipotent and multiform, who originally looked like a tiger but transformed himself into the first Sun that lit the world. But the other gods were unhappy under his rule and created a race of giants who were to destroy Tezcatlipoca. These giants were nomads and lived off the land, not delving in agriculture or animal husbandry. There were also humans at that time, but they were wild, living by instinct alone. Eventually it was another god, *Quetzalcoatl*, the benevolent god of the wind, founder of agriculture, industry, and the arts, who rose up in arms against Tezcatlipoca and deposed him. Tezcatlipoca then resumed his tiger form and, in the sunless darkness, devoured all the giants and humans.

Quetzalcoatl then became the second Sun over a spiritual age of ghosts and transparent beings. He ruled until the day Tezcatlipoca took revenge: he reached up with his paw and pulled Quetzalcoatl down. In falling he caused a great hurricane that destroyed the second world. A few humans of the time did survive, but they could not understand the gods and so were transformed into monkeys.

After this quid pro quo the rest of the gods banished Quetzalcoatl and Tezcatlipoca and asked *Tlaloc*, the god of rain and heavenly fire, to rule over the third world. But Quetzalcoatl was not to be so easily disposed of: he caused a devastating rain of fire to fall upon the land. The rivers dried and most humans perished; the survivors were transformed into birds. Quetzalcoatl then made *Chalchiuitlicue* ("she of the jade-green skirts") and made her the fourth Sun. But Tezcatlipoca was jealous and sent a great deluge that yet again destroyed the world and almost all the humans living in it. The few survivors turned into fish. This was such a terrible flood that all the land was covered, and remained so until two gods lifted it so as to begin the fifth world. But the fourth Sun had been destroyed and all was in darkness, so all the gods gathered in Teotihuacan to do penance and ask *Onteotl*, the father of all the gods, for light. In the end two gods offered to immolate themselves in order to create the fifth Sun: *Tecuhciztecatl*, rich and powerful, and *Ranahuantzin*, poor and modest; the former offered rich presents to Onteotl, the latter only moss and maguey leaves, but the latter were wet with his blood. For four days these gods made sacrifices; on the fifth day, a big fire was made to where one of them could be purified and rise as the new Sun. Three times Tecuhciztecatl came up to the fire, and three times he hesitated and drew back, then Ranahuantzin, closing his eyes, hurled himself into the blaze whence a great flame leaped up to the sky and became our Sun. Tecuhciztecatl was then ashamed and also jumped in, but the fire had died and he was slowly consumed, rising then as the Moon. This angered the gods who threw a rabbit at it, leaving a mark we still can see today. Even then the Moon's troubles were not over, soon afterwards the stars rebelled against it and had to be subdued; this being eventually accomplished the world finally knew a measure of peace.

China

As in all such rich cultures there are many Chinese creation myths. In this version the Universe began in chaos where heaven and earth were as one. The Universe was like a black egg carrying *Pan Gu* inside. After 18 millennia Pan Gu woke, and feeling hemmed in, cracked this cosmic egg with a broad axe. The top part of the shell was light and clear, and rose to make the heavens, while the lower turbid part stayed below and formed the earth. The heavens and the earth then began to grow, and Pan Gu grew with them. After another 18 millennia, Pan Gu was 9 million li in height, and he placed himself between heaven and earth to insure they never joined again.

When Pan Gu died (apparently around 2,229,000 BC), his breath became the wind and the clouds, his voice the thunder. One eye became the Sun and the other the Moon. His body became five large mountains and his blood the roaring water. His veins became far-stretching roads and his muscles fertile land. The stars in the sky came from his beard and hair and from his skin and fine hairs came trees and plants. His marrow turned into precious stones such as jade and pearls, and his sweat flowed like the good rain and dew to nurture all things on Earth . . . in one version, his fleas and lice became the ancestors of mankind.

Notes

1. Since the middle of the twenty-third century BC.
2. To gauge the antiquity of these original investigations one has but to remember that the first Egyptian Empire was as ancient to Herodotus as he is to us.
3. It is worth pointing out, however, that at the time it was not customary for philosophers to write and publish manuscripts and to insure authorship rights, so most of these results are "attributed" to one or other historical personage, with no guarantee that he was indeed responsible for the work, or even for the inspiration.
4. Pythagoras' theorem is valid for triangles drawn on flat surfaces, it does *not* hold for triangles drawn on curved surfaces. Similar statements hold for Euclid's postulate that parallel lines never meet.
5. Some relate this to the origin of the decimal system, but it seems to me more reasonable to associate the decimal system to our having ten fingers. Also, we cannot avoid remarking (tongue in cheek) that certain string theories also place great importance on the number ten, in this case in connection with the dimensions of space and time.
6. This model had been proposed earlier by Phiolaus.
7. This was also proposed by Indian philosophers; possibly related to Democritus' ideas.
8. We will see below that in Aristotle's view objects in a vacuum would move at infinite speeds.
9. Phrase from W. Durant's *The Age of Faith* (Chap. XXXVI, Sect. II).
10. Even though it is sometimes hard to know if a given work is indeed the creation of those whose name appears in the front page, or is the result of the collective effort of a series of interlopers that have shaped and added to it over the more than 20 centuries that separate us from the assumed author.
11. The belief was that "nothing" had a real existence.
12. In other versions of this myth, the fates are the daughters of Zeus with the nymph *Thamis*, but the Pythian princess in an oracle confessed that this version is incorrect.

3
From the Middle Ages to heliocentrism

Felix qui potuit rerum cognoscere causas.
(Happy he who has been able to know the cause of things.) Virgil

The look of the time

The Middle Ages is often described as a fearful drama of plague and famine, decorated with religious fanaticism and war motifs; so bleak a picture is presented that one can understand the absence of mass suicide only by the promise of eternal punishment in hell. Such stark landscape, as one might imagine, is but a caricature of an epoch that in reality was much more complex. The collapse of the Western Roman Empire cannot (and should not) be underemphasized, with it disappeared the complex economic and cultural system that supported a vast array of activities, leaving people bereft of the underlying security of a cohesive government and a structured social system. The chaos and poverty that followed this collapse took centuries to remedy. During this time the Catholic Church played the pivotal role of a stable and secure institution that provided the moral code that framed medieval Europe. With such dramatic political changes, new directions of thought and economic and social activities were developed, often through a process of trial and error, where errors often carried enormous pain and suffering to the population.

Despite often appalling surrounding circumstances medieval thought slowly achieved the monumental results that led into the Renaissance, for it was during the Middle Ages that the seeds of modern Western European civilization were planted and first cared for. It is unfortunate that this age is often criticized for lacking a genteel approach towards the life and happiness of the inhabitants, without giving it credit for having provided the foundations for our civilization; none but a few like to see sausage being made, but many welcome the finished product.[1] One of the achievements of the Middle Ages was the construction of a philosophy that coexisted (almost) peacefully with Christian theology, and which opened the way for the development of modern science. It is also from these misnamed "Dark Ages" that we inherited the universities and hence the medium to store and develop knowledge. The intellectual climate fostered by these institutions led to a rediscovery of classical culture, made learning respectable (again) and then indispensable;

the medieval universities were midwives, milkmaids and governesses of the Renaissance.

It is also important to remember that the rest of the world did not partake of the medieval miseries endured by Europe between the fifth and ninth centuries. This time also witnessed the apex of the Islamic Empire, and the heyday of Islamic thought. The stability and prosperity reached during the reign of the Ommayad dynasty (eighth century AD) and during the Abbasid Caliphate (ninth–tenth centuries) fostered the growth of philosopher-scientists who preserved and expanded the knowledge so carefully gathered by the Greeks. The works of Aristotle, Archimedes, and Euclid, originally introduced into the Islamic world through Syria, soon diffused throughout the Empire, and fostered significant developments in almost all areas of science, from astronomy to zoology. Eventually, as commerce propitiated intellectual exchange, and also with the crusades and less benign interactions, these ideas were reintroduced into Western Europe and were received, like prodigal children, with a mixture of elation and mistrust. Whether considered good or evil, the ideas from Classical Greece could not be ignored, and their hybridization with Christian philosophy eventually led to the contradictions that gave rise to Copernicus.

The late Middle Ages witnessed the momentous development of mechanized printing. This was an idea of Chinese origins (1041) that was apparently developed in the Netherlands by the introduction of movable type (1430), and culminated with the famous and beautiful "Gutenberg Bible" (Mainz, 1456). This was an enormously important advance with innumerable ramifications. Printing made books accessible to almost all that could read, and these were complete and faithful copies, not subject to the whims, inaccuracies or proclivities of a copyist. Printing made possible the success of Luther's appeal, for it

Box 3.1

Magellan's trek

The most impressive adventure in the second generation of transoceanic exploration was the circling of the globe. Fernão de Magalhães (Fernando Magellan) persuaded Charles I (V) to finance an expedition that would seek a southwest passage to Asia. Strapped for funds, the King allotted Magellan five weather-beaten and hardly seaworthy ships (the largest displaced 120 tons, the smallest 75). On September 20, 1519 the fleet sailed, and returned after three years on September 8, 1522 having completed the circumnavigation of the globe. The trip was a long history of bone-wrenching hardships and death, including Magellan's; of the initial 280 men only 18 returned.

allowed him to disseminate his ideas rapidly and accurately, and the general population could understand his complaints since copies of the Bible were by then generally available. The introduction of printing ended the Church's monopoly on learning, and it facilitated scientific cooperation; it fostered vernacular literature, making authors receptive to the tastes of the middle class, on whose purse they depended. Printing allowed the dissemination of the ideas that led to the way for the Enlightenment, to the American and French revolutions, and from these to democracy.

There were (and are) also darker sides to this discovery: both Church and State recognized printing as a powerful vehicle for the dissemination of subversive ideas; this paved the way for future censorship and more violent suppression of ideas. And, from its inception, printing became the most powerful instrument for the dissemination of nonsense (except perhaps for the internet) and spread it even more vigorously than knowledge.

By the end of the fifteenth century, a variety of circumstances forced a dramatic restructuring of the European culture, economy, politics and philosophy. The failure of the crusades, the influence of Islamic culture, the rediscovery of the classic pagan ideas, the expansion of commerce generated by the discoveries of Columbus and Vasco da Gama, the rise of a business class that was to finance the budding monarchies, the development of national states that were to challenge the authority of the Vatican, the attrition of the Church's power due to Luther's schism, the fall of Constantinople in 1492; they all produced a widening of the cultural horizons and a diversification of ideas that were incompatible with medieval society. They also generated economic and political structures that could not grow within the constraints of the Middle Ages; and so, Europe evolved into the Renaissance.

The passing of the Middle Ages, just as Christianity was giving ground in the adolescent West, also marked the waning of Islamic science and philosophy, smothered under the weight of religion. The highest honors now went to theologians, dervishes, fakirs, and saints, while scientists aimed to absorb the findings of their predecessors rather than look Nature freshly in the face. Yet Moslem astronomy died out slowly, producing useful astronomic tables even as it was disappearing; ironically it was Arab tables, maps and navigators that piloted the historic voyage of Vasco da Gama from Africa to India, a trip that ended the economic ascendancy of Islam.

The Renaissance was enamored of literature and style, somewhat interested in philosophy, almost indifferent to science. At that time (sixteenth century) the popes were not initially hostile to science; indeed, Leo X and Clement VII listened with open minds to Copernican ideas, and Paul III even accepted the dedication of Copernicus' world-shaking *Book of Revolutions*. But this was not to last: the reactionary policies of Paul IV, the development of the Inquisition in Italy, and the dogmatic decrees of the Council of Trent made scientific studies increasingly difficult and dangerous after 1555.

Box 3.2

Pierre Abélard (1079–1142)

Born in Brittany, he surrendered his property rights in order to pursue his studies. He was a student of Roscelin, and was involved in the controversy concerning "universals." In the middle of this controversy Abélard was forced back home to deal with personal matters (his parents had decided to enter religious orders). He then returned to Paris where he faced a glowing future. At this time he fell in love with Héloïse, the 16-year-old daughter of canon Fulbert; soon she was with child and they married in secret (at her insistence for it would limit his prospects). Fulbert cruelly punished his niece for her pregnancy, so Abélard sent her to a convent for protection; enraged, Fulbert stole into Abélard's quarters one night and castrated him. After this tragedy Abélard convinced Héloïse to take the veil, and he became a monk, but continued his intellectual activities. He published *Sic et non* where he discussed questions of faith, including the nature of the Holy Trinity and the unity of God. Condemned by the Church, the book was burned and he was confined to a monastery for one year. After this he spent three years in a modest hermitage, lecturing and publishing, favoring mystic Christian doctrines. This brought on renewed criticisms and, probably tired of controversy, he retired as abbot of St. Gildas (*c.* 1125), where he wrote his autobiography. Legend has it that Héloïse, having read this manuscript, sent him a series of astonishingly beautiful and passionate letters, which Abélard was morally and physically incapable of answering in kind; his responses are chaste, moderate, even tepid. The authenticity of Héloïse's letters is very much uncertain, but they have become an integral part of French romantic literature, and in this capacity their provenance is irrelevant. Abélard's belief that all of faith could be rationally understood brought on what was to be his last battle: St. Bernard condemned him, was found guilty and forced into the Cluny monastery, where he died. Héloïse joined her lover and partner in 1164. In 1817 their bodies were rejoined in the Pére Lachaise Cemetery in Paris where even now one might see on bright summer days, men and women adorning their tombs with flowers.

As Copernicus was to expose the astronomic insignificance of the Earth, so the exploration expeditions of the fifteenth and sixteenth centuries, and the ensuing expansion of commerce revealed vast realms ignorant of and indifferent to the Christian world. Even the authority of Aristotle and the other Greeks suffered when it became apparent how little of the planet they had known. The Renaissance idolatry of the Greeks declined, modern science and philosophy rose, and undertook the epochal task of re-conceiving the world; but such changes did not occur smoothly. The discovery of America and the exploration of Asia, the demands of industry and the extension of commerce, did turn up

Box 3.3

Thomas Aquinas (1225–74)

Born in his father's castle at Roccasecca, of noble progeny, Thomas renounced his lordly inheritance to follow his theological and philosophical pursuits. He received his first schooling at the abbey of Monte Cassino, at 14 he entered the University of Naples, where he was first exposed to the ideas of Aristotle. Though invited by Frederick II himself to join his court, Thomas joined instead the Dominican Order (1244) and was soon sent to Paris to study theology. On the way there he was kidnapped by his family and kept for a year hoping to convince him to return to the laity; they were unsuccessful and he escaped. At Paris he studied under Albertus Magnus whom he followed to Cologne. In 1252 Thomas returned to Paris where he taught his Christianized version of Aristotle. In 1259 he left Paris to teach at the *studium* of the papal court now in Anagni. He returned to the University of Paris to counter the Averroistic revolution, where he had to also fend off attacks from Franciscan monks for his efforts to reconcile Aristotle with the Christian doctrine. In 1272 he was asked by Charles of Anjou to reorganize the University of Naples. In his final years he ceased writing, through weariness or disillusionment with his arguments. In 1274 Gregory X summoned him to attend the Council of Lyons; on the way he grew weak, took to his bed in the monastery of Fossanuova, and there, still but 49, he died. He was canonized in 1323.

knowledge that often contradicted traditional beliefs and encouraged fresh thought. But many travelers were liars (Figure 3.1) or careless reporters of their discoveries, and many of the technological or scientific claims proved to be exaggerations or unscrupulous quackery.

During the Renaissance (and even after this period), the classic Greek scientific texts, such Apollonius' *Conics* and Archimedes' works, were translated from the Arabic and became generally available. This provided a measured stimulus for the growth of mathematics and physics, but these developments were hampered by the almost primitive scientific instruments then available (microscopes, telescopes, thermometers, and barometers were still in the future). Clocks, however, did not suffer such delays: by the late thirteenth century the growth of commerce generated the need of better mechanisms for measuring time; the old method of dividing daylight into the same number of periods in all seasons proved too inaccurate. This led to the invention of the mechanical clock; it used a weight slowly turning a wheel, whose revolution was checked by an escapement tooth that allowed a turn by one cog in a given interval of time. By the middle fourteenth century clocks of this type were available (Figure 3.2) that showed hours and minutes, the tides and motions of the Sun and Moon.

Figure 3.1 Illustration of Manderville's race of men with heads in their chests. (from: Livre des Meerevilles, French, fifteenth century)

Figure 3.2 Lady Wisdom and H. Suso examine an astrolabe and a variety of fourteenth century clocks (from a book by Suso; French, *c.* 1334). (from: Henri Suso, *L'Horloge de sapience Bruxelles, Bibliothèque royale de Belgique*, ms. IV III, fol, 13v)

The influence of the Catholic Church

The Catholic Church has been often portrayed as having been an obstacle to the development of knowledge, especially during the dark confusion following the fall of Rome; but in fact it played a much more ambiguous and complex role. On

Box 3.4

William of Ockham (*c.* 1280–1349)

Probably born at Ockham in Surrey, he entered the Franciscan order, and when 12 entered Oxford, expected to become a shining light in the Church. There he learned (and opposed) the realism of Duns Scotus (another Franciscan); but he carried Scotus' rationalist critique of philosophy and theology to a skepticism that would dissolve alike religious dogmas and scientific laws. Taught at Oxford for six years, and perhaps at Paris; while in his 20s wrote commentaries on Aristotle and Lombard, and his important treatise *Simma totius logicae*. He applied his logic, and in particular the "razor," to a variety of philosophical and theological arguments; and though he advocated differentiating logical and theological truths, his arguments could not be accepted by the Church. John XXII ordered an ecclesiastical inquiry into his "abominable heresies," and summoned him to the papal court in Avignon. Ockham came, and was imprisoned (1328) but escaped (and was excommunicated) to Aiguesmortes, there he embarked in a small boat, was picked up by a galley and taken to the Emperor Louis of Bavaria at Pisa, who protected him. William went with Louis to Munich, lived in an anti-papal Franciscan monastery, and issued from it a torrent of publications against the power and heresies of the popes, advocating a more democratic, humble and simple faith. Of Ockham's last years little is known except his having felt isolated and banished into an intellectual wilderness. He died of the Black Plague.

one hand, during the Middle Ages university instruction was overwhelmingly carried out by priests, with a curriculum that carefully covered lay philosophy and science (subject, of course, to the available texts). So the Church made it its business to preserve knowledge; and by the thirteenth century it was more intent on assimilating intellectual developments than suppressing them. On the other hand, worried about the indiscriminate dissemination of new ideas that could easily lead to a diversification and weakening of the faith, the Church actively countered any such perceived threat. Particularly worrisome during the early Middle Ages were the non-standard interpretations of the Bible: at this time books were scarce and expensive, most copies of the Bible accessible to private ownership were incomplete and often inaccurate, and such defects often led to misinterpretations of the Scriptures. On this basis the Church forbad people from even privately *owning* such a copy of the Scriptures. Later the Church's concerns turned to subtle philosophical interpretations of Christian dogma and the intellectualization of the faith.

After the poverty and confusion of the fifth to the ninth centuries the circumstances in Europe steadily improved, allowing the leisure for intellectual pursuits. This activity was fostered by an improvement in the economic situation, it was fed by the waves of Islamic translations of Greek philosopher-scientists that reached Europe in the twelfth and thirteenth centuries, and it was prodded by the dissemination of Islamic philosophy (notably, the writings of the Islamic philosopher Averroes that became fashionable in the thirteenth century) that threatened to depose its incipient Christian counterpart. In them we find the notions that natural laws rule the Universe, that the Universe is co-eternal with God, that there is one universal soul of which the individual one is but a temporary "blip," and that heaven and hell are but stories made to terrorize the naïve. All these ideas directly contradicted Church doctrine and became dangerously popular, and so were vigorously attacked by Christian philosophers. Many of the most strenuous endeavors of Aquinas in favor of classical philosophy were made not out of love for Greek philosophy, but out of fear of Islam.

Meager as were the medieval scientific achievements, the Church and the Inquisition participated actively in them. Though they supervised university instruction, professors enjoyed considerable latitude, and so the Dominican and Franciscan orders tolerated the studies of Albertus Magnus and Roger Bacon. It is true that zealous objectors to these pursuits were not uncommon, some as weighty as St. Bernard himself, but the condemnation of scientific interests was not the policy of the Church. The noblest of legacies of the Middle Ages, the universities, appeared in the late twelfth century, and for almost half a millennium thereafter clergyman shouldered most of the joy and burden of instruction. Lectures were often held in schoolrooms or auditoriums scattered about the town, though sometimes in colleges such as those maintained by Benedictine, Franciscan and Dominican monks at Oxford. From these academies came some of the most brilliant men of the fourteenth century; among them William of Ockham, who was to be so troublesome for orthodox theology.

The role later played by the Protestant Church was quite different. Luther, being mainly concerned with a purist interpretation of the Scriptures, rejected the involvement of the Church in most intellectual secular pursuits. In particular, Protestantism could not favor science, believing in an infallible Bible. Hence, Luther's rejection of the Copernican theory because the Bible told how the Sun's motion, not the Earth's, was arrested during the take of Jericho.[2] Melanchthon was more favorable to science, even lecturing on ancient mathematics, but was overwhelmed by Luther's personality and by a narrowing Lutheranism after Luther's death. Calvin had little use for science, Knox none.[3]

The thought of the time

The late Middle Ages witnessed an enlargement of horizons brought by the crusades, the creation of the universities and the consequent revival of Roman secular law, and the first tentative re-awakenings of scientific investigation. This revival was enlivened by the reintroduction of Aristotle's ideas (through the translations of Arabic texts), which though initially perceived

Box 3.5

Petrus Ramus (Pierre de la Rameé) (1516–72)

Born near Noyon in Picardy was an avid learner. Twice he tried, but failed, to walk to Paris avid to enter one of its colleges; in 1528 he finally succeeded. He studied by night and worked by day and despite many hardships completed the curriculum in the Arts faculty after eight very difficult years. He admired Plato and Socrates, but considered useless Aristotle's logic. He advocated skepticism stating that "unbelief is the beginning of knowledge"; this brought on accusations as an enemy of the faith and he was tried and punished (1544), forbidding him from lecturing or publishing. One year later Henry II repealed this sentence and Remus soon became the most famous teacher in Paris. In 1561 he announced his adherence to Protestantism, and when civil war broke out was forced to leave Paris under the protection of the Médicis; he returned on the signing of peace. In 1567 the civil war between Catholics and Protestants resumed and Remus left the city again; the government excluded Protestants from teaching positions and he retired to private life (supported by the Médicis). In July 1572 the Bishop of Valence invited Remus to join in an embassy to Poland, perhaps foreseeing the massacre of St. Bartholomew, but Remus refused. On August 26, 1572 two armed men came into Remus' study (apparently directed by an unscrupulous colleague of Remus), stabbed and shot him, then hurled him out the window. Students and ragamuffins dragged the still-living body and threw him in the Seine; others recovered and hacked it to pieces.

as a dangerous threat to orthodoxy, were soon accepted in the universities' curriculums. For example, Aristotle's *Metaphysics* was forbidden at the University of Paris in 1210, yet by 1255 it had become a required reading at that institution, with other universities soon following suit. It is notable that the Church allowed and even fostered this enlightened attitude of assimilating such new ideas instead of repressing them, an attitude that was not to be followed during and after the Renaissance. By the twelfth century it became apparent that Aristotle's philosophy was often inconsistent with the teachings of the Church and immune to its authority, yet its popularity required that such inconsistencies be resolved. And so, just as the Moslem philosophers of the ninth century strove to meld Aristotle into Islamic theology, Christian philosophers faced a similar task 300 years later. This led medieval Western philosophers, beginning with Roscelin and Abelard and culminating with St. Thomas Aquinas and Albertus Magnus, to investigate the logical structure of the Scriptures, and to search for a rational interpretation of the Bible and the Christian dogma. One can imagine how Aristotle would wonder at seeing such major faiths struggling with his writings, a struggle whose result would affect millions of lives!

As an example of the subtle intellectual pursuits that rocked the theology and philosophy of the time let me mention the controversy concerning "universals," that is, the question of whether collective names (such as "men") represented real objects, or were mere idealizations. The issue was that if only individuals had real existence while universals were abstractions, then the Holy Trinity, and even the Church itself, could be but abstractions. Roscelin claimed that universals *were* abstractions, and was of course punished by the Church. Anselm argued for the reality of universals and was accordingly praised and supported by the clergy. Abélard took a middle course stating a universal is real in as much as its individual components are real, but it has no existence separate from their parts. The Church rejected his ideas, condemning him to a life of strife and controversy. This and similar issues, which are so monumentally unimportant today, were of paramount relevance during the Middle Ages; men were ruined and dragged in the mud, or rose among universal praise, according to their position in such matters. *Sic transit gloria mundi.*

The concern of medieval thinkers with philosophy and religion lead to a relative stagnation in scientific pursuits. But this lack of activity was not due to stupidity or monomaniacal concern with religion, but because, on the one hand, there were no economic demands for dramatic technological improvements, and on the other hand, because the most fundamental questions addressed by modern science were then thought to have been definitively answered by theology: religion provided a description of the origin, fate, and constitution of the Universe (Figure 3.3), and a rationale for the behavior of all objects, animate or inanimate. With such satisfactory (for that period) explanations, and with the unconditional and almost universal acceptance of this paradigm, the mind naturally became concerned with improving our position in the universe, with providing an accurate interpretation of the Scriptures as the source of these momentous answers, and with the drive to have a complete intellectual structure that combined the Bible with the solid intellectual reasoning represented by Aristotelian philosophy. This relatively stable state of affairs was not to last: the economic developments of the following centuries, the enlargement of horizons brought by exploration, and natural human curiosity would generate questions and needs that assailed the comfortable medieval world picture and eventually destroyed it.

Medieval skepticism: from Ockham to Remus

The continual re-examination of the Christian doctrine, the attempts to make it consistent with logic and philosophy, and the eventual realization that some of the basic religious precepts could not be proved through reason also led to a movement aimed at clarifying the logical structure of religious dogma; its members came to be known as skeptics. William of Ockham and, later, Petrus Remus were the most notable representatives of this group.

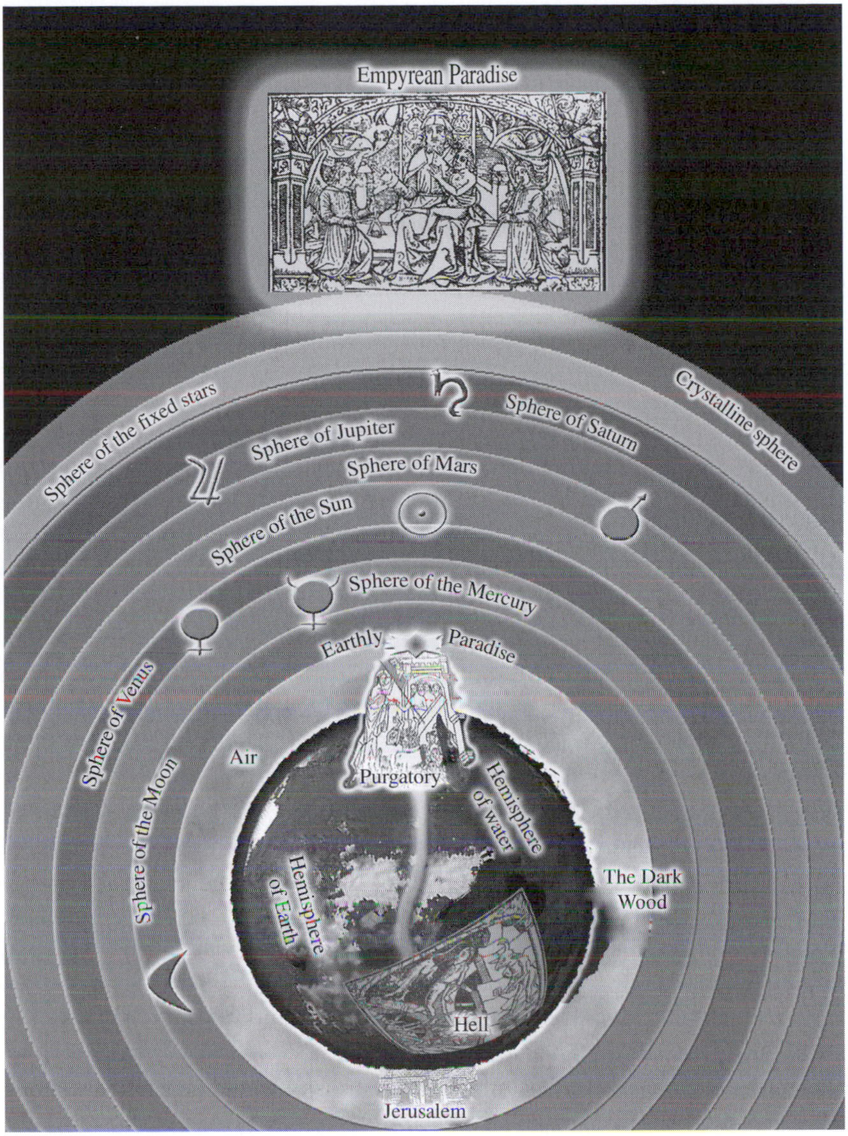

Figure 3.3 A summary of the Medieval universe.

William of Ockham lived a life of controversy. He was recognized as the most forceful thinker of the age, and the universities shook with disputes over his ideas. In philosophy he admitted no authority but experience and reason, and this led him to question many accepted tenets of the faith. His occasional sallies into physical science – his perception of a law of inertia, his doctrine of action at a distance – stimulated thinkers from Jean Buridan to Isaac Newton; but his most momentous contributions concerned the logical consistency of ideas. Using a precise (though often dry) terminology, Ockham managed to avoid the inaccuracies of the terms often present in the gothic edifice of medieval philosophical abstraction. This clarity of ideas led him to the famous formula, traditionally called "Ockham's razor": *entia non sunt multiplicanda praeter necessitate* – entities are not to be multiplied beyond need.[4] This postulate was not new; Aquinas had accepted it, Scotus had used it; but it was in Ockham's hands that it became a deadly weapon, cutting away occult fancies and grandiose abstractions in secular and religious philosophy. William argued that conclusions reached by reasoning have meaning only in so far as they refer to experience that involves individual entities or acts, otherwise they are but vain and perhaps deceptive abstractions. Our ways and means of perception mold and limit our knowledge, which should then never be thought to be the objective or ultimate truth about anything.

The general effect of these arguments was to undermine the basic (Scholastic) assumption of thinkers, such as Abélard and Aquinas, that medieval Christian dogma could be proved by reason; Scholasticism survived, but never recovered from these blows. The "razor" also led Ockham to reject several of Aristotle's arguments. For example Aristotle argued for the existence of a "prime mover," whose action was supposed to generate the motion of all celestial objects, without itself being affected. Ockham noted that the existence of this prime mover could not be justified by the observation that celestial bodies move.

The razor also took William along more perilous paths: he argued that since nothing can be known save through direct perception, we can never have any clear knowledge that God exists. He then concluded that monotheism and polytheism are equally rational; that there may be many worlds, and more gods to govern them. Facing these conclusions, and aware of the ruin of theology to which they could lead, Ockham sacrificed reason to faith. He argued that, though it could not be proved, it was probable that God exists, and he endowed each of us with an immortal soul. Reverting to Scotus' Scholasticism, he also argued that we must distinguish between theological truth and philosophical truth, and despite reason's doubts, accept in faith the basic tenets of the Christian religion; philosophical and religious truth must be distinct and distinguished. Nonetheless Ockham did apply his "razor" to the dogmas and rites that the Church had added to early Christianity, noted many of them led logically to intolerable absurdities,[5] and demanded a return to the simpler creed and worship of the New Testament.

Due to its simplicity and its universality Ockham's razor has survived as a rigorous test for a variety of intellectual constructions. It provides now, just as it did 650 years ago, a way of separating what is assumed for convenience or prejudice from what is required from observation and experimentation. This is particularly useful when faced with new data leading to hypotheses, for then the old prejudices should be treated differently from the old data: the new hypotheses cannot ignore the data, but are quite free to dispose of (or modify or replace) any or all preconceptions as needed. Yet, despite the cleanliness of Ockham's arguments, and their fiery revival by Petrus Ramus, Scholasticism (the attempt to rationally interpret all issues of faith) and Aristotelian philosophy and logic, though damaged by the skeptics' arguments, eventually regained their ascendancy. By the sixteenth century almost all anti-Aristotelian pursuits came to a dead end, and philosophy reached a plateau. Most thinkers were concerned with theology, Luther condemned skepticism and rejected reason since it might lead to atheism; the Inquisition gained power with a corresponding increase of repression of philosophical pursuits.[6]

In other areas, however, new ideas were forming: a spirit of doubt appeared in politics in the form of attacks on the divine right and inviolability of kings (usually Protestant thinkers uncomfortable under Catholic rulers, or Catholic thinkers smarting under the triumph of the State), while in science the field was ripe for the Copernican revolution.

The magic of the time

In parallel with the lofty thoughts of philosophers and theologians, magic, alchemy and astrology flourished and touched all human activities. Both science and religion (though for very different reasons) condemned and attacked these pursuits, and still they prospered. Knowledge could be allegedly obtained by magic, and there were many accepted methods of divination. Medicine contended at every step with astrology, theology, and quackery, and was regarded almost as a branch of theology with illness believed to result from evil influences (a belief supported by Luther and Aquinas); seldom has the belief in demonic possessions been so widespread.

These ideas not only impeded scientific growth, but were also threatening to the authority of the Church, and consequently were severely condemned. Witchcraft was at first tolerated by the Church hoping and trusting it would die out, but it grew and spread instead. So in 1298 the Inquisition changed tactics and started suppressing witches with fire. For example, Cecco d'Ascoli, an astrology professor at the University of Bologna, dared trust astrology so completely that he cast Christ's horoscope stating that it made the crucifixion inevitable, and was burned at the stake (1327) for denying the freedom of will.

Even after the Renaissance the life of the mind had to fight for air in a jungle of superstition, intolerance, and fear. Famine, plague, and war laid waste on the population, and led people to seek explanations for such miseries in occult forces or take a mystical escape from so harsh a reality. Demons lurked everywhere, especially in one's bed. Anyone might inadvertently cross the threshold into the realm of magic; every natural object had supernatural qualities. Most Europeans believed in sorcery (including Thomas More and Erasmus), a belief from which the so-called Dark Ages had been comparatively free.[7] But the emphasis laid by preachers upon the reality of hell and the wiles of Satan strengthened popular

Figure 3.4 An alchemist's laboratory. (Print by Brueghel the Elder, engraved by Cock, sixteenth century)

belief in the ubiquitous presence of devils; and many a diseased mind or desperate soul harbored the idea of summoning such evil to its aid. This pandemonium of occultism continued to surround, confuse, and even – as in the case of the sixteenth-century alchemist Paracelsus – threaten the sanity of the would-be scientist. The ecclesiastical response strengthened accordingly, reaching appalling ferocity by the sixteenth century, under Catholic and Protestant auspices alike; in 1554 an officer of the Inquisition boasted that in the preceding 150 years the Holy Office had burned at least 30 000 witches.

Alchemy, with its claims of instant riches through elemental transmutation was assiduously pursued from late Roman times. But the many unsubstantiated claims and unfulfilled promises eventually led to its losing some of its reputation. Still the adepts continued to carry both honest research and chicanery through the sixteenth century, protected by the inherent secrecy of their "art" and supported by human greed. They were denounced by royal and papal edicts, yet used by some kings in an attempt for a facile replenishing of exhausted treasuries (Figure 3.4).

Astrology is far from being a medieval invention, having by then an ancestry of at least 40 centuries, but it reached an enviable popularity during this time. Our age, despite its efforts, has not matched the Middle Ages in the universality of the belief that the future is written in the stars, and that we can devise the means to decipher it. If heavenly bodies affect the climate on Earth, why should not such bodies affect growth, nature, illness, the fate and character of men and states? The stars, following their predictable paths supposedly guided by angels, were studied as a guide to navigation and to date religious festivals, and also to forecast terrestrial occurrences and personal destinies. The influence of Sun and Moon on climate,

Box 3.6

Robert Grosseteste (*c.* 1168–1253)

Born in Suffolk of humble parentage, served as a cleric in the households of the bishops of Lincoln and Hereford; graduated from Oxford and took his divinity degree in 1189, being the first of a thousand brilliant minds whose achievements created the magnificent prestige of Oxford in the educational and intellectual world. He was chosen as Oxford's first chancellor; while in this office was also elected Bishop of Lincoln (1235), and superintended the completion of the great cathedral. Grosseteste was an ardent reformer, committed to his pastoral duties; he energetically promoted the study of Greek and Hebrew, with a view to converting the Jews (whom he protected from the frequent attacks of the populace). He was a lawyer and a physician as well as a theologian, a scientist and an active social reformer; loyal to the Church, but daring to send Pope Innocent IV a memorial in which he ascribed the shortcomings of the Church to the practices of the Papal Curia. He wrote French poetry and a treatise on husbandry; he shared in the heroic effort of the thirteenth century mind to reconcile Aristotle's philosophy with the Christian faith. He worked for a reform of the calendar; he understood the principles of the microscope and the telescope, and opened many paths for Roger Bacon in mathematics and physics. He insisted on the need for experimental verification of scientific claims, and on the notion that all science must be based upon mathematics.

season, and tide, the subservience of agriculture to the whims of the weather, even the lunar periodicity of women, seemed to justify the claims of astrology that the heavens of today forecast the events of tomorrow. And so, any household that could afford it had an astrologer, and astrology was taught at all respectable universities, encompassing and feeding on astronomy. Predictions of the heavenly influences were regularly published and reached a wide and avid audience.

The most illustrious (and often the most illustrated) dared not take any important action without assurance from the astrologers that the stars favored the enterprise; there were 30 000 astrologers in Paris in the sixteenth century (Figure 3.5). Even thinkers of the quality of Aquinas believed that angels guided the celestial bodies in their motions, and allowed much truth to the belief that celestial processes influence or determine human affairs. He wrote:

> The movements of bodies here below . . . must be referred to the movements of the heavenly bodies as their cause . . . That astrologers not infrequently forecast the truth by observing the stars may be explained in two ways. First because a great number of men follow their bodily passions, so that their actions are for the most part disposed in accordance with the inclination of the heavenly bodies; while there are a few – namely, the wise alone – who moderate these inclinations by their reason . . . Secondly, because of the interference of demons.[8]

Figure 3.5 Fifteenth-century astrologers. (Astrologers of the fifteenth century (Tarot card said to have belonged to Charles VI, Bibliothèque Nationale, Cabinet des Estampes))

It was during this period that Nostradamus wrote his book of prophecies written in such a cunningly ambiguous way that any section could be (and has been) applied to almost any event in later history. Illness was also a stellar responsibility, and most physicians related prognosis and cause of a disease to the constellation under which the sufferer had been born or taken ill; the very name influenza indicates a belief in its celestial origin. Some astrologers

Box 3.7

Roger Bacon (*c.* 1214–92)

Doctor mirabilis, Bacon was an experimentalist who urged the need for observation in science, and believed that intellectual comprehension came from God alone. Born in Ilchester, Somerset, to a knightly family, he probably studied at Oxford and pursued his studies at Paris where he was one of the first to lecture on Aristotle's natural philosophy, which had hitherto been banned. He entered the Franciscans around 1256, but his opinions were seemingly denounced by the Master General of the order. Bacon considered theology the mistress of the sciences, and regarded scientific study as a tool for illuminating theological truths. He fervently believed in the unity of knowledge and maintained that an understanding of the Scriptures and Nature would lead to comprehension of God. Inspired by Augustine's *City of God*, he conceived a Christian world united by faith. Bacon wrote a variety of works in a number of genres including treatises on medicine, commentaries, polemic and scientific explanation. In the latter he was heavily influenced by Robert Grosseteste. In 1266 Clement instructed him to present a summary of his thoughts; he responded by writing a seven-part encyclopedic *Opus Maius*, followed by the *Opus Minus* and *Opus Tretium*. Bacon's image as a prescient scientist dates from the Elizabethan era; during this time his works (especially those dealing with science) were rediscovered and popularized.

predicted a second Deluge for February 11, 1524 (when Jupiter and Saturn coincided in Pisces); thereupon Toulouse built an ark, and cautious families stored food on mountaintops, only to be (hopefully pleasantly) disappointed. Despite these failures magic flourished.

Reason tiptoed among sorcery, witchcraft, necromancy, palmistry, phrenology, numerology, divination, prophecies, fateful stellar conjunctions, chemical transmutations, miraculous cures, and occult powers in all of Nature, marvels that remain with us to this day, apparently deathless, yet not with the preponderance reached between the tenth and sixteenth centuries. This bonanza, perhaps fueled by a need to feel connected with a Universe that was becoming increasingly complicated and impersonal, carried with it a suppression of other pursuits, especially those scientific. Still the deficiencies, unreliability and mystery inherent to occultism served and motivated thinkers such as Bacon and Grosseteste to fertilize the ground for the ideas of Copernicus, Galileo and Newton.

The science of the time

Though Rome had valued science, it apparently forgot the pure approach of the Greeks; already in the works of the elder Pliny we find superstitions on every other page. This indifference of Romans and early Christians toward scientific progress managed to almost dry up the progress in this area long before the Barbarians took over. What remained of Greek science in Europe was buried in the libraries of Constantinople, and these suffered terribly during the sack of 1204. Fortunately Greek science was not lost: having migrated into the Islamic world in the ninth century, it stirred Moslem thought and helped in one of the most remarkable intellectual awakenings in history. While Christian Europe struggled to lift itself from barbarism and superstition, Islamic philosophers received this classical knowledge, preserved and expanded it (though without breaking the ferrets of the Aristotelian assumptions). This enlightened era lasted 300 years; by the tenth century Islamic science had stagnated and after the twelfth century Islam waned. It was at this turning of the tides that the West became re-acquainted with classical philosophy and science, partly through the cross-pollination fostered by Jewish thinkers who were forced to emigrate from the Islamic world by growing insecurity, poverty, instability, and a climate hostile to intellectual pursuits.[9] These philosophers carried their knowledge to Christendom and then dispersed it throughout Europe, as they were persecuted and expelled from one country to the next. Of particular note is the astronomer Levi ben Gerson (1288–1344) who argued that, "The Torah cannot prevent us from considering to be true that which our reason urges us to believe." Among his many writings he discussed, and rejected, the heliocentric hypothesis. His followers believed that the Cosmos was created out of an inert, formless mass that had existed from eternity.

There is a tendency to dismiss the writings of medieval thinkers pertaining to scientific issues since they are almost always tinted with the theology and philosophy of the time, and this philosophy often hindered the development of the subject. But ignoring the results of people of the stature of Grosseteste and Roger Bacon produces the misleading impression that the creativity of Copernicus, Galileo, and Newton occurred spontaneously in an intellectual vacuum; it also presents a picture of complete medieval obscurantism. These (and similar) perceptions are quite inaccurate. Dismissing medieval thinkers ignores their achievements, reached *despite* the constraining ferrets of the pervading theology and philosophy and the constant occultist attacks on reason. It also creates the misleading impression that after the sixteenth century scientists became immune to their surroundings, pursuing their high trade in the impersonal vacuum of pure thought.

The infallible researcher concerned only with abstractions is, of course, unrelated to the reality of scientific endeavor, and implicitly suggests a false superiority and infallibility of our current theories. We can only imagine the smiling condescendence with which future researchers will discuss our most pressing questions, noting the simple misleading errors that lead us into dead alleys with such assurance. Our advantage over Bacon is not our being divorced from society and its effects, but the availability of the scientific method that allows us to separate the results from such influences. The topics of research, the questions we ask, the areas we select, are influenced and often determined by philosophical prejudices and practical needs; this was as true in Bacon's time as is today.

Notable among medieval scientists was Robert Grosseteste, the first in an illustrious line of Oxford alumni of the thirteenth century. In relation to science Grosseteste noted the need of discovery through experimentation and the fact that Nature is amenable to a mathematical interpretation. These ideas were accepted and made more precise by Roger Bacon; he also made few practical contributions to science except in optics and in devising a new more accurate calendar, but his ideas concerning the ways and method of science opened the door to our current approach, which has proved so fruitful. More clearly than Francis Bacon, Grosseteste stated that, in science, experimentation is the means, method, and verification of scientific ideas; it is the *only* way of obtaining incontrovertible proof. Prior to experimental verification all claims can be doubted: only a burned finger decides whether fire burns. Once obtained, no conclusions are rigorous unless cast in mathematical form; and, with an echo of Pythagorean ideas, ultimately all natural science is mathematics. Non-spiritual phenomena are results of matter and forces, which act uniformly and consistently and can be expressed in terms of mathematical formulas.

Though the Aristotelian picture of the Universe (as interpreted by the Church) still ruled supreme during this period, the first cracks in its armor began to appear by the end of the Middle Ages, and these developments paved the way for the scientific revolutions of the sixteenth century. For example, Nicole Oresme (*c.* 1323–82) and Nicolas of Cusa (1401–64) argued that the speed of a falling body increases regularly (to be proved later by Galileo), and also noted that the heliocentric theory was quite viable, despite the old Ptolemaic objections; Cusa also championed the importance of experimentation in scientific investigations and opened the possibility that celestial objects do not follow circular orbits. Albert of Saxony rejected the Aristotelian objections to the existence of a vacuum, and Jean Buridan (*c.* 1299–1359) argued that given an initial impetus (presumably of divine responsibility), the daily rotation of the Earth explains the motion of celestial bodies, whose behavior can be described by the same mechanical laws that operate on Earth. Buridan also anticipated Galileo, Descartes, and Newton by stating what is now known as the law of inertia: a body continues its state of motion or rest until affected by a force. These propositions, now so trite, were then revolutionary and damaging to the prevailing medieval world view.

Nonetheless, all astronomical investigations assumed implicitly the accuracy of Ptolemy's *Almagest* and most astronomers up to the sixteenth century believed in a geocentric system with celestial bodies fixed on crystalline spheres steered by divine intelligences. At the center of it all was man, that despicable worm tainted with sin and mostly condemned to the hell of the theologians. The alternative heliocentric model of Aristarchus and others was preserved through the works Macrobius and John Scotus Erigena, but was not shown to be viable until Copernicus.

One must not, however, use these glimpses of modernity to disparage the achievements of the sixteenth and seventeenth centuries, and label them but reflections of medieval ideas. Cusa, Buridan, and even Bacon merely *stated* their belief that the classical interpretations or descriptions of Nature might be misleading or wrong, but in supporting these claims provided no proof (nor the means for obtaining it); they merely noted that it was not logically inconsistent for there to exist alternative explanations to the Aristotelian view. This did open the way for future investigations, but did not supersede them. It required the genius of Newton and Galileo, of Copernicus and Kepler, to create an edifice which would withstand experimental tests; they did not provide possibilities, they *proved* them.

Medieval epilogue

The road from the Middle Ages through the Renaissance and the Reformation provides a remarkable testament of the evolution of human thought. During the Middle Ages religion and philosophy learned the subtle art of peaceful coexistence, and thus provided a stable intellectual environment that proved so fertile to new pursuits; and, indeed, the advances in science during the fifteenth century, the time of Copernicus and Vesalius, were epochal. This was also the age of exploration during which the known Earth was doubled, and as a result the world view was changed as never before in recorded history. Knowledge was growing rapidly in scope and spread; the use of the vernacular in science and philosophy was extending to the middle classes the instruction and ideas formerly confined to scholars and priests. The mold of belief and the hold of authority had been broken and humanity had left the comfortable and limited medieval universe and started its first timid steps into the mysterious, exciting and often frightening world of modern ideas.

The heliocentric revolution

The closing of the sixteenth century marked an expansion in commerce, a stabilization of monarchies and other governments and an increase in wealth. With this came a reinvigoration of technological and scientific pursuits. Better and more reliable ships were designed and built, more accurate time-keeping devices constructed, and such developments required a better and deeper understanding of the workings of Nature, which Aristotelian physics was incapable of providing.

The Church continued its oversight of human activities, but its power was waning. The freedom that gushed from the Renaissance and Luther's schism permanently weakened its grip on society. It was in this cautiously optimistic climate that Copernicus reintroduced the heliocentric hypothesis that had languished for centuries under the weight of Aristotle's opposition.

Box 3.8

Nicolaus Copernicus (Mikolai Kopernik, Niklas Koppernigk; 1473–1543)

Was born at Thorn on the Vistula, West Prussia, which then belonged to Poland. His mother came of a wealthy Prussian family, his father hailed from Cracow, and dealt in copper. When the father died (1483), the mother's brother, Lucas Watzelrode, Prince Bishop of Ermland, took charge of the children. When 18, Nicolaus was sent to the University of Cracow to prepare for the priesthood. Not liking the Scholasticism that had there suppressed Humanism, he persuaded his uncle to let him study in Italy. Acquiescing, the uncle had him appointed a canon of the cathedral at Frauenburg in Polish East Prussia, and gave him leave of absence for three years. At the University of Bologna (1497–1500) Copernicus studied mathematics, physics, and astronomy. There he learned the intricacies of the Ptolemaic system and was also introduced to the works of ancient Greek astronomers such as Aristarchus of Samos who had questioned geocentrism. At 27 Copernicus went to Rome where he lectured, and it is said that he tentatively proposed the motion of the Earth. He then returned to his duties as canon in Frauenburg; but his mind kept circling the heliocentric idea. He begged (and obtained) permission to resume his studies in Italy, now in medicine and canon law – which to his superiors seemed more appropriate than astronomy. He received a degree in law at Ferrara (1503), but apparently took no degree in medicine, and again reconciled himself to Frauenburg. In 1506 his uncle, probably to give him time for further study, appointed him his secretary and physician; and for six years Copernicus lived in the Episcopal Castle at Heilsberg. There he worked in manuscript the basic mathematics of his theory, which he felt could explain the data more compactly. When his kind uncle died, Copernicus again returned to Frauenburg and practiced medicine, treating the poor without charge. He represented the cathedral chapter on diplomatic missions, and prepared for King Sigismund I of Poland a plan for reforming the Prussian currency; and he continued his astronomic researches. In 1514 he published his *Little Commentary*, favored by Leo X who asked for a demonstration of this thesis; but Luther and Calvin rejected it, noting that it contradicted the Scriptures. Copernicus did not pursue his astronomical research. Instead he proceeded with his duties and even delved into politics; in his 60s he was accused of having a mistress. In 1540 Copernicus' assistant, Georg Rheticus, published a simplified version of the heliocentric system; Copernicus completed a full version in 1542, only shortly before his death one year later.

The need for a new theory

With the end of the Middle Ages came a series of mathematical advances (the introduction of the plus, minus, and equal signs, new treatises in trigonometry, the adoption of algebra as a tool for calculations, etc.) that allowed an increasing precision in astronomical calculations. Astronomical instrumentation, however, did not keep pace and was still limited to medieval devices (celestial and terrestrial spheres, Jacob's staff, astrolabe, armillary sphere, quadrant, clock, compass, etc). With modern mathematics and primitive equipment Copernicus moved the Earth.

The heliocentric hypothesis of Aristarchus had been dismissed for it could not compete in accuracy with Ptolemy's geocentric one. Still Ptolemy's model was exceedingly complex and had been repeatedly questioned during the Middle Ages (Nicole Oresme, Nicholas of Cusa); and during the Renaissance by Leonardo da Vinci (who wrote, "The sun does not move . . . The earth is not in the center of the circle of the sun, nor in the center of the universe."). But these statements were not supported by calculations, let alone predictions. Copernicus is famed not for what he said, but for what he could *prove*.[10]

A bit of history

Copernicus spent a life alternating between scientific studies, priestly duties, medical practice, and even diplomacy and high finance. But his heart was in astronomy; his research and readings led him to the belief that the heliocentric theory could explain the observed phenomena more compactly than the Ptolemaic view. His location was not propitious for these pursuits: he lived in Frauenburg, near the Baltic, which was often shrouded in mists or clouds; no wonder he was so fond of the Sun! His astronomical observations were neither numerous nor precise, but neither were they vital, for he could use Ptolemy's own data (obtained in the Egyptian climate Copernicus envied), and that of his followers, to gauge the efficacy of the heliocentric view.

About 1514 Copernicus published the gist of his ideas in his *Little Commentary*. Curiously this revolutionary publication by a respected astronomer, containing supporting evidence against the time-honored and Church-sanctioned geocentric model, was mostly ignored. Copernicus, discouraged by this reaction, withheld the publication of an extended version of his theory, and dedicated himself to more rewarding pursuits. In 1540, however, the 67-year-old Copernicus allowed G. Rheticus, his admirer and assistant, to publish a simplified version of the heliocentric ideas, and he sent a copy to Melanchthon, who was not convinced; again, this publication had little impact. Still, when Rheticus went to teach at Wittenberg and praised the Copernican hypothesis during a lecture, he was "ordered," he says, to discuss instead other topics; and yet, despite these worrisome signs, he (and others) repeatedly begged Copernicus to publish his *magnum opus*.

Finally, and perhaps feeling that his life was coming to a close, Copernicus agreed. He made some final additions, and in 1542 allowed Rheticus (who assumed all costs and risks) to send the manuscript to a printer in Nuremberg. By then Rheticus had left to teach in Leipzig, so he asked his friend A. Osiander, a Lutheran minister at Nuremberg, to see the book through the press. Osiander had already written to Copernicus and Rheticus suggesting that the new view should be presented as a hypothesis rather than proven theory (an approach followed in some of Copernicus' own writings), hoping by this device to appease both Aristotelians and theologians. It is unclear whether Copernicus acquiesced to this idea, for his position on the matter was ambiguous, referring to his ideas as a hypothesis, but also stating that they are supported by "the most transparent proofs." Nevertheless Osiander did append an unsigned preface to Copernicus' book emphasizing his proposed tentative view.

The book appeared at last, in the spring of 1543, Copernicus' *annus mirabilis*, with the title *Nicolai Copernici revolutionum liber primus (First Book of Revolutions)*; later the book came to be known as *De revolutionibus orbium coelestium (On the Revolutions of the Celestial Orbs)*. One of the first copies reached Copernicus on May 24, 1543 as he was on his deathbed. Legend has it that when he read the title page, he smiled, and in the same hour died. The dedication to Pope Paul III was itself an effort to disarm resistance against his hypothesis, which he well knew directly contradicted the letter of the Scriptures. He professed to have avoided theories altogether foreign to orthodoxy; and to have published only at the insistence of various learned churchmen (among whom he wisely avoided mentioning Rheticus, who was Lutheran). He acknowledged his debt to Greek astronomers, but unaccountably omitted mentioning Aristarchus by name. He notes his belief in the need of a better theory than the Ptolemaic, which had trouble calculating the length of the year accurately. And he appeals to the Pope, as a man of learning, to protect him against "slanderers" who, without adequate mathematical knowledge, would judge him or would "attack this theory of mine because of some passage of Scripture . . ." In any case Osiander, without appending his own name, prefaced the book as follows:

> To the reader, concerning the hypotheses of this work.
>
> Many scientists, in view of the already widespread reputation of these new hypotheses, will doubtless be greatly shocked by the theories of this book . . . However . . . the master's . . . hypotheses are not necessarily true; they need not even be probable. It is completely sufficient if they lead to a computation that is in accordance with the astronomical observations . . . The astronomer will most readily follow those hypotheses which are most easily understood. The philosopher will perhaps demand greater probability; but neither of the two will be able to discover anything certain . . . unless it has been made known to him by divine revelations. Therefore let us grant that the following new hypotheses take their place beside the old ones which are not any more probable. Moreover, these are really admirable and easy to grasp, and in addition we shall find here a great treasure of the most learned observations. For the rest let no one expect certainty from astronomy as regards hypotheses. It cannot give this certainty. He who takes

> everything that is worked out for other purposes, as truth, would leave this science probably more ignorant than when he came to it . . .

This preface has often been condemned as an insolent interpolation. Copernicus may have resented it, for the old man, having lived with this idea for 30 years, became convinced of the reality of the heliocentric ideas. But the preface reduced the natural resistance of many minds to a disturbing and revolutionary idea, it allowed an easier dissemination of the book; and it is a good reminder that our descriptions of the Universe are the fallible.

In 1581 Bishop Kromer raised a monument to Copernicus against the inner wall of Frauenburg Cathedral, next to the canon's grave. In 1746 the monument was removed to make a place for a statue of Bishop Szembek. Who was he? Who knows . . .?[11]

The Copernican system

The fundamental ideas of Copernicus' heliocentric model, as published in his *Little Commentary*, were:

- There is not one center of all the celestial circles or spheres.
- The Earth is not the center of the Universe, but only of gravity and of the lunar sphere.
- All the planets move in circles with the Sun as their midpoint; the Sun is the center of the Universe.
- The distance from the Earth to the stars is much larger than that to the Sun.
- The motions in the firmament arise from the Earth's motion. The Earth, together with its circumjacent elements, completes a daily rotation on its fixed poles, the firmament and highest heaven abide unchanged.
- The motions of the Sun arise not from its motion but from the motion of the Earth and our sphere, with which we revolve around the Sun like any other planet.
- The retrograde and direct motion of the planets arises not from their motion but from the Earth's. The motion of the Earth alone, therefore, suffices to explain the heavens' apparent inequalities.

These ideas are repeated in *De Revolutionibus* (Figure 3.6). There the exposition begins with postulates: (1) the Universe is spherical; (2) the Earth is spherical – for matter, left to itself, gravitates toward a center and arranges itself into a spherical form; (3) the motions of the celestial bodies are uniform circular motions, or are composed of such motions – for the circle is the "most perfect form," and "the intellect shrinks with horror" from the supposition that the celestial motions are not uniform. He sums up his system in a compact paragraph:

> First and above all lies the sphere of the fixed stars, containing itself and all things, for that very reason immovable . . . Of the moving bodies [planets] first comes

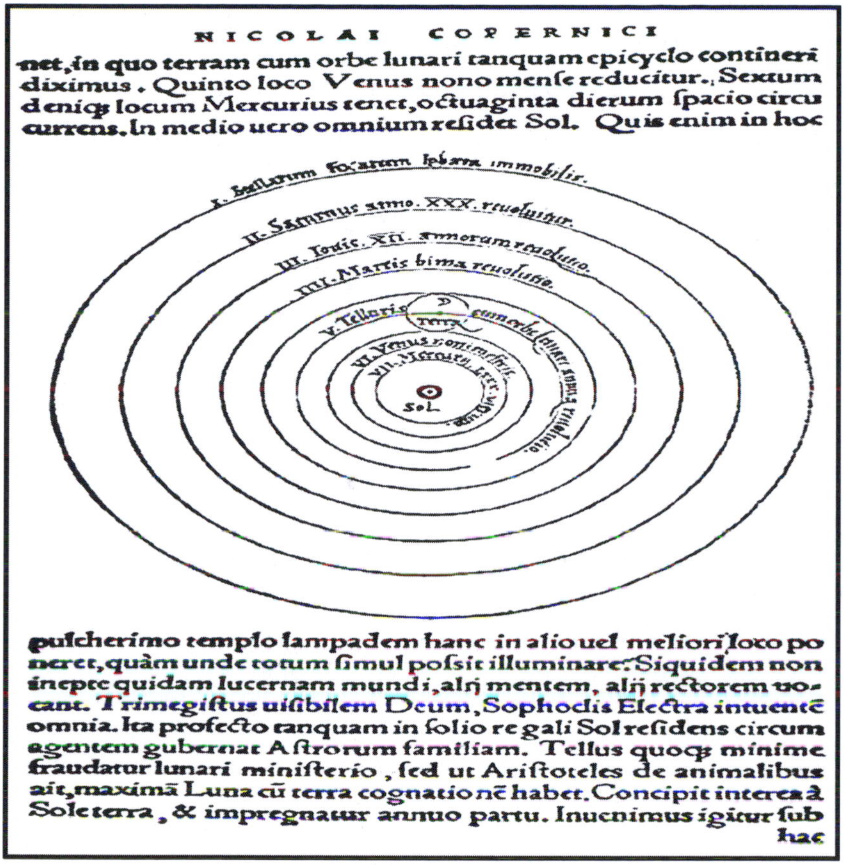

NICOLAI COPERNICI

net, in quo terram cum orbe lunari tanquam epicyclo contineri diximus. Quinto loco Venus nono menſe reducitur. Sextum denique locum Mercurius tenet, octuaginta dierum ſpacio circu currens. In medio uero omnium reſidet Sol. Quis enim in hoc

pulcherrimo templo lampadem hanc in alio uel meliori loco poneret, quàm unde totum ſimul poſsit illuminare? Siquidem non inepte quidam lucernam mundi, alij mentem, alij rectorem uocant. Trimegiſtus uiſibilem Deum, Sophoclis Electra intuentē omnia. Ita profecto tanquam in ſolio regali Sol reſidens circum agentem gubernat Aſtrorum familiam. Tellus quoque minime fraudatur lunari miniſterio, ſed ut Ariſtoteles de animalibus ait, maximā Luna cū terra cognationē habet. Concipit interea à Sole terra, & impregnatur annuo partu. Inuenimus igitur ſub hac

Figure 3.6 Illustration of Copernicus' heliocentric hypothesis (from Nicholas Copernicus *De Revolutionibus Orbium coelestium*, facsimile reprint of the first edition of 1543 (New York: Johnson Reprint, 1965))

> Saturn, who completes his circuit in thirty years. After him Jupiter, moving in a twelve-year evolution. Then Mars, who revolves biennially. Fourth in order, an annual cycle takes place, in which . . . is contained the Earth, with the lunar orbit as an epicycle. In the fifth place Venus is carried round in nine months. Then Mercury holds the sixth place, circulating in the space of eighty days. In the middle of all dwells the sun . . . Not ineptly some call it the lamp of the universe, others its mind, and others again its ruler . . . rightly, inasmuch as the sun, sitting on a royal throne, governs the circumambient family of the stars . . . We find, therefore, under this orderly arrangement, a wonderful symmetry in the universe, and a definite relation of harmony in the motion and magnitude of the orbs, of a kind it is not possible to obtain in any other way . . .

One of the most important features of the Copernican theory is its ability to explain the retrograde motion of the planets. This effect consists of the observation that in reference to the fixed stars planets such as Mars move sometimes faster, sometimes slower, and sometimes even in the opposite direction. Epicycles were invented for the sole purpose of explaining this behavior within

Box 3.9

Giordano Bruno (1548–1600)

Born in Nola, near Naples, he became a Dominican monk and learned Aristotelian philosophy, being attracted to "unorthodox" streams of thought (e.g. Plato). His ideas forced him to leave Naples (1576) and later Rome (1577) to escape the Inquisition. He lived in France until 1583 and in London until 1585. In 1584 he published *The Ash Wednesday Supper* where he defended the heliocentric theory (though he was confused on several points), and *On the Infinite Universe and Worlds* where he argued that the Universe was infinite, strewn with worlds inhabited by intelligent beings. Wherever he went, Bruno's passionate statements led to opposition; he lived off the munificence of patrons, whom he sooner or later outraged. In 1591 he moved to Venice where he was arrested by the Inquisition and tried; eventually being sent to Rome. Despite many opportunities he refused to recant; he was kept imprisoned and repeatedly interrogated until 1600 when he was declared a heretic, guilty of denying the Trinity, the Incarnation and the Transubstantiation. Naked and tongue-tied, he was publicly burned at the stake (for public edification). Bruno's philosophy dealt with many of the cosmological questions of his time; it advanced the possibility of an infinity of inhabited worlds, as well as the astrological influence of the heavenly bodies on human affairs. Scientists such as Galileo and Kepler were not sympathetic to Bruno in their writings, for his frenzied and uncompromising defense of the heliocentric theory produced more opposition than conversions.

the geocentric model, so explaining it using the heliocentric hypothesis was a "must." Copernicus' answer was simple: the effect is due not to the motion on epicycles as Ptolemy claimed, but to the fact that planets move around the Sun at different speeds (Figure 3.6, Figure 3.7).

Effects

The Copernican ideas were certainly not new: they are essentially the same as the ones proposed by Aristarchus (and which brought so much trouble upon his head) preserved and repeated by several luminaries throughout the centuries, including Leonardo himself. So one might rightfully ask why it is that Copernicus is praised as a discoverer of the heliocentric system, and not as one who merely showed its consistency with the data and then popularized it. To understand this it is important to recall that Aristarchus' hypothesis was dismissed for not being able to answer a series of problems, stated by Aristotle and emphasized by Ptolemy, concerning the motion of bodies lying on a planet, such as the Earth, when that planet itself rotates, while moving about the Sun. The main points, together with Copernicus' responses are shown in Table 3.1

Table 3.1. *Copernicus' responses to the problems with Aristarchus' hypothesis*

Ptolemy's complaints	Copernicus' responses
Clouds and surface objects on a rotating Earth would fly off.	This same objection exists, and is more problematic, in the geocentric model where distant planets orbit the Earth at great speeds.
An object propelled directly upward from a rotating Earth would not fall back to its point of origin but be left behind.	Such objects, like the clouds and air, are parts of the Earth and carried along in its rotation.
The annual revolution of the Earth around the Sun should evince a change in the position of the "fixed" stars when observed at opposite ends of the Earth's orbit.	There *is* such a movement, but the great distance to these stars makes it imperceptible (given the accuracy of the data available to both Copernicus and Ptolemy; this motion is now observable).

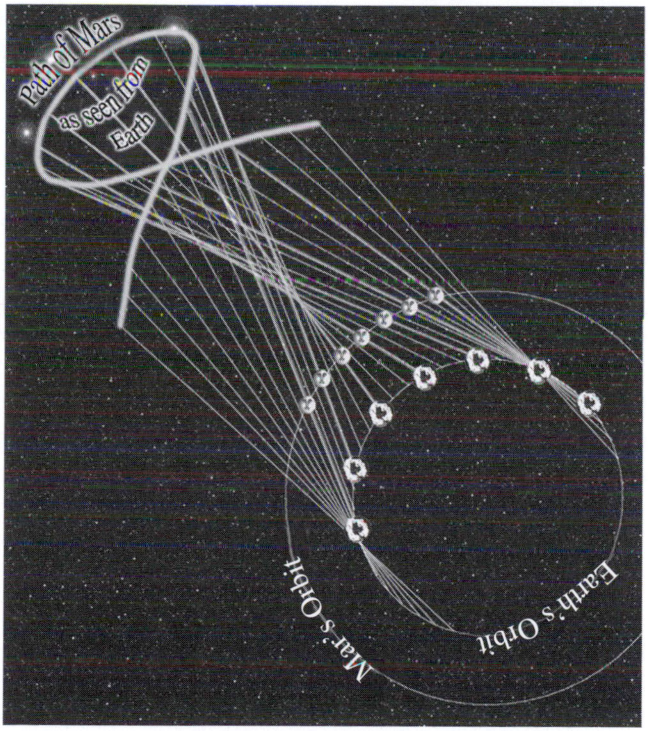

Figure 3.7 Retrograde motion within the Copernican theory. Earth and Mars move at different speeds hence Earth-bound observations of Mars will trace the indicated curve against the background of the fixed stars.

In order to provide these answers to such venerable complaints Copernicus had to re-investigate the properties of motion, to propose new ideas, and then to prove that these new hypotheses were consistent among themselves and with the heliocentric view of the world. Anticipating Galileo, Copernicus notes the relativity of motion: "All change in position which is seen is due to motion either of the observer or of the thing looked at, or to changes in the position of both, provided that these are different. For when things are moved equally relatively to the same things, no motion is perceived as between the object seen and the observer." So the daily rotation of the planets about the Earth could be explained as due to a daily rotation of the Earth on its axis; and the annual movement of the Sun around the Earth could be explained by supposing the Earth to move annually around the Sun.

Copernicus showed that his hypothesis was supported by data, logically consistent, and able to make verifiable predictions; in modern language, it was a falsifiable theory. It also contained the promise of a simplified model of the world which would allow for a more accurate calendar, and so it had its positive practical aspects also. It is not unusual for new hypotheses to carry with them some remnants of the theory displaced; Copernicus based his conceptions on observations handed down by Ptolemy, and he retained some of the Ptolemaic celestial machinery.[12] Later, with the observations of Brahe and Kepler, these constructions would prove unable to explain the increasingly precise data, but the changes required are, to a certain extent, technical: Copernicus adamantly held on to the Aristotelian belief that the motion of celestial bodies must be circular, and this, though a good approximation, is in fact incorrect. There is no preference toward circles in the motion of heavenly bodies, and planets follow a different geometrical figure called an ellipse (and even this orbit is perturbed by the interaction among planets).

The Copernican system moved the Earth to the humble position of a satellite of the Sun, which was endowed with the honor of being at the center of the Universe. With more data and thought the Sun was also to lose this exalted position. Later we will see that the Universe is much more democratic, it has no preferred location, and one may choose *any* point as a reference, be it the center of the Sun, the Earth or our own head. That point is chosen for convenience when investigating one phenomenon and will serve as the center of the Universe for such purposes. Of course the same results can be obtained (though perhaps with more difficulty) for any other choice of reference point.

Knowing that such central parts of Copernicus' theory – that planets "must" move in circles around an absolutely immovable Sun – have proved wrong, we might again ask why he has received so much praise. To understand this one should not forget that the data available to the man was quite consistent with a fixed Sun surrounded by a retinue of planets moving in circular orbits. In fact Earth's orbit differs from a circle only by 1.7 percent,[13] while the Sun ponderously moves about the center of the Milky Way at the rate of 200 million years per orbit; so the heliocentric model of 1543 is, in fact, quite accurate. In addition, and using modern language, Copernicus showed the enormous advantage of studying a system using different reference frames, for there might be one where its behavior could be understood most easily. In later years, as we learned that the Sun itself moves about in our galaxy, and then that the Milky Way itself moves with respect to near-by galaxies, it became natural to ask where is the center of the Universe, about which all these behemoths each containing a billion suns move, or, more suggestively, whether there *is* such a point at all, and even whether such questions make any sense. It would take the work of Newton and Galileo to phrase these questions precisely, and the genius of Einstein to answer them. Copernicus did not travel far along this route, nor were his initial steps infallible, but his is the credit for having found a most fruitful of ideas; like Moses he was to look onto this new land, but not to dwell in it.

The Ptolemaic system, honored by tradition, and consistent with naïve observation and theology had reached its limit of usefulness; its replacement had arrived. But the heliocentric ideas reached much farther than the accurate calculation of astronomical quantities. Instead of being the very center of the Universe, the new system had the Earth as one more planet in an average stellar system, one more speck in a vast Universe. What is the meaning now of the Christian dogma of creation? Where is Heaven, and which direction will the damned take to Hell, now that the traditional locations of these regions as simply being "up" or "down" change as the Earth moves? And why would an all-powerful creator be as concerned with such a piddling world as ours to send his son to die in order to redeem its inhabitants?

The heliocentric astronomy compelled men to re-conceive God in less provincial, less anthropomorphic terms; it gave (Western) theology the strongest challenge in its history, before which the difference between Catholic and Protestant become trivial. Not until Darwin, three centuries later, would religious thought face such controversy.

Followers and detractors

The Copernican system was dismissed or rejected by many, but this was not done out of spite, pigheadedness or ignorance. For example, the new system required the Earth to move at frightening speeds around the Sun, in apparent blaring contradiction to the evidence of the senses. In addition the full theory was hardly simpler than Ptolemy's and gave only slightly more accurate results, so there was no practical reason to make this momentous change. The full impact of the heliocentric idea had to wait for the modifications provided by Kepler, Galileo, and Newton to demonstrate its extreme usefulness and accuracy. Nonetheless there were far-sighted astronomers, such as Rheticus, Osiander, and Digges, who immediately took to (and extended) the Copernican idea. In Digges' version of the theory the "fixed" stars are no longer attached to a distant sphere, but strewn over the Universe; and Giordano Bruno raised the possibility of these being suns like our own, surrounded by inhabited worlds like ours.

The Catholic Church raised no objections so long as it was represented as a hypothesis; but the Inquisition struck back mercilessly when Giordano Bruno stridently asserted this hypothesis was, in fact, a certainty, and made explicit its consequences for religion. In 1616 the Congregation of the Index forbade the reading of *De Revolutionibus* "until corrected." In 1620 it allowed Catholics to read editions from which nine sentences that represented the theory to be a fact had been removed. The book disappeared from the revised Index of 1758, but the prohibition was not explicitly rescinded until 285 years after its publication.

Tycho Brahe, the foremost astronomer of the time, rejected the Copernican hypothesis arguing that it had failed to answer Ptolemy's objections. One of his arguments was roughly the following: if the Earth moves in circles around the Sun, nearby stars will appear in different positions at different times of the year. Since the stars are fixed they must be very far away, but then they should be enormous, and this is "unreasonable" (of course they only need to be enormously bright . . . and they *are*).

Box 3.10

Tycho Brahe (1546–1601)

Born in Denmark, Tycho had a romantic childhood, being abducted by a wealthy and childless uncle who raised him at his castle in Tostrup, Scania, and financed his education. He began studying law at the University of Copenhagen, but soon turned to astronomy, which he pursued assiduously after inheriting his father's and uncle's rich estates (and also supported by the Danish crown). In 1572 he noted the appearance of a supernova (now carrying his name) and showed it to be a star; this discovery and the publication of his observations in *De nova stella* marked his becoming an astronomer with a European reputation; in 1577 he observed a comet and showed that its orbit lay beyond Venus'. In 1576 Frederick II of Denmark granted him the title to the island of Ven, where Tycho established Uraniborg, the best observatory of the time (before the invention of the telescope). He published a comprehensive study of the Solar System and accurately plotted the positions of 777 stars. Tycho was an artist as well as a scientist and craftsman; he established a printing shop to produce his manuscripts, imported craftsmen to construct the finest astronomical instruments; he also invented a sanitary lavatory system. In 1573 he married a peasant's daughter whom he seldom mentions in his extant correspondence; they had eight children, six of whom survived him. Frederick II died in 1588, and under his son, Christian IV, Tycho's star waned: most of his income was stopped, due to lack of funds, to Tycho's unreasonable demands, and his disinclination to carry out certain civic duties. Brahe left Ven and went to Prague as Imperial Mathematician (1599) supported by Emperor Rudolph II; there he first met J. Kepler, who became his assistant. Tycho continued his observations, but his spirit was broken, and he died two years later of a burst bladder, leaving all his observational data to Kepler.

Brahe, however, could not fully ignore the heliocentric arguments, so he produced a "compromise" hypothesis where the Earth does revolve around the Sun, but all the other planets remain faithfully circling the Earth. This, as any other attempt to rescue the Ptolemaic system, was doomed to fail: the observational evidence soon favored overwhelmingly the heliocentric system. The cosmic preponderance of the Earth was but an illusion.

Aristotle in the sixteenth century

On November 11, 1572, Tycho had the fortune to observe in the constellation Cassiopeia what he called a "new star": a star which had not been there before, but which appeared on or around that date with a brightness that surpassed Venus'. He carefully observed the new star and showed that it did not change in position with respect to the other stars, so, accepting as he did the Aristotelian ideas, he concluded this new star must also inhabit the sphere of the fixed stars, a realm of supposedly perfect heavenly bodies. This was an extremely disquieting discovery, for this perfection required the eternal permanence of everything there. In particular, it forbad stars from suddenly appearing and later vanishing (as Tycho's new star was soon to do) without any visible trace. In addition the disharmony implied by such fickle stellar behavior, compounded with the publication of the Copernican hypothesis, shook the confidence in the immutable laws of antiquity and suggested that the chaos and imperfections of Earth were reflected in the heavens. It is hardly surprising that the event was regarded as an ominous one!

The same type of problems arose from Tycho's observations of a comet that appeared in 1577, for he could determine that this object was farther than Venus, again contradicting the Aristotelian idea that the Universe beyond the Moon was perfect, eternal and unchanging.

The contradictions between Tycho's data and the Ptolemaic system provides a good example of a situation where new and more accurate observations proved fatal to the accepted theory, and provided the foundation that would eventually support a better, more precise description of the Universe. The great strength of the scientific framework is to insure, and even demand, a continuous criticism and reexamination in the face of the data.

By the time Tycho made these important observations, most of the medieval approach to physics had also been shed, though not completely. For example, the motion of a projectile was thought to be composed of an initial violent part (when thrown) and a subsequent natural part (which returned it to the ground). Still it was during this time that the importance of velocity and force in determining the motion of objects was realized. The birth of new theories is not easy. In this case it was not until the late seventeenth century that a complete heliocentric view of the Universe was polished and could be used as a tool for investigating Nature; only then was the Aristotelian doctrine finally set aside. The first step in this long road was taken by Copernicus, the next by Johannes Kepler in his description of the motion of the planets, and then by Galileo through his investigations on the nature of motion and of the Solar System; finally, Newton produced his "system of the world" based on his laws of motion and the universal law of gravitation.

Kepler

The Copernican model was slowly disseminated throughout Europe under the watchful eye of both the Catholic and Protestant churches. This prevented a completely open discussion of the merits of this newcomer, but, on the other hand, it forced its advocates to amass overwhelming evidence in order to gain its acceptance. The most prominent of these supporters was Johannes Kepler, whose modification of the original Copernican idea turned the heliocentric hypothesis into a theory whose accuracy Ptolemy could never hope to achieve. But even with the strength of these successes Kepler was cautious: he hid his discoveries by burying them in almost impenetrable Latin prose which he published as a series of works that did not circulate widely.

It was Kepler's belief that Nature was the book where God's plan is written, in such a fashion that divine ideas correspond to geometrical objects (a Platonic concept), and that the human mind is ideally created to understand this structure. This belief was suffused with mystical overtones, for example, he became attached to the image of the Holy Trinity as symbolized by a sphere (God the Father being the center, Christ the Son the surface, the Holy Ghost the intervening space) of which the world as we perceive is (literally) but a reflection. The presence of God as a dynamic and creative being was exemplified by our Sun whose influence generated the motion of the planets. Despite these ideas (or perhaps because of them) Kepler had the perseverance to collect data with the necessary accuracy, the integrity to accept these observations even when they contradicted his initial hypotheses, and the genius to modify the latter to accommodate the former.

Kepler's researches were not restricted to pure astronomy but spanned "the science of the stars," which had mathematical and physical components. The mathematical aspect was rooted in the old Pythagorean and Aristotelian belief that the working of the heavens could be explained using geometry. The physical aspect had an astronomical component (creating tables and constructing instruments), an astrological component (producing forecasts for individuals, cities, states, the weather, etc.), an optical component (understanding light propagation and perception), and a musical component of Pythagorean progeny. Taking an approach perfectly consistent with Ockham's razor he separated all his mystic–religious beliefs from the study of the properties and regularities of planetary motion, even when he felt pressed by the most weighty philosophical arguments he did not ignore the data in favor of a facile acceptance of his preconceptions. A particularly good example of this attitude is given by his development and treatment of his first planetary hypothesis.

Box 3.11

Johannes Kepler (1571–1630)

Was born Weil der Stadt, Germany, died Regensburg, Germany. A renowned astronomer, Kepler also studied astrology, optics, music and geometry; best known as the discoverer of the three laws of planetary motion that provided a solid observational basis for Copernicus' heliocentric system, and defined the path for the astronomical investigations of the seventeenth century that culminated with Newton's discoveries. In Kepler's time, European education was still controlled by the Catholic and Protestant churches; local rulers often provided scholarships to poor boys as means of increasing loyalty and Kepler was one such beneficiary, allowing him to enter the University of Tübingen (1589), aiming to become a theologian. Yet, "Divine Providence guided him to the study of the stars": having learned of Copernicus' theory, Kepler decided to rigorously demonstrate its accuracy. At this time (1600) he became Tycho Brahe's assistant (at miserly wages) and when Tycho died (1601), Kepler succeeded him as Imperial Mathematician to the Holy Roman Emperor Rudolf II. His first publication in this capacity criticized Ptolemaic astrology (though not astrology per se). Tycho's data, which he had refused to share while living, had come down to Kepler (after a squabble with Tycho's heirs) and described Mars' position with an accuracy of 2 arc-minutes. During the years 1601–5 Kepler used this information to devise his first two planetary laws (published in 1609); he also wrote treatises on the nature of light and on the appearance of a new star (the 1606 supernova) which led him to much astrological speculation (e.g. the collapse of Islam and the return of Christ) for it coincided with a conjunction of Jupiter and Saturn. Later (1611) he also published his observations of the Jovian satellites that would support Galileo's momentous discoveries. In 1612 Kepler's life took a turn for the worse. His wife became ill, his three children contracted smallpox and one died; Emperor Rudolf abdicated his throne, and the theology faculty of Tübingen rejected his appointment (his desire for a peaceful coexistence of Calvinists and Catholics had gained him many enemies). He did retain the position of Imperial Mathematician under the new emperor Matthias, and was appointed District Mathematician in Linz. Here he lived until 1626, but his personal tragedies continued relentlessly: he was excommunicated from the Lutheran church; he remarried but five of his seven children from that marriage died in childhood; his mother was accused of witchcraft and only exonerated shortly before her death; in the Counter-reformation of 1625, Catholic authorities removed his library and ordered his children to attend mass. Still he pursued his research, publishing the first textbook of Copernican astronomy modified by his rules of planetary motion and which also contained his third law. In 1627 he published the Rudolphine Tables (named after Emperor Rudolph II) where, using the heliocentric system, he was able to provide the positions of 1005 stars with an accuracy that could not be matched within the geocentric scheme. In this year he left for Sagan in Silesia, supported by the Imperial

Box 3.11 (cont.)

General Albrecht von Wallenstein whose murder in 1634 he forecast in a horoscope (though he was less successful in collecting his back-wages). In 1630 Wallenstein lost his position and Kepler left for Regensburg in hopes of improving his financial situation; soon after arriving he became seriously ill with fever, and on November 15 he died. His grave was swept away in the Thirty Years' War, but the epitaph that he composed for himself survived:

> I used to measure the heavens, now I shall measure the shadows of the earth. Although my soul was from heaven, the shadow of my body lies here.

Kepler's geometric hypothesis of planetary motion

In 1595, while teaching a class in a high school in Graz, Austria, Kepler experienced a moment of illumination. It struck him suddenly that the spacing among the six Copernican planets (Uranus, Neptune, and Pluto would not be discovered for almost 300 years) might be explained by the following purely geometric construction.

He assumed all planets move on the surface of spheres (Aristotle cast a long shadow indeed), then he imagined these six planetary spheres being inscribed and circumscribed by regular polyhedrons, of which he knew there were *precisely* five (as proved by Euclid long ago): cube, tetrahedron, octahedron, icosahedron, and dodecahedron. Explicitly his idea was to (Figure 3.8):

- Put an octahedron around Mercury's sphere; surround this octahedron by a new sphere; this second sphere is Venus'.
- Put an icosahedron around Venus' sphere; surround this icosahedron by a new sphere; this third sphere is the Earth's.
- Put a dodecahedron around Earth's sphere; surround this dodecahedron by a new sphere; this fourth sphere is Mars'.
- Put a tetrahedron around Mars' sphere; surround this tetrahedron by a new sphere; this fifth sphere is Jupiter's.
- Put a cube around Jupiter's sphere; surround this cube by a new sphere; this sixth sphere is Saturn's.

This hypothesis predicted that planets move in circles, it gave numerical predictions for the ratios of planetary orbits, and it also predicted the *number* of planets; in addition it complied with Kepler's deeply rooted belief that Nature should be described in terms of elegant geometrical arguments. To his initial delight he found agreement up to a 5 percent error (except for Jupiter, which he presumably ascribed to inaccurate measurements due to the planet's distance form Earth). But this optimism was premature: the data he later obtained from Tycho could not be reconciled with his hypothesis, nor could the discrepancies

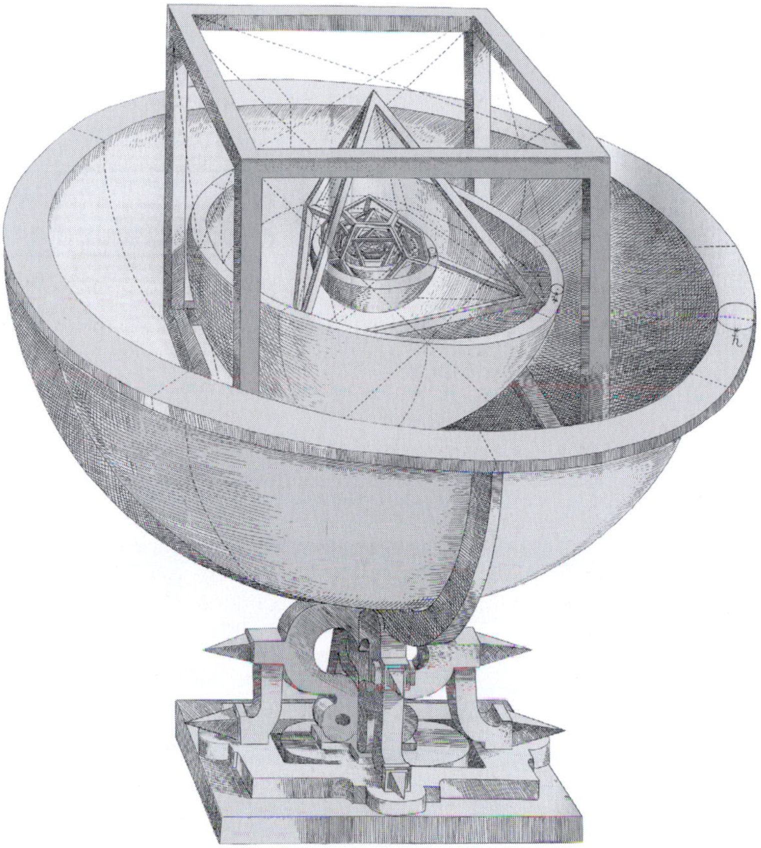

Figure 3.8 Kepler's geometrical model of the Solar System. (from Johanes Kepler, *Gesammelte werke* (Monich: C. H. Beck, 1937) vol. I)

be dismissed, for the data was very accurate. Kepler spent almost two decades trying to make this hypothesis work . . . and failed: the predictions would not match the observations. Faced with this reality he took the very difficult step of dropping this line of investigation, dismissing his cherished geometrical idea as unrelated to the workings of Nature. This work, however, was of some use: because of it Kepler was recognized as "someone" and, in 1600, was hired by Tycho Brahe (then in Prague) as an assistant.

The laws of planetary motion

Even while studying his geometric hypothesis, and especially after he discarded it, Kepler realized the crucial importance of determining the dynamical principle responsible for planetary motion. He was not satisfied with the assumption that celestial beings were responsible for it, but was convinced that the motion of all planets was the result of the Sun's dynamical influence, a consequence of his mystic assumptions concerning the underlying structure of the Universe. In the early 1600s, having read W. Gilbert's *De Magnete* (then recently published),

Table 3.2. *Period and average distances to the Sun for the innermost five planets: a plot of the last two columns gives a straight line as claimed by Kepler's third law (an astronomical unit – AU – is a measure of distance approximately equal to* 1.5×10^8 *km)*

Planet	Period (years)	Avg. distance to the Sun (AU)	$(\text{Period})^2$	$(\text{Average distance})^3$
Mercury	0.24	0.39	0.06	0.06
Venus	0.62	0.72	0.39	0.37
Earth	1.00	1.00	1.00	1.00
Mars	1.88	1.52	3.53	3.51
Jupiter	11.9	5.20	142	141
Saturn	29.5	9.54	870	868

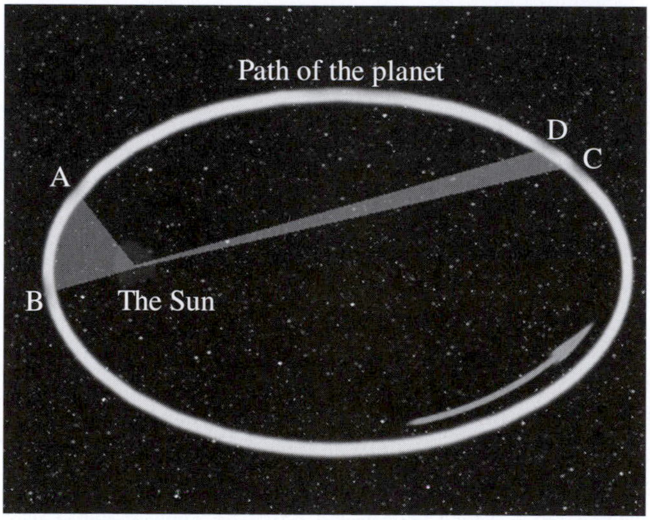

Figure 3.9 Kepler's first two laws of planetary motion. (1) Planets move in ellipses around the Sun and (2) planets sweep equal areas in equal times (that is, if the planet takes the same time to go from **A** to **B** as from **C** to **D**, then the shaded areas are equal).

Kepler became convinced that this Solar influence was of magnetic origin. Though he never managed to turn this idea into a quantitative theory with solid predictions, it still led him to the conclusion that he could extract the nature of the Solar influence from its effects. Accordingly he set out to determine the shape of the planetary orbits with extremely high accuracy, first concentrating on Mars, then on the other planets. From this arduous investigation he elucidated the simple and subtle laws that describe the motion of planets; first describing the orbit of each planet (first and second laws), then the relationship between these orbits (third law). But the nature of the Solar influence remained a mystery that only Newton would decipher. The three laws obtained by Kepler are:

1. Planets move in ellipses with the Sun at one focus (Figure 3.9).
2. Planets sweep equal areas in equal times in their motion around the Sun (Figure 3.9).
3. The average distance to the Sun cubed is proportional to the period squared; see Table 3.2 for the data which led Kepler to this conclusion.

The transformation from the Copernican to the Keplerian universe presents a good example of the evolution of a scientific theory. The data required Kepler to modify the original Copernican hypothesis (planets move in circles with the Sun at the center) to a new hypothesis (planets move in ellipses with the Sun at one focus). He showed that this was the case for Mars, and then checked whether it was also true for the other planets (it was).

Kepler's results divested the heliocentric hypothesis from the prejudices that favored circular motion, the last vestiges of Aristotelian baggage. While he did not know *why* planets followed the regularities he observed (for he did not refer to the above three rules as "laws" at all, but only as a set of celestial harmonies that reflected God's design for the Universe), his results provided Newton with a guiding light. The fact that Newton's predictions agreed with Kepler's laws was instrumental in the acceptance of Newton's theories of motion and of gravitation. In modern language, Kepler's rules for planetary motion are the result of the gravitational force exerted by the Sun on the planets, this force is attractive and directed along the line from the planet to the Sun, and decreases as the square of the distance. While keeping his religious faith fervent and undiminished, he revealed a Universe in which the same mathematical laws ruled the Earth and the stars. "My wish," he said, "is that I may perceive the God whom I find everywhere in the external world in like manner within me."

Notes

1. Paraphrasing Bismarck.
2. This comes from a quote in Luther's *Tabletalk*. One must be aware, however, that this is a compilation of informal after-dinner discussions; Luther's quick and facile dismissal of the Copernican ideas must be understood within this context. In fact, the complaints against Copernicus had a more complicated and diversified basis.
3. W. and A. Durant, *The History of Civilisation* (New York: Simon and Schuster, 1963).
4. Though this precise phrasing does not appear in his (extant) works, equivalent statements do: *pluralitas non est ponenda sine necessitate* – a plurality of entities, causes or factors is not to be posited or assumed without necessity; and *frustra fit per plura quod potest fieri per pauciora* – it is vain to seek to accomplish or explain by assuming several entities or causes what can be explained by fewer.
5. If, for example, Mary is the mother of God, and God is father of us all, Mary is the mother of her father.
6. For example, a Dutch priest was burned in The Hague in 1512 for denying creation, immortality, and the divinity of Christ.

7. Saint Boniface and Saint Agobard had denounced it as sinful and ridiculous; Charlemagne made it a capital crime to execute anyone on a charge of witchcraft; and Pope Gregory VII Hildebrand forbade inquisition to be made for sorcerers as the cause of storms or plagues.
8. Thomas Aquinas: *Summa Theologica*, lcxv 3; xcv 5.
9. For example, in 1305 Solomon ben Abraham ben Adret, a respected and influential rabbi of Barcelona, forbad under penalty of excommunication, the teaching of science or philosophy to any Jew under the age of 25, arguing that such instruction might damage religious faith.
10. Paraphrasing J. Moore.
11. W. and A. Durant, *The History of Civilisation* (New York: Simon and Schuster, 1963).
12. In addition Copernicus believed that in order to explain the precession of the equinoxes the Earth had an additional motion, distinct from its rotation about its axis and around the Sun.
13. For the other planets known to Copernicus the deviations from a circular orbit are: 5.6 percent for Saturn, 4.8 percent for Jupiter, 9.3 percent for Mars, 0.7 percent for Venus and 20.6 percent for Mercury. These effects were obscured by the errors in the data.

4
Galileo and Newton

Europe began the seventeenth century with the weighty legacy of the Copernican revolution and the Protestant schism. Over the following two centuries most of the continent would shed its religious ferrets and give birth to modern science. The paradigm presented by Copernicus required a re-evaluation of the way we perceived the Universe and the way we understood motion. A moving Earth implied that all medieval ideas about a heaven above with its hellish counterpart below could not be literally true, for "up" was now a changing direction and even the place God looked down from became uncertain. Accepting the motion of our world also required new theories of motion that could accommodate ever-changing reference frames; only then could the behavior of falling bodies be reconciled with a rotating Earth.

This period also marked one of the most creative periods in the history of science. A myriad of instruments, from telescopes to air pumps were devised or improved, and then used to prod Nature. The insights of Galileo's and Newton's theories provided an elegant and useful theory of motion, and as a result the behavior of bodies and their interactions could be predicted with unprecedented accuracy. Newton also discovered some of the fundamental properties of that most universal of forces, gravity, and so divested the mechanics of the heavens of its aura of mystery, as the motion of celestial bodies could now be anticipated through routine calculations. The mechanistic universe was born, where natural laws regulate Nature without the need of divine intervention. Science rose to a place of honor in human activities, it was to be the bringer of Utopia and, as today, the savior and hope of mankind. Everything advanced, except, perhaps, man.

The way in which the new theories of motion and gravity were developed changed the way we do science to this day; the understanding gained by Galileo and Newton on the workings of Nature survived with little modification for more than three centuries, and many of their discoveries and insights are as valid today as they were 400 years ago.

The scientific environment

Science in the seventeenth century did not grow in a pleasing and propitious environment. Popular superstitions were as endemic as during the Middle

Ages: elves, hobgoblins, ghosts, witches, and demons patronized every forest, house, and human body;[1] talismans and amulets were endowed with miraculous powers; and kings would cure illnesses by simply touching the afflicted. Superstition eventually abated, but survived, and the majority of people continued to believe in a world populated by good or evil spirits that made our planet their battleground. Natural phenomena and the behavior of all living things could be affected by these beings, who acted without any regard to law or regularity; human reality was a helpless island in a supernatural sea. Astrology, though increasingly repudiated by the learned, was still ubiquitous (horary, judicial, and natural versions are all found in Shakespeare – though this does not prove he subscribed to any of them), and most people used horoscopes in trying to understand past misfortunes and prepare for future ones. Leprosy had mostly disappeared, but epidemics of typhus, typhoid fever, malaria, diphtheria, scurvy, influenza, smallpox, and dysentery scourged the land (two-fifths of all children died before their third year), and the religious and/or astrological interpretations of these calamities efficiently undermined the development of rational medicine.

Belief in magic was not as prevalent as previously, with the exception of witchery, which was believed to be almost epidemic. Witches were assumed to have the power to haunt houses, to engender love or hate, to kill by hurting a wax effigy, and to raise devastating storms (all ably described in King James VI's *Demonologie*, one of the horrors of literature). Though sometimes the officers of the Inquisition professed disbelief in the accusations, and despite condemnation from enlightened popes (such as Urban VIII) and monarchs (such as Maximilian II), and the protest and appeals of men like Wier, Scot, and Montaigne, witches' trials multiplied at a pace matched only by the spread of rumors: this was the heyday of judicial murders for witchcraft. Frightened by the religious implications of the practice of witchery, and despite the regular extraction of confessions by torture, the populace demanded severity and haste in executions. Catholics and Protestants competed in sending witches to the stake, reaching a demented frenzy of burning: about 100 000 executions for witchcraft were carried out in Germany in the seventeenth century, while between 1560 and 1600 some 8000 women were burned in Scotland, almost 1 percent of the population. Still even this mania waned, the persecution of those accused of witchery declined and the executions stopped in most countries in Europe (in England by 1712, in France six years later); though the Scottish clergy continued to burn witches and the practice spread to the American Colonies.[2]

The hold of religion on the passions of men weakened as the Thirty Years War became openly political, and not a struggle between Catholicism and Protestantism. Instruction started to spread, fostered by the popularization of printing, the rise of compulsory elementary education and the creation of new universities[3] (though the great majority of the people in Western Europe remained illiterate). Men of all walks of life were now willing (and cautiously unafraid) to test and experiment with new ideas. All over Europe scientists measured and tested, mixed chemicals, looked into cells and charted the heavens until the laws regulating these phenomena could be glimpsed, and so the whole edifice of science was restructured and our view of the world revolutionized.

The passion for new knowledge would not be curtailed. When religious censorship on universities increased and academic freedom was limited by state and religious pressures, and by a rigid and anachronistic curriculum, the intellectuals of the time responded by creating private academies (such as the Accademia dei Lincei, founded in 1603, and the Royal Society of London, founded in 1662) where the new ideas of the time could be discussed. A variety of publications overcame national boundaries despite religious and secular censorship. Libraries, such as the Bodleian at Oxford and the *Bibliothèque Royale* (now *Nationale*) in Paris, began collecting all published books; journals such as the *Philosophical Transactions of the Royal Society* (published from 1665 to this date), the *Giornale de' letterati*, and the *Acta eruditorum* disseminated scientific developments and technological discoveries. The spread of knowledge would not be quelled, and the prestige of the supernatural dropped.

Alchemy, after a thousand disappointments, declined and began to be slowly replaced by chemistry. In 1661 R. Boyle published his *Sceptical Chymist* rejecting all mystical explanations and occult qualities and abandoned the division of matter into the four elements of Empedocles arguing instead (without proof) that all matter is composed of small parts quite similar to Democritus' atoms.

Fueled by the changing political panorama and compelled by the necessities of commerce and industry, the demand grew for a mental procedure that would deal with facts and quantities as well as with theories and ideas; and so science was revived, and with this resurgence came the need of a deeper understanding of the workings of Nature. This could be obtained only through more precise scientific instruments, for our organs are not subtle enough for the task; and so, one by one, microscopes, telescopes, thermometers, barometers, and chronometers were created to aid scientists. Better ways of calculating were also needed; and so mathematics also bloomed: the decimal system was proposed and adopted; Napier and Burgi invented logarithms, with the slide rule soon following for efficient and accurate calculations; Descartes developed analytical geometry; and Newton and Leibniz invented calculus.

Despite all these changes the shift from the stifling scientific environment of the Middle Ages to the enlightened curiosity of the eighteenth century was neither sudden nor easy. The influence of the Christian dogma waned but slowly and was felt to a greater or lesser degree in most investigations of the time. The competing Catholic and Protestant faiths agreed on intolerance, denouncing and punishing heresy (as defined by the churches themselves). But their impact steadily decreased; scientific and technological developments revealed inconsistencies with the literal interpretation of the Bible, and soon made this uncompromising position untenable (for example, R. Hooke pointed out the impossibility of reconciling the chronology in the Book of Genesis with the antiquity of fossil remains); geographical discoveries also played a role in eroding the preeminence of Christian theology and philosophy through the inevitable contact with such a range of faiths as those in the Americas and East Asia. In addition a variety of luminaries such as Locke, Milton, and Spinoza pleaded for freedom from censorship and freedom of worship, even when this involved ideas traditionally considered heretical. By the end of the seventeenth century no church would have dared to do what had been done to Bruno, or to Galileo.

The relaxation of religious constraints combined with the prodding of natural curiosity and the necessities of industry and commerce produced an explosion of new discoveries and hypotheses seldom seen in history. The meticulous study of the heavens started by Tycho Brahe continued, but now with the help of powerful telescopes; Gregory of Edinburgh designed the first reflecting telescope, a design which Newton later improved. Extensive star catalogs were compiled and new discoveries were made, such as the first double-star system, and the fact that the "fixed" stars in fact vary their positions albeit very slowly. The monarchies realized the paramount importance of astronomy in navigation and hence in commerce, and so Louis XIV funded the Paris Observatory and Charles II built the observatory at Greenwich; there followed very precise measurements of the Earth and the realization that it is not perfectly spherical, which was explained by centrifugal effects that also explained the flattening of Jupiter. Even the principles of the evolution of the Earth's crust appeared at this time (N. Stensen, 1669).

The main exponents in the exact sciences of the time were unquestionably Galileo and Newton, but they worked in a rather fertile environment. At this time René Descartes stated the law of inertia, to be later expanded by Galileo and Newton themselves; Gassendi measured the speed of sound (accurate to 35 percent); Palissy revived the view that fossils were the petrified remains of dead organisms (and was denounced as a heretic, but not burned); Harvey described the circulation of blood and demonstrated the absence of spontaneous generation of life (*omne vivum ex ovo*, everything living comes from an egg), and led to a critical re-evaluation of Galen. Science began to secede from philosophy, and it would soon shed its Aristotelian moorings and develop new methods for investigation. Tradition and ancestral authority would be dismissed in favor of "lowly" facts; and whatever "logic" might say, theories were to be accepted only when demonstrated by experiment and after their mathematical predictions were verified. This approach was advocated by Francis Bacon and Descartes in their brave attempt to revamp the way by which scientific knowledge was obtained, dismissing the deductive approach used by Aristotle (and in vogue since antiquity), and replacing it with an inductive method that resembles the one used today.

Light became the target of intense scrutiny and its speed was measured for the first time; Snell formulated the laws of refraction; the phenomenon of diffraction was observed and led Hooke and Huygens to the wave theory of light (rejected by Newton), and from this the idea of the ether as the medium in which light travels (much as air for sound waves in our atmosphere) was proposed.[4] These researches were fueled by a natural curiosity about the nature of this phenomenon, and yet the knowledge gained through such an "impractical" occupation would later allow the construction of extremely sensitive optical devices: microscopes to probe the smallest components of cells and track down infectious bacilli, telescopes to discern the bowels of the Universe and even our modern spy satellites that siphon delicate and dangerous earthy secrets to their safe orbital homes.

The machinery of science developed rapidly due to the natural feedback from science to industry. Instruments such as the barometer, the thermometer, the air pump, the telescope, and the compound microscope were significantly improved. In addition mathematics was recognized as essential for future scientific progress and its technological progeny. Fermat created the theory of numbers, the calculus of probabilities, and analytic geometry (independently from Descartes). Calculus, originally glimpsed by Archimedes, approached by Kepler, discovered but left unpublished by Fermat, was completed and applied by Leibnitz and Newton. Statistics was established by J. Grant's studies on the occurrence of disease, and was immediately appropriated by political pollsters and insurance analysts.

Heat was understood to be the result of the random motion of infinitesimal parts of the object (a view first proposed by F. Bacon and adopted by Boyle and Hooke) and this insight led to a deeper understanding of thermodynamics; soon practical applications began to appear and in 1712 J. Cawley created a steam engine, an invention that would soon beget the machinery of the Industrial Revolution that would change the face of the world. The first efficient air pump was devised by O. van Guericke (1665) and his experiments disposed of the idea that Nature would not allow a vacuum.[5]

Francis Bacon

Bacon has been sometimes described as a prescient thinker who invented the scientific method, while in other occasions he is presented (almost patronizingly) as a well-intended philosopher of science whose recipe for extracting knowledge from experiments was doomed to sterility and failure. Both of these descriptions, while accurate to a degree, paint only a partial picture of the man's impact on the development of science. Bacon emphasized the need to dismiss all prejudices (or, as he called them, "idols") before beginning to inquire into the nature of things, and that such inquiries should be pursued following a coherent, repeatable and unprejudiced method. He dismissed the Aristotelian deductive approach and the empiric experimental method followed by the alchemists, noting that they were both encumbered by such "idols," and this clouded the types of experiments that were carried out and the interpretation of the results. He specified that the knowledge obtained though scientific endeavors should

Box 4.1

Francis Bacon (1561–1626)

Philosopher, essayist, and statesman, born in London, educated at Trinity College, Cambridge, and at Gray's Inn. His father, Sir Nicholas Bacon, Lord Keeper to Queen Elizabeth I, died when Francis was 18 leaving him penniless. From his youth Bacon was interested in politics; in 1584 he was elected to Parliament, where he served for 19 years, gaining a reputation for learning and clear striking speech. He opposed the Queen's tax program and this limited his prospects, but was accepted as an unofficial member of the Queen's Learned Council through the efforts of the Earl of Essex. Despite this Bacon participated in Essex's prosecution for treason (1601) – and was condemned for this; yet one could argue that Bacon merely put loyalty to the Crown before personal friendship. Throughout his life Bacon was divided between a passion for learning and a consuming desire for political and monetary gains. He left no children and had little use for women; yet when 45 he married Alice Barnham, perhaps because she augmented his income. With the ascension of James I, Bacon's star rose: he gained a government post and used this position and his lawyer's skill to further his fortune. He was knighted in 1603, became Solicitor General (1607), Attorney General (1613), Lord Keeper (1617), Lord Chancellor and Baron Verulam (1618), and Viscount St. Albans in 1621. By then he had everything except an honorable reputation: he sacrificed principle for expediency or convenience, used his influence to obtain favorable judicial verdicts, defended and protected the most oppressive monopolies, and as a judge he accepted substantial presents from wealthy plaintiffs. His conduct was ourageous even by the loose customs of his age; he was impeached by Parliament in 1621 and was charged with corruption. Following James' advice Bacon confessed, was fined £ 40 000, banished from the court (neither was acted upon), disqualified from holding office, and sentenced to the Tower of London (which he graced for four days). He spent the rest of his life in retirement writing about philosophy and science. His main writings were a major philosophical work, the *Instauratio Magna*, and various works on the philosophy and organization of science and the application of the inductive method: *The Advancement of Learning* (1605), the *Novum Organum* (1620), *De Augmentis Scientiarum* (1623) and *The New Atlantis* (1627) – all left incomplete to a larger or lesser extent. In the nineteenth century Macaulay initiated a movement to restore Bacon's prestige as a scientist and today his contributions are widely recognized.

not have as its purpose the solace of airy intellectuals divorced from the toil of lesser human beings, but was to be applied for the betterment of the population at large, with the government and other institutions providing funds for scientific enterprises and laboratories. The clarity and charm with which he expressed these ideas, combined with his visibility as a public figure insured that his arguments would have a deep impact in the development of science. Bacon might not have invented the phrase "knowledge is power," but he certainly embodies one of its best illustrations.

His achievements in modifying the purpose and scope of natural philosophy (as science was known at that time) were not evenly matched by the procedure he devised for obtaining knowledge (an incipient scientific method). In Bacon's mind the aim of scientific knowledge is the understanding of the causes of a phenomenon – in a curious agreement with Aristotle. He went on to assert that these causes could be obtained by the simple expedient of first listing the properties of a variety of objects or situations that exhibit this phenomenon in greater or lesser degree, and then noting the commonalities among these lists. He believed from this approach one would be able to automatically infer the correct hypothesis, which would be verified by future experiments. Thus, for example, to understand the nature of heat, one should observe hot and cold bodies and list the properties that are prominent in the former and almost absent in the latter; these properties can then be ascribed to the phenomenon of heat.

This approach, however, does not usually work: a mindless enumeration of properties generates enormous lists that lack a guiding principle that can be used to eliminate those properties that are irrelevant to the phenomenon at hand from the ones central to it. For the case of heat, for example, is color a relevant property of an object? What is more relevant, the temperature or the amount of time the body is exposed to the Sun? etc.

Bacon's writings had the virtue of increasing awareness of the importance of science, and emphasized the urgency for devising a reliable procedure for pursuing scientific endeavors. His detailed and incomplete program for rebuilding the whole of science was both grandiose and impracticable, yet many of the visionary social improvements he demanded as goals of the scientific effort have been carried out: the health and longevity of the population have improved markedly (though to a degree that shamefully depends on location and social class), psychic phenomena have been closely scrutinized and exposed, a measure of knowledge of social psychology has been obtained, governments support laboratories and research institutes, and a new "method of science" has been devised and used to great profit.

Galileo Galilei

Introduction

Only rarely are we fortunate enough to witness the flourishing of a mind as keen and fertile as that of Galileo, a man of infinite resource and sagacity, who longed to do everything and lacked only the time to accomplish it. From the seed of his discoveries grew our modern scientific understanding of the Universe, his thought represents the watershed between science and philosophy. Grotius called him "the greatest mind of all time," but for Galileo, as for Newton and Einstein, such superlatives provide poor and incomplete descriptions of the man. His faults – pride, temper, vanity – were the price of his qualities: persistence, courage, and originality. Though he certainly surpassed many of his contemporaries and predecessors, he seldom found the grace to give credit to, or learn from, their achievements. Though Galileo shares with Kepler the honor of winning the acceptance of the heliocentric hypothesis, and demonstrating the regularities obeyed by heavenly bodies, he never (at least openly) recognized the importance of Kepler's calculations.

Galileo could have been any kind of great man, but in boyhood his interest was turned towards playing with machines, and this developed into a passion for understanding the physical properties of the Universe. His investigations revealed in greater measure the frightful immensity of the Cosmos, but did so without falling into fantastic speculations about the world around us. He realized that mathematics provides the best language for this task, and he set to learn and develop it, noting its superiority over Aristotelian or Scholastic philosophy for describing the laws of Nature. He avoided philosophical conjectures concerning the infinity of the Universe and the existence of alien-inhabited worlds, and concentrated on investigating hypotheses that could be tested by experiments, described in mathematical form, and whose predictions could be verified; thus he helped cement the usefulness and applicability of the scientific method. His influence suffused Europe and raised the status of science in the north, though his condemnation suppressed it in the south (and even persuaded Descartes to withhold publication of his manuscripts, lest he suffer a fate similar to Galileo's). In his wake European natural philosophy became a Protestant monopoly.

To Galileo we owe our current notions about motion, the law of inertia, the concepts of velocity and acceleration, and the correct description of projectile motion; he made dynamics a full-fledged science. His astronomical observations generated a sense of wonderment about the fantastic and surprising variety of Nature. From such discoveries we learned not only the practical tools for obtaining information about Nature, and the necessity of severing all ties with Aristotle's view of science, but also a sobering realization of our modest role in the Universe. No-one since Archimedes had done so much for physics. Being hampered by deficient instruments he proceeded to improve them or devise new ones. To him we owe the use of the telescope in astronomical observations and of the pendulum in time-measuring devices; he invented a thermometer, a hydrostatic balance, a water pump, and various instruments of military use; and, like a good son of the Renaissance, he wrote in the best Italian prose of the time.

Many of Galileo's hypotheses have stood the test of time and, in fact, provide the foundation of modern exact sciences. Others have been replaced by more accurate descriptions of Nature or have been proven wrong. Many of the views for which he was condemned and persecuted are not the ones currently accepted. Like most martyrs, Galileo suffered for the right to be wrong.

Box 4.2

Galileo Galilei (1564–1642)

Was born near Pisa, Italy on the day of Michelangelo's death; died near Florence, Italy. His father was a cultured Florentine who taught him Greek, Latin, mathematics and music (his perennial consolation). The family moved to Florence (*c.* 1571) where he attended the near-by monastery school at Vallombrosa (*c.* 1579). In 1581 he entered the University of Pisa to study medicine and philosophy; there he discovered Euclid and took clandestine lessons on the *Elements*, a year later he described correctly the properties of the pendulum. In 1585 he returned to Florence without a degree but followed his passion for mathematics under the guidance of court mathematician O. Ricci while giving private lessons on this subject. In 1586 he invented a new hydrostatic balance and published his results in *The Little Balance*; at this time he began his life-long studies on the properties of motion, and published an essay on the center of gravity of solid bodies. In 1588 he was forced to apply for a teaching position as his father's financial support ran out; he was rejected by the Universities of Padua, Pisa and Florence as being too young; disheartened he thought of seeking his fortunes in the Near East, but was saved by obtaining a three-year position in the mathematics faculty at Pisa (1589) at starvation wages. Soon thereafter he completed his manuscript *On Motion* where he described his anti-Aristotelian ideas and perhaps for this reason his appointment was not renewed in 1592 (the legend of his experiment from the Leaning Tower of Pisa refers to this time). He was then offered a much better position at the University of Padua with a generous salary; yet, being burdened with the full financial support of his family (his father having died in 1591), he was forced to supplement his income by taking boarding students and by selling scientific instruments of his construction. Galileo never married, but he took a mistress, Marina Gamba, a Venetian woman with whom he had two daughters and a son; the daughters were placed in a convent (he could not provide adequate dowries) and he eventually managed to have his son legitimated. In 1613 Marina married Giovanni Bartoluzzi, and Galileo apparently kept cordial relations with the couple. At Padua he studied the properties of motion and determined the rules describing the trajectories of projectiles and the acceleration of falling bodies (both contradicting Aristotle's claims); he published his results in the *Dialogues Concerning Two New Sciences*. By then Galileo was convinced of the veracity of the Copernican hypothesis and lectured on the subject (Pisa, 1604). In 1609 he observed the Moon using telescopes of his construction, a year later he discovered Venus' phases and four of Jupiter's satellites; yet his colleagues refused to believe him or even look through his telescope! Tiring of Padua and hoping for a more enlightened intellectual environment he moved to Florence in 1610 (at double the salary but without Marina); he soon published *Siderus Nuncius* summarizing his astronomical observations, which he dedicated to Cosimo II of Medicis and was accordingly appointed (for life) "Chief Mathematician of the University of Pisa and

Box 4.2 (cont.)

Philosopher and Mathematician to the Grand Duke of Tuscany." In 1611 he was admitted to the Lycean Academy, and was lauded by the Society of Jesus as the greatest astronomer of all times; Pope Paul V assured him of his favor. Yet the friendship with the Jesuits was soon to cool over a controversy over the discovery of sunspots between him and Fabricius – the honor belongs to both. Galileo published his observations in *Letters on Sunspots* (1613). In 1615 he faced his first accusation before the Inquisition, and defended himself in the *Letter to the Grand Duchess Christina*. In 1616 the Copernican doctrine was declared heretical and Galileo was warned not to support it. In 1623 Galileo published *The Assayer*, where he supported heliocentric ideas; he dedicated it to his friend Pope Urban VIII who allowed its publication, provided heliocentrism was presented as a convenient hypothesis and not a fact. In 1630 he completed his book *Dialogue Concerning the Two Chief World Systems* contrasting the Ptolemaic and Copernican models. The book was printed in 1632 but Urban VIII stopped its distribution; the case was referred to the Inquisition and Galileo was summoned to Rome despite his physical infirmities. In 1633 Galileo was formally interrogated by the Inquisition and he recanted his support of the Copernican theory, he was placed under house arrest, spent six months in the palace of the archbishop of Sienna and then moved to his villa in Arcetri where he remained until his death. Despite these tribulations he continued his studies on motion and in mathematics; in 1633 he began writing his *Discourse on Two New Sciences* that appeared in 1638 after clandestine negotiation with a Leiden publisher. His health deteriorated steadily: in 1634 he suffered a painful hernia, in 1637 with his eyesight failing he made his last astronomical discovery (the librations of the Moon). By 1638 he was totally blind: "This universe," he said, "that I have extended a thousand times . . . has now shrunk to the narrow confines of my own body. Thus God likes it; so I too must like it." In 1639 he was granted special permission to visit a physician in Florence and to hear Mass. For three more years he continued his researches; he died January 8, 1642 surrounded by his disciples.

Galileo on motion

One of Galileo's life-long interests concerned the properties of moving bodies. It was Aristotle's claim that all unimpeded motion was determined by the very nature of the object, and he provided an immovable Earth as a universal reference from which these motions could be determined. With the advent of the heliocentric hypothesis these ideas were no longer sustainable, and the nature and properties of motion again came into question. The simplest situations were analyzed and correctly interpreted by Galileo, and these results proved to be the basis of the Newtonian theory of this subject, which would stand uncorrected for more than two centuries.

Galileo carefully defined the concepts of *velocity* and *acceleration* that have proved very useful in describing the motion of all objects. Since colloquially these two words are often misused we will provide their "technical" definitions:

Velocity: The rate of change of *position* with time.

Acceleration: The rate of change of *velocity* with time.

Note that velocity is characterized by both speed and direction; if the speed remains constant but the direction changes, the body *is* accelerating. This is the case for an object moving in a circle.

Galileo remained interested in the properties of motion throughout his life; he devised mathematical descriptions of his observations and derived practical applications from them. For example, he was the first to note that as a pendulum swings back and forth the time for each swing remains the same even though their size decreases steadily[6] (legend has it that he discovered this during Mass at the Cathedral of Pisa by observing the swing of a lamp as it was pushed by gusts of wind), and he used it to construct the first of a venerable line of grandfather clocks.

The motion of free bodies and Galilean relativity

Imagine a person inside a ship that is sailing on a perfectly smooth lake at constant speed. This passenger is in the ship's windowless hull and, despite it being a fine day, is engaged in performing mechanical experiments (such as studying the behavior of pendulums and the trajectories of falling bodies). A simple question one can ask of this researcher is whether s/he can detect the motion of the ship (with respect to the lake shore) *without going on deck or looking out a porthole*.

Before going into answering this question I'd like to point out that saying that this is a question one "can ask" does not mean this is a relevant or fruitful exercise. It is a mark of Galileo's genius that he saw the importance of asking and answering such a query. As is often the case, it is asking the right questions that leads to fruitful results, Galileo had the wit to do so and the intellect to provide the answers.

Back to the ship. Since the craft is moving at constant speed and direction, the researcher will not *feel* its motion. This is the same situation as when flying on a plane: one cannot tell, without looking out one of the windows, that the plane is moving once it reaches cruising altitude (at which point the plane is flying at constant speed and direction). Still one might wonder whether this is due to a limitation of our senses and that some precise experiment done in the ship's hull would give some indication of its motion. The amazing *experimental* result first obtained by Galileo is that this is *not* the case: an experiment completely contained in the ship's hull would give precisely the same result as when performed on shore, no matter how precise the measuring devices. Generalizing these results Galileo postulated his *relativity hypothesis*:[7]

Any two observers moving at constant speed and direction with respect to one another will obtain the same results for all mechanical experiments.

It is important to note that this is true *only* if the ship is sailing at constant speed and direction. Should it speed up, slow down, or turn, the researcher inside *can* tell that the ship is moving; for example, if the ship turns one can see all things hanging from the roof (such as a lamp) tilting with respect to the floor.

Since one cannot determine uniform motion without referring to some point or object (like the shore for the boat) Galileo concluded that, contrary to Aristotle's claim, *velocity is relative* and can only be determined with respect to some reference observer. Other observers may measure a completely different velocity. For example, if the researcher chooses a corner in the ship's hull as reference, then all his/her apparatuses are at rest (zero velocity) with respect to him/her. If s/he chooses to sit at a quay by the shore, then s/he finds that his/her apparatuses move at the same speed as the boat with respect to *him/her*. The inevitable conclusion that the question "are we moving?" has no meaning, unless we specify a reference point ("are we moving with respect to that rock?" *is meaningful*). This fact, formulated in the 1600s remains very true today and is one of the cornerstones of Einstein's theories of relativity.

A concept associated with these ideas is the one of a "frame of reference." We intuitively know that the position of a small body relative to a reference point is determined by three numbers: if we imagine three long rods at 90° to one another, then the position of an object with respect to the point where the rods meet is uniquely determined by the distance one must travel parallel to each rod in order to get from the point where they join to the object (Figure 4.1). Using a frame of

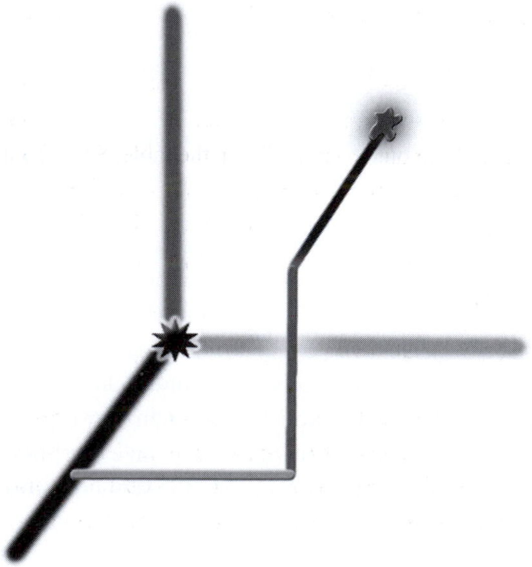

Figure 4.1 A reference frame (the length of the three thin lines determines the position of the five-point star with respect to the ten-point one).

Box 4.3

Turbulence (from *Relativity and its Roots*, by B. Hoffmann)

Although this question will seem silly, consider it anyway: Why do the flight attendants on an airplane not serve meals when the air is turbulent but wait until the turbulence has passed?

The reason is obvious. If you tried to drink a cup of coffee during a turbulent flight, you would probably spill it all over the place.

The question may seem utterly inane. But even so, let us not be satisfied with only a partial answer. The question has a second part: Why is it all right for the flight attendants to serve meals when the turbulence has passed?

Again the reason is obvious. When the plane is in smooth flight, we can eat and drink in it as easily as we could if it were at rest on the ground.

Yes indeed! And *that* is a most remarkable fact of experience. Think of it.

reference and a clock anyone can determine the position and motion of any object. The set of rods and clock(s) is called a *reference frame*. In short: a reference frame determines the where and when of anything with respect to a reference point.

The researcher on the ship can choose a reference frame where s/he is at rest; if another investigator on shore does the same, the ship reference frame will be moving at constant velocity with respect to the shore reference frame, *and vice versa*. Galileo's relativity hypothesis implies that experiments using *either* frame as a reference will give equivalent results. He concluded that there are no preferred reference frames, and, in particular, that the preponderance of the Earth in Aristotle's ideas about motion must be wrong. Being at rest on the Earth's surface provides just one reference frame, which might be convenient, but is certainly not special.

Now imagine that the researcher is in the boat and places some object like a banana on a table inside the hull. In his/her frame of reference the banana is not affected by any forces[8] and it does nothing but rest placidly on the table. Seen from the shore the banana is *not* at rest but is moving at the same speed as the ship, so in *that* reference frame the banana is moving at constant speed in a straight line; this change in behavior is due to our changing reference frame, nothing has been done to the banana. Since both reference frames are equivalent:

Objects not affected by forces are either at rest or move at constant speed in a straight line.

This is the so-called *law of inertia* later adopted by Newton as his first law of motion. The observer will measure a different speed and direction when observing from a different reference frame; the velocity of bodies is determined only relative to the observer, there is no absolute velocity. This is the essence of Galileo's theory of relativity.

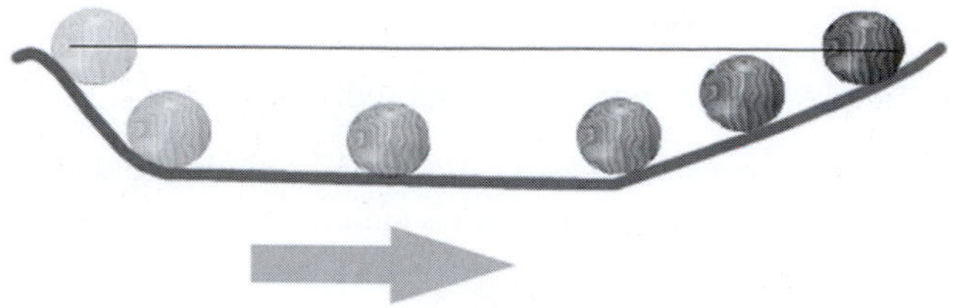

Figure 4.2 Illustration of Galileo's experiments with friction (fainter images give the position of the ball at earlier times). A ball released from the left will roll down and rise to the same height on the right provided friction is negligible. These experiments also showed that when friction is negligible the ball will move in the horizontal section at constant speed irrespective of how long this section is and despite Aristotle's claims that this type of motion is unnatural to the ball.

Galilean relativity predicts that free motion is in a straight line at constant speed, an important conclusion that cannot be accepted without experimental evidence. Though everyday experience seems to contradict this conclusion (for example, if we kick a ball, it will eventually stop), Galileo realized that this is due to the fact that most objects are *not* left alone while moving: they are affected by friction. He then devised and carried out a series of experiments where he determined that frictionless motion would indeed be in a straight line at constant speed.

Consider a ball rolling in a smooth track placed on a table (Figure 4.2). The ball rolls from its release point to the opposite end, rising to a certain height above the table (and then back). As the surfaces of the track and ball are made smoother the height reached after rolling approaches the initial height; in addition he found that when the track is level the ball moves at almost constant speed. He concluded that in the limit of zero friction the ball would endlessly go back and forth in this bowl moving at constant speed in all horizontal stretches of the track. Following this reasoning and "abstracting away" frictional effects he concluded that it is indeed true that force-free motion is characterized by motion at constant velocity.

This directly contradicts the then standard Aristotelian ideas about motion that assigned a natural motion to all objects. A ball, being "earthy," has a unique natural motion downwards; if it is forced to move horizontally, Aristotle would say, it will slow down and stop since that type of motion is just not natural to it. Galileo's experiments showed that there is no "natural" motion, and that the observed slowing down is due to the influence of friction.

An interesting sideline about Galilean relativity is the following. Up to that time the perennial question had been: what keeps a body moving? Galileo realized that this was the *wrong question*, since uniform motion in a straight line is not an absolute concept: the same object would present different motions to different observers. The right question is: what keeps a body from moving uniformly in a straight line? The answer to that is "forces," which were to be so ably described and used by Newton. This illustrates an important issue in physics: we have, in principle, the answers to all questions, for Nature is before us to observe; but only when the right *questions* are asked does the regularity of these answers become apparent. Three centuries after Galileo, Einstein revisited these same old observations, and by asking a different set of questions was led

Figure 4.3 According to Aristotle a 2 lb rock falls faster than a 1 lb rock . . . but what if we have two 1 lb rocks just touching? This argument demonstrated the inconsistency of Aristotle's hypothesis.

to what are perhaps the most beautiful insights into the workings of Nature that have been reached.

The motion of falling bodies

As with horizontal motion, Aristotle believed that the motion of freely falling bodies was determined by the objects' nature (Figure 4.3). This belief was held by most people, some as keen as Lucretius[9] and Leonardo.

The force of gravity is quite unique in that it affects all objects in the same way, so that, for example, all bodies dropped from the same height take the same time to hit the ground. For other forces the details of the motion depend on the weight and/or shape of the object. For example, if you kick two rocks (thus applying a force to them) while on the shore of an icy lake, the heavier one will move more slowly, and so the lighter one will reach the opposite shore first. This unique property of gravity was one of the most important motivations for Einstein's General Theory of Relativity.

From his observations of falling bodies Galileo also determined that the acceleration of these bodies is constant. He demonstrated that an object released from a height starts with zero velocity and increases its speed with time (before him it was thought that bodies when released acquired instantaneously a velocity which remained constant but was larger the heavier the object was). Experimenting with rolling balls down inclined planes, Galileo discovered the mathematical expression of the law of falling bodies: the speed increases uniformly with time while the distance traveled increases as the *square* of the time.[10]

The motion of projectiles

The art of accurately aiming weapons for optimal reach and damage has been practiced since the invention of human aggression. By understanding the mathematical rules that describe this type of motion Galileo turned this art into a science.

The rules that determine projectile trajectories are quite simple: the motion can be decomposed into horizontal and vertical components that occur independently of each other. If a projectile is fired horizontally (and air friction is ignored) it will move in the horizontal direction with constant speed, so that the horizontal distance will increase uniformly with time; in the vertical direction it will experience the pull of gravity and will undergo free fall, and so the vertical distance fallen will increase as the square of the time. As a result the ball follows a curved trajectory technically known as a section of a parabola.

In order to show that horizontal and vertical motions are indeed independent one can imagine an experiment where one ball is thrown horizontally as described above, while at the same time a similar ball is dropped. The first ball has horizontal and vertical motions, the second one only a vertical one, yet they reach the ground simultaneously since the horizontal motion of the first ball in no way affects the rate at which it approaches the ground (Figure 4.4).

Astronomy

Throughout his life Galileo provided some of the most compelling arguments in favor of the heliocentric hypothesis, and though his criticisms of the Ptolemaic system brought him endless trouble in his lifetime, he was ultimately vindicated. Galileo had the persistence to observe the heavenly bodies and the intelligence to discern the regularities in their behavior; his results told also of the impending demise of the medieval system of the world and Galileo was strongly resented for being a harbinger of this change.

The beginnings of Galileo's astronomical studies were quite dramatic. On October 9, 1604 astronomers witnessed the appearance of a new star, as red and bright as Mars, in the constellation *Ophiuchus* (the serpent bearer). But this newcomer had only an ephemeral existence for it vanished a few days later (it is now known that this was a supernova, a star that destroyed itself in a tremendous and brilliant explosion). Galileo demonstrated that this object must lie beyond the Moon, and concluded this region could host stars that came into being and soon disappeared, a characteristic that directly contradicted the Aristotelian doctrine that claimed that everything beyond the Moon was perfect and unchanging.

A few years later, in 1609, Galileo heard of an instrument made by Hans Lippershey (a spectacle maker of Middelburg) that allowed the observer to see distant objects with (relative) clarity. Realizing its astronomical potential, Galileo quickly reproduced the instrument and improved it until it reached a magnification of 1000 times; he then used it to probe the heavens. This

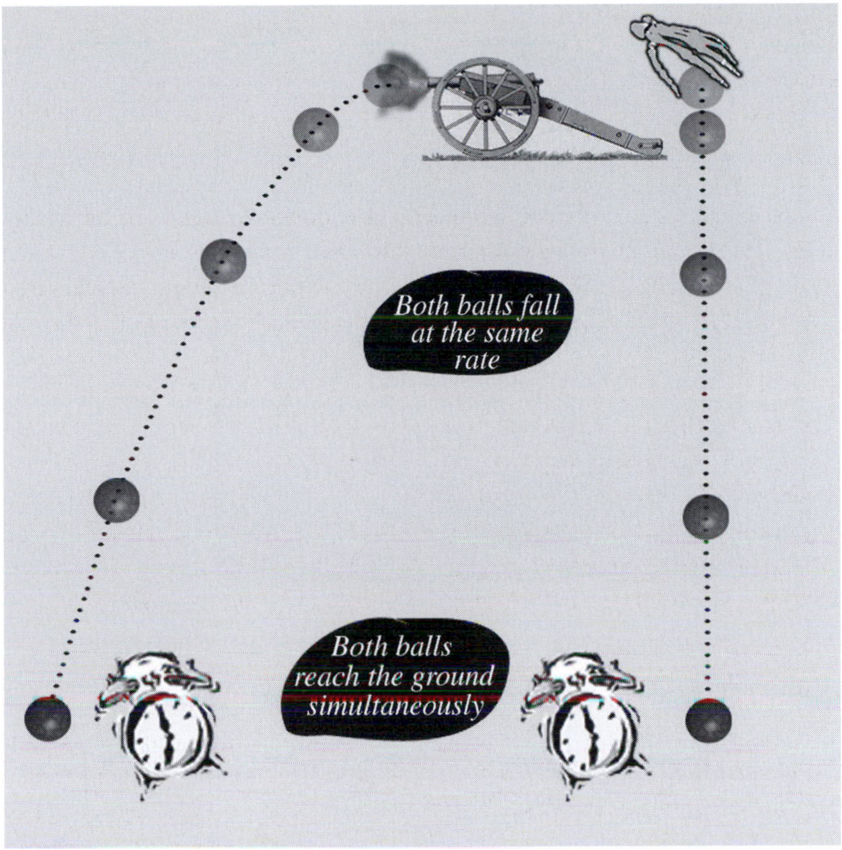

Figure 4.4 An experiment that illustrates the independence of the horizontal and vertical components in projectile motion. The cannon fires its ball *horizontally* and simultaneously the hand drops the other ball; they both hit the ground at the same time (fainter images give the position of the ball at earlier times).

instrument resolved the blurriness of the Milky Way into a swath of thousands upon thousands of stars; it also revealed that the familiar constellations contained very many members that were too dim to be seen with the naked eye. Under Galileo's gaze the population of the Universe exploded.

The first object that he studied in detail was the Moon, of which he made many drawings (Figure 4.5) some of which are quite accurate even by today's standards. When observing the Moon's surface at high magnification he found that it was heavily scarred, and identified some of the dark features he observed as shadows; the Moon was not exactly spherical and hardly perfect. He also guessed correctly that the vague illumination of the dark side of the Moon is due partially to sunlight reflected by the Earth. Directing his attention to other celestial bodies, Galileo discovered the rings of Saturn and the presence of spots on the Sun's surface, he published his solar observations in 1613, in a book where he correctly interpreted the motion of these dark features as a consequence of the Sun's rotation, but erroneously believed them to be clouds.

Soon after the publication of Copernicus' *De Revolutionibus*, detractors of the heliocentric hypothesis noted that if all the planets move around the Sun, then

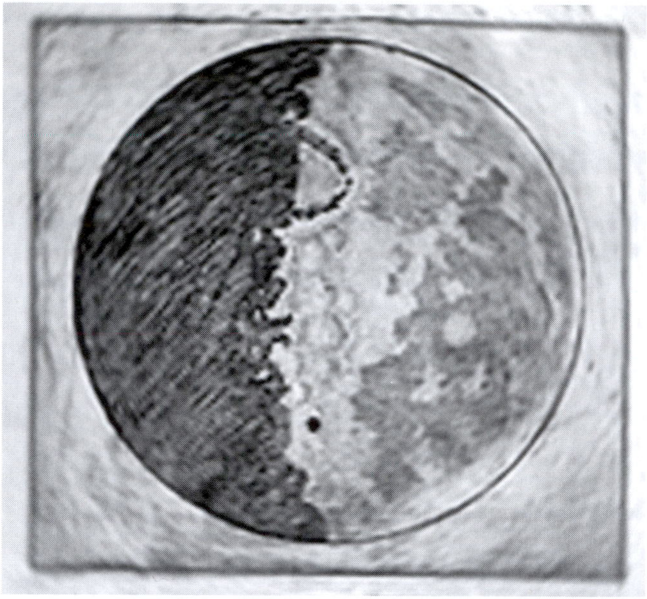

Figure 4.5 An example of Galileo's drawings of the Moon. (Courtesy of the *Biblioteca Nazionale Centrale* in Florence)

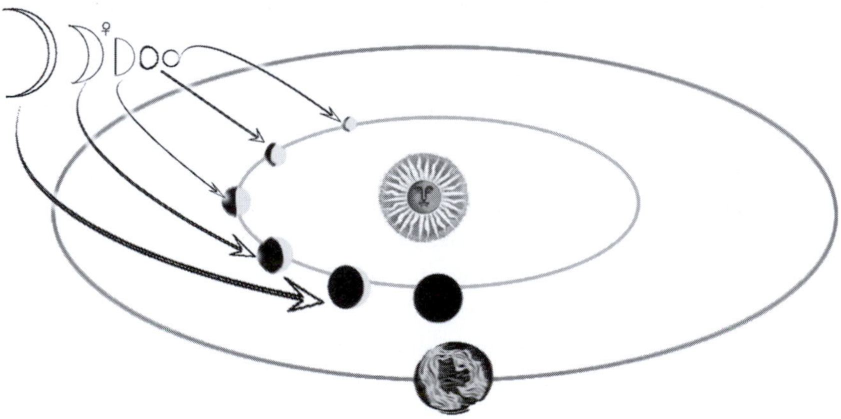

Figure 4.6 Galileo's drawings of the phases of Venus (upper left) (*Instituto e Museo di Storia della Scienza*), and their explanation within the heliocentric theory.

Venus, like the Moon, should exhibit periodic phases due to its changing position relative to the Earth. Naked-eye observations, however, failed to provide confirmation of this effect and provided an argument against heliocentrism. This changed dramatically when, in December 1610, Galileo made a series of observations of Venus using his telescope and verified the Copernican prediction (Figure 4.6); our eyes are simply not sensitive enough to perceive this effect without using magnifying apparatuses. These results were later verified by the Jesuit mathematicians of the *Collegio Romano* (although they did not necessarily agree with Galileo's interpretation!).

But the most dramatic of Galileo's astronomical discoveries was that of Jupiter's satellites.[11] In 1610 he observed four "stars" in the vicinity of Jupiter that moved about the planet, sometimes approaching it and disappearing for a short time, but soon reappearing on the other side. He correctly explained this by hypothesizing that these were not stars, but satellites that circled Jupiter, just as the Moon circles the Earth; and that their vanishings were due to their being eclipsed by the planet when they moved behind it (as seen from Earth). "These new bodies," he wrote, "moved around another very great star, in the same way as Mercury and Venus, and peradventure the other known planets, move around the Sun."[12] Seeing a series of bodies that circle Jupiter provided for Galileo a miniature model of the Solar System as described by Copernicus, and, in addition, the observations indicated that these satellites undoubtedly did *not* circle the Earth. All this was in blatant contradiction of the Aristotelian model; any remaining doubts which he might have had in his belief of the heliocentric model vanished.

After publishing his observations of Jupiter, Galileo's astronomical writings and research centered on the heliocentric theory, which he defended almost openly, providing some of the most cogent arguments supporting it (and against the Church-held geocentric hypothesis).[13] Though there were certainly other advocates of the Copernican hypothesis, Galileo's prestige, his compelling arguments, and the fact that he published them in Italian (making them more readily accessible to the public at large), made him the lightning-rod of the Church's rebuke. In 1616 Pope Pius V declared the Earth to be at rest and dubbed the heliocentric model heretical, Copernicus' *magnum opus* was black-listed (where it so remained for more than two centuries), and Galileo was eventually forced to recant his belief in these ideas. And yet, despite these efforts, the geocentric view was doomed; within 50 years all but the most medieval scholars had rejected it.

Galileo's impact

There are two aspects to Galileo's enormous influence in the subsequent development of science. On the one hand he opened new avenues of research providing fresh interpretations that proved extremely fertile. He exemplified the need for tempering imagination and philosophical inclination with accurate observations, and showed the importance of verifying hypotheses with data; more through action than by specific statements Galileo advocated the scientific method.

On the other hand, Galileo brought science out of the seclusion of a natural philosopher's room into the forefront of public debate. He popularized and publicized his discoveries, and defended them without timid subservience to the mandates of religion. This vocal support, coupled to his influence and visibility, helped science shed its medieval constraints and made him both herald and midwife of the rational interpretation of Nature that has proved so fruitful to this day.

Isaac Newton

On Christmas Day 1642, in the manor house of Woolsthorpe, a weak child was born, and christened Isaac; he was destined to be the most influential scientist of his time, and one who completely dominated the development of the exact sciences for the next two and a half centuries, and whose pre-eminence is matched only by Einstein. Newton provided a simple set of rules (or "laws") that describe the manner in which objects interact, and the magnitude of this achievement lies precisely in its simplicity and universality, for these rules are of such elegance, their applicability so wide, and their accuracy so outstanding, that they produced a qualitative change in the way scientists, and even the public at large, viewed Nature. In the words of Alexander Pope:

Nature and Nature's laws lay hid in night;
God said, "Let Newton be!" and all was light.

This new perception made us realize that we inhabit a Universe ruled by laws, where a known action will produce predictable effects, and not a world filled with careless and whimsical (at best) spirits whose passions dangle us as leaves in the wind. Not only did Newton devise a set of practical laws for understanding Nature, but he also provided a prescription for gathering information and inferring conclusions from it. This procedure summarizes the medieval and Renaissance thought on this matter and echoes of Ockham, Bacon and Galileo; but it is in Newton that the fruitfulness of the approach first became apparent. With characteristic elegance he described this procedure in terms of the following four simple rules:

***RULE 1*:** We are to admit no more causes of natural things than such as are both true and sufficient to explain their appearances.

***RULE 2*:** Therefore to the same natural effects we must, as far as possible, assign the same causes.

***RULE 3*:** The qualities of bodies that admit neither intention nor remission of degrees and which are found to belong to all bodies within the reach of our experiments, are to be esteemed the universal qualities of all bodies whatsoever.

***RULE 4*:** In experimental philosophy we are to look upon propositions collected by general induction from phenomena as accurately if very nearly true, notwithstanding any contrary hypotheses that might be imagined, till such a time as other phenomena occur by which they may either be made more accurate, or liable to exceptions.

At a practical level Newton's laws allowed him and all subsequent researchers to accurately predict the behavior of all (inanimate) objects under all circumstances present in our everyday experience. Newton's equations describe with exquisite accuracy the motion of celestial bodies, whose positions and motions for the next 1000 centuries can be predicted. These equations, first published in 1687, were the ones used by NASA scientists to plot the trajectory of the Apollo 11 spacecraft in 1969, and insure a safe passage from the Earth to the Moon and back.

Newton's equations are also used in designing engines, bridges, and buildings, in insuring the safe working of elevators, rollercoasters, and transatlantic ships. The dangers of turbulence in jet airliners are evaluated using Newton's laws, which also determine the necessary design modifications for avoiding tragedy; even billiard balls faithfully obey these rules.

So successful was Newton's description of Nature that by the end of the nineteenth century it was believed that,[14] when supplemented by a few other ideas such as the laws of thermodynamics, it was the tool with which humanity would soon reach a complete understanding of Nature, and hence (following F. Bacon), a complete mastery over it. But this was not to be: during the first quarter of the twentieth century it become clear that Newton's theory could not have such universal applicability: it fails dramatically when describing behavior at the atomic and subatomic level (where quantum effects come into play), or when describing motion close to the speed of light (where the Special Theory of Relativity must be used), or when considering the physics of very massive and compact objects (for which the General Theory of Relativity is needed). This led to a less fantastic assessment of Newton's theory: it was no longer supposed to provide a complete description of Nature, but was understood to be an accurate tool that is useful in a very wide – albeit limited – set of circumstances. This change of perception generated enormous turmoil and great excitement, once again a feeling of security was lost, but we gained a new sense of wonderment about the world around us.

Half a century ago one could safely assert that despite its failings, Newton's theory of motion was applicable in all everyday circumstances; but this is no longer true today. The miracle of electronics, the ubiquitous presence of lasers, the menace of atomic weapons are all technological consequences stemming from quantum mechanics and the Special Theory of Relativity, theories that replaced Newton's. It is true that there is little in our life that pertains to the General Theory of Relativity, which describes the interactions between very compact objects such as neutron stars and black holes, but in view of the significant changes that have occurred over the last 50 years it would be perhaps more prudent and wise to say that General Relativity has no effects on our daily life . . . *yet*.

Mechanics

During the plague of 1665 Newton was forced to leave Cambridge for his home in Lincolnshire; and during this sojourn he formulated what was to become his remarkably successful theory of motion. Newton's theory is based on three laws that provide the means for calculating the way in which bodies react to forces, as well as the interpretation of computations and observations. He created a language for describing how objects behave as well as the tools for predicting this behavior. Newton's laws (as they are now called) appeared in 1687 in his *Philosophiae Naturalis Principia Mathematica*, or the *Principia* as it is commonly known, one of the greatest scientific books ever written. These laws have an immense range of applicability, failing only at very small distances (below 10^{-8} cm), for very strong gravitational fields (about 10^8 stronger than the Sun's), or for very large speeds (near $10^8\,\mathrm{m\,s^{-1}}$).

The first of Newton's laws addresses the motion of free bodies and is a restatement of a result obtained by Galileo. The second law states quantitatively how a motion deviates from free motion, while the third law states the effect experienced by a body when exerting a force on another object. In Newton's own words:

> ***First Law***: Every body perseveres in its state of rest, or uniform motion in a right line, unless it is compelled to change that state by forces impressed thereon.
>
> ***Second Law***: The alteration of motion is ever proportional to the motive force impressed; and is made in the direction of the right line in which that force is impressed.
>
> ***Third Law***: To every action there is always opposed an equal reaction; or the mutual actions of two bodies upon each other are always equal, and directed to contrary parts.

In particular Newton constructed his mechanics to comply with Galilean relativity: an observer in uniform motion with respect to another cannot, without looking outside his laboratory, determine whether they are at rest or not. And even if they look outside, they cannot decide whether they are in motion or the other observer is. In fact for two inertial observers moving relative to each other the question, "which of us is moving?" is un-answerable and meaningless. The only thing to be said is that they have a certain relative velocity.

These three laws, when complemented by some basic assumptions about space and time, summarize Newton's basic hypotheses; the rest of the *Principia* consists of his derivation of consequences of these laws for various special but very interesting cases. It is noteworthy that these laws are stated to be valid for all bodies and under all circumstances; in particular, the very same rules apply on Earth and on (and for) all heavenly bodies; the acceptance of Newton's ideas marks the final passing of Aristotelian physics. All experimental evidence of the time (and for the next two centuries) was to support these hypotheses; Newton's theory became *the* theory of Nature, some of whose features I will now discuss.

Newtonian space and time

The *Principia* is carefully constructed by first providing a few basic definitions and postulates, and then deriving a variety of consequences from them. The main topic of the book is the study of motion, and since motion implies change of location with time, Newton needed to define the properties of space and time in order to clarify where and how objects move. Hence the preface to the *Principia* contains the following statements:

> Absolute space, in its own nature, without regard to anything external, remains always similar and immovable.
> Absolute true and mathematical time, of itself, and from its own nature, flows equably without regard to anything external, and by another name is called duration.

That is, *space and time are absolute*, an arena where the play of Nature unfolds, they are both featureless objects that serve to define a universal and preferred reference frame (Figure 4.1) and their existence is quite unaffected by the objects they host. Since any distance represents the separation between two immovable points in eternal space, its magnitude will be agreed upon by *all* observers. In the same way, a time interval marks two notches on eternal time and will also be universally agreed upon. Within Newtonian theory distances and time intervals are *absolute*, that is, their values are completely independent of the observer and his/her state of motion.

These basic assumptions about space and time are the foundation of Newton's theory of Nature, and were accepted because they seemed self-evident and because the results derived from this theory were enormously successful. By the nineteenth century, however, experimental results concerning electric and magnetic phenomena clashed with these assumptions about space and time.[15] These problems did not refer to some detail of calculation that might require a technical modification, but to a fundamental defect concerning Newton's basic initial postulates. This led Einstein to propose a complete revision of our understanding of space and time, as embodied in the Special and General Theories of Relativity. Only after this dramatic paradigm shift did Newton's pre-eminence end. Do we know that the current theories of space and time are *the* truth? The answer is, of course, no; we *do* know that the current theories (including the one explained by Newton and more) explain all the data, but we cannot determine whether they represent the ultimate and complete description of Nature. But in the seventeenth century there was no inkling of these problems and very few scientists questioned Newton's hypotheses.

There is one final aspect of Newtonian space. Consider a pail filled with water and hanging from a tightly twisted rope. Initially the water is at rest with respect to the bucket and its surface is flat. When released the pail will start to rotate and will soon transmit its rotation to the water whose surface will then become concave; at this point the water is again at rest with respect to the bucket. We then have two situations in which pail and water are at rest with respect to one another and which can be differentiated by the shape of the water surface. Newton interprets this difference by asserting that in the first case pail and water are at rest with respect to absolute space, while in the second they are accelerating with respect to it:[16] though there are only relative velocities, there *are* absolute accelerations.

The First Law

Newton's first law tells of the behavior of objects that are not affected by forces, and yet it is not as trivial as it may sound. Imagine, for example, a falling rock and, say, a coyote both falling from a very high cliff. In their trip to the ground they will fall at the same rate (as Galileo noticed) so that, *from the point of view of the coyote*, the rock is at rest. Should the coyote be cultured enough to know of Newton's laws he could conclude that the rock is a free object ... only to be sadly disappointed a few seconds later when they both hit the ground. An observer on the ground would disagree with the coyote even before his demise, this observer would see that the rock accelerates as it falls, and behaves in this way because it is affected by the force of gravity.

The point of this vignette is that different observers may observe different behaviors for a given object, even to the point that the object will be at rest for one observer even though it is under the influence of forces. What the first law implies is that there exists a privileged class of observers, whom we call *inertial observers*, for whom free bodies do indeed move at constant speeds in straight lines, or else are at rest. A frame of reference used by an inertial observer will be called *an inertial reference frame*; the First Law, in reality, asserts the existence of inertial reference frames, and tells how free bodies move as seen from such vantage points. The laws of physics devised by Newton take a particularly simple form when expressed in terms of quantities measured by an inertial observer (such as positions, velocities, etc.).

The Second Law

The Second Law quantifies the conditions under which bodies do not move at constant speeds in straight lines (as seen from an inertial reference frame); using this law one can predict the trajectory of a given body once the actions of other bodies on it are known and so it is of great practical importance. If a body of mass m is acted upon by a force F then the body will acquire an acceleration (denoted by a) equal to F/m, that is:

$$F = ma$$

If the mathematical expression for the force is known (e.g. its dependence on the position and the velocity of the body) one can use this formula together with the machinery of calculus to obtain the path followed by the body. If the expression for F is not known one can still obtain it by measuring the acceleration of a body of known mass. Once F is known the motion of *any* body can be predicted; for the case of gravity we can determine the motion of Jupiter if we know how the Sun affects the motion of a pebble weighing, say, 1 g.

In all these considerations the only property of the body that affects its response to a force is its mass m, two bodies of the same mass will exhibit

Box 4.4

Isaac Newton (1643–1727)

Was born to a family of farmers in the manor house of Woolsthorpe, near Grantham in Lincolnshire on Christmas Day 1642, the year Galileo died. He was a small weak child whose father died before he was born; his mother remarried, moved to a nearby village, and left him in the care of his grandmother. Upon the death of his stepfather in 1656, Newton's mother removed him from grammar school in Grantham where he had shown little promise in academic work. Legend has it that one day the student just ahead of him in class kicked him in the stomach, Newton won the fight but he also decided to get ahead of this student in class ranking; he succeeded admirably. An uncle decided that he should be prepared for university, and he entered Trinity College, Cambridge (1661), aiming at a law degree. Though instruction was dominated by the philosophy of Aristotle, he also studied the philosophy and analytical geometry of Descartes, Boyle's works, and the mechanics of the heliocentric astronomy of Galileo. He took his degree in 1665 and was soon elected fellow of the college. This same year the plague reached Cambridge and Newton returned home, where he stayed until 1669. This was to be was a seminal year: his ideas about calculus began to take shape, and he hypothesized the main properties of gravity (the anecdote about the apple dates from this period, but it is likely to be a legend). The dependence of the gravitational force on distance had been proposed earlier by Bouillard (1645), Borelli (1666), and Hooke (1674), but only Newton could derive Kepler's rules of planetary motion . . . and yet he did not publish. At this time he also began his optics experiments that led to his constructing a reflecting telescope that avoided many optical aberrations present in refracting telescopes (like Galileo's); he presented this instrument to the Royal Society in 1671 and was elected a member one year later. After a series of experiments with prisms he published his *New Theory of Light and Colors* (1672) where he elucidated the composition of white light. Hooke had done a similar set of experiments, but failed to reach any useful conclusions, yet he felt slighted by Newton's failure to give him credit; this controversy lasted three years. Another point of contention was the nature of light itself: Newton believed it was made of a stream of particles and hence could travel through empty space; Huygens and Hooke considered light to be a wave which would – so they argued – need a medium, called the "ether," to undulate in. Possibly because of these controversies Newton delayed the publication of *Opticks* until after Hooke's death.

In 1669 Newton's mathematics teacher I. Barrow resigned his chair and recommended that Newton be appointed his successor, and his judgment was sound; Newton held this chair for 34 years. He soon built the sole experimental laboratory in Cambridge University dedicated mainly to alchemy, a subject he studied assiduously for the next 15 years (probably looking for some underlying unity in the structure of matter rather than for commercial gain). He was an unsuccessful teacher with frugal habits and a preference for walking in his garden; many legends grew out of his

Box 4.4 (cont.)

absent-mindedness. As early as 1669 Newton wrote to Barrow describing his "method of fluxions," an early though completely workable form of differential calculus, which he did not publish until 1704 and then only in an appendix to his *Opticks* (this was characteristic of Newton, for example, he derived his binomial theorem in 1665 but waited 11 years before publishing it).

This deferment led to a rather disgraceful squabble. In 1671 Leibnitz published a rival (though equivalent) form of differential calculus on which he carried a polite correspondence with Newton; accordingly, the first edition of the *Principia* contained an acknowledgement of Leibnitz's discovery. But in 1699 a Swiss mathematician claimed that Leibnitz had borrowed his idea from Newton; Leibnitz in an anonymous review of Newton's *Opticks* claimed that it was Newton who had done the copying. In 1712 the Royal Society investigated the matter and gave the priority to Newton. Leibnitz died in 1716 and the third edition of the *Principia* (1726) did not contain any comments on Leibnitz's contribution to the discovery of calculus.

The *Principia* was both acclaimed and strongly criticized. Hooke (of course) claimed to have anticipated Newton's force of gravitation, Newton retorted that Borelli had anticipated Hooke, but still credited him with Wren and Halley for having guessed the law of inverse squares; Voltaire praised the *Principia* without restraint (1738), but the absence of a description of the underlying nature of gravity was criticized by Cassini and Leibniz. French philosophers complained that the theory was not sufficiently mechanical, while in England the complaints were that Newton's system was so mechanistic it left no place for God. Newton addressed some of these criticisms in the second edition, again with mixed results. The absence of religion in the *Principia* did not extend to his personal life; he was a life-long member of the Church of England, studied the Bible avidly and in his old age became quite orthodox and a mystic.

In 1687 Newton was elected to protest before James II this king's attempt to have a monk admitted to Cambridge without taking the usual oaths; the mission failed but the university was pleased with Newton and he was chosen Member for Parliament in 1689, served until it was dissolved in 1690 and then was reelected in 1701, but left no memorable mark. Newton suffered a nervous breakdown in 1692; thereafter he feared to have lost "the former consistency of his mind" (yet his active and visible lifestyle belie this). Newton retired from Cambridge and moved to live with his niece in London. At his friends' suggestion Lord Halifax appointed him Warden (1696), and then Master (1699) of the Royal Mint; these were no charity appointments for the government wished to use Newton's chemical and metallurgical knowledge in minting new coinage.

His old age should have been happy: he was recognized as the foremost scientist of the time; he had a good salary and fruitful investments and lived in comfortable luxury without being ostentatious. In 1703 he was elected president of the Royal Society and was reelected each year until his death; in 1705 he was knighted by Queen Anne, the first scientist to be so honored for his work. And yet he was moody, suspicious and irritable, timid but proud, resentful of all criticisms; he cherished his

Box 4.4 (cont.)

privacy and did not make friends easily. In his seventy-ninth year began a series of old-age diseases (gout, bladder stones, and hemorrhoids) that would end only with his death in 1727. He was buried in Westminster Abbey after a funeral attended by all the luminaries of the time; his tomb is inscribed with these words: "Mortals! Rejoice at so great an ornament to the human race!"

identical accelerations when in the presence of the same force, while a heavier body will suffer a correspondingly smaller acceleration. The mass in the above formula measures how strongly a body responds to a given force (the larger m is the less it will be accelerated): m *measures the inertia of the body*.

The Third Law

The Third Law, colloquially speaking, states that no action remains unanswered: if, for example, an astronaut floating in space at rest with respect to a rocket throws a wrench, they will find themselves moving in the opposite direction (again with the rocket as reference). The action here is the impulse given by the astronaut to the wrench; Newton's Third Law then predicts that the astronaut will receive an impulse equal in magnitude but in the opposite direction . . . and they do. Of course, since the astronaut is more massive than the wrench they will move slower, but move they will.

Despite this illustration it seems that here on Earth this law is violated regularly. For example, if I kick a ball I do not start moving opposite to the ball, or when you sneeze you are not thrown backwards. The reason for the discrepancy between the experiences on Earth and those of an astronaut in space is that in space there is no friction, while on Earth this is a very difficult force to avoid; it is friction that prevents your backward motion when you sneeze. Still there are some cases where friction is very small and the predictions of the Third Law can be observed directly. Imagine, for example, two hockey players standing in the middle of a rink, they are about to fight and one pushes the other, the pushed player moves in one direction while the pusher actually slides in the other direction (as can be verified in just about any game).

Gravitation

Another of Newton's greatest achievements was in the field of celestial mechanics where he produced the first synthesis in the theories describing Nature: he realized that the same force that makes things fall, gravity, is responsible for the motion of the Moon around the Earth and the planets around the Sun.

He reasoned (more or less) as follows. Suppose we let an apple fall from a very high tower; it will take, say, t seconds to reach the ground. Imagine now that we

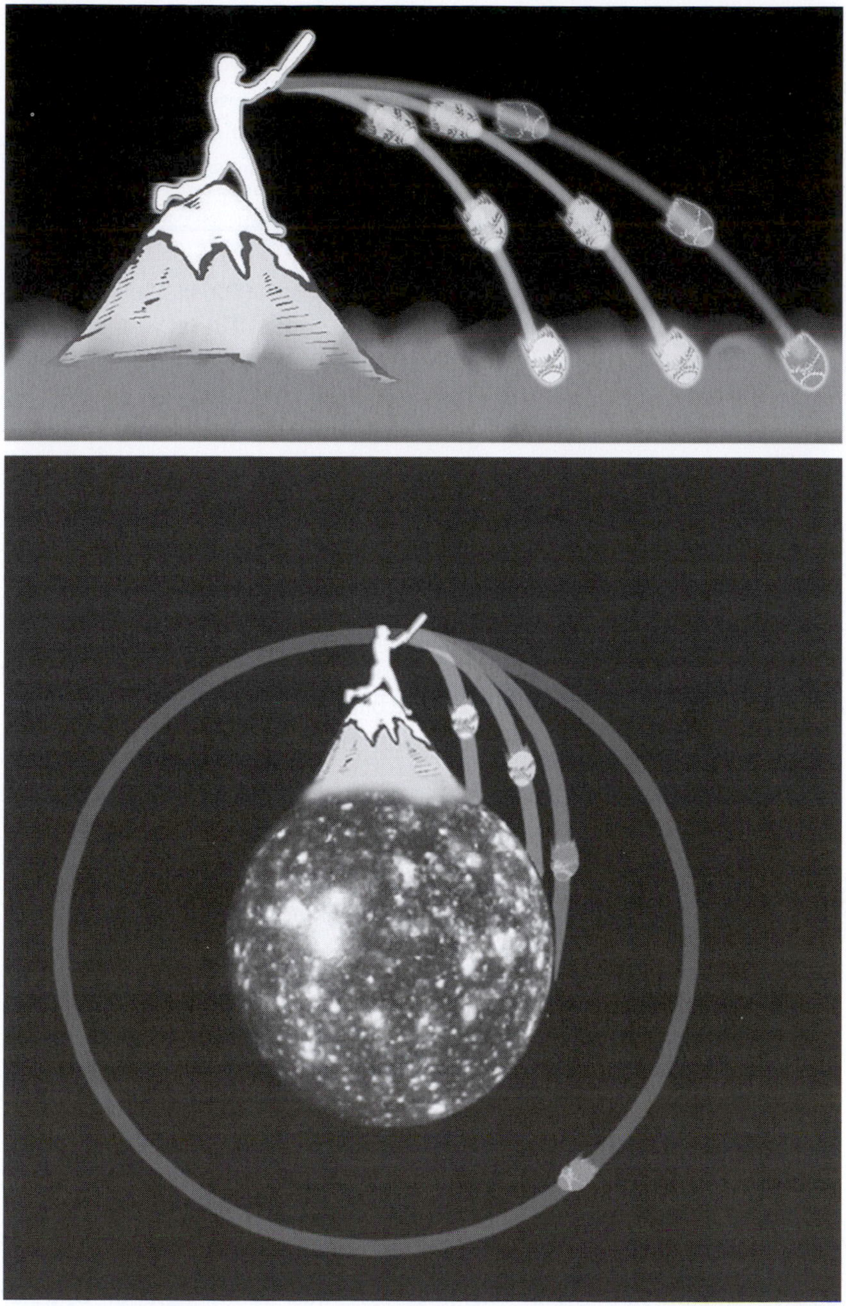

Figure 4.7 Newton's explanation of the equivalence between the force making baseballs fall and the one responsible for the Moon orbiting the Earth.

throw the apple horizontally with great vigor, it will now travel some distance horizontally, but it will still take t seconds to reach the ground, *provided* we assume the Earth is flat (Figure 4.7, top). The Earth, however, *isn't* flat and will curve beneath the falling apple; hence it will take longer to hit the ground. The further the apple travels horizontally the more the Earth curves under it (Figure 4.7, bottom); by throwing the apple with increasing force one eventually reaches a point where the apple never hits the ground since the distance it falls equals the distance the Earth has curved under it: the apple is in orbit.[17] With this thought experiment Newton convincingly argued that an apple can behave in the same way as the Moon, that it is the very same force – gravity – that makes the apple fall, satellites orbit the planets, and planets orbit the Sun. This reasoning provided the first "unification" in modern physics: effects that were originally though to be unrelated were found to have a common origin and explanation. This argument also contains the hypothesis that *all* objects are affected by gravity (the apple was chosen for illustrative purposes only, one could have used elephants or rocks with identical results). The universal character of gravity is supported by a host of experiments involving all sorts of objects, from galaxies to subatomic particles.

Having reached this unification, Newton naturally searched for the mathematical expression for the force of gravity with the intention of deriving the orbits of the planets through the application of his Second Law, and then comparing this with the data. After intense labor he showed, in a major *tour de force*,[18] that the laws of planetary motion inferred by Kepler from the observations can indeed be *derived* from his laws of motion provided that the force of gravity between two bodies of masses m and M separated by a distance r is given by

$$F_{\text{grav}} = mMG/r^2$$

where G is a universal constant (known as Newton's constant). This force is always attractive and acts along a line from one body to the other; in Newton's words:

- That the forces by which primary planets are continually drawn off from rectilinear motions, and are retained in their proper orbits, tend to the Sun; and are reciprocally as the squares of the distances of the places of those planets from the Sun's centre.
- That there is a power of gravity tending to all bodies, proportional to the several quantities of matter which they contain.

Translated this means, "any two bodies attract with a force proportional to the product of their masses and inversely proportional to the distance between them" which is precisely what the above equation states.

Having discovered this Newton was able to explain a wider range of phenomena, previously thought to be unrelated: the eccentric orbits of comets, the tides and their variations, the precession of the Earth's axis, and motion of the Moon as perturbed by the gravity of the Sun. With this theory we can predict the positions of the planets for thousands of years, foretell eclipses, and calculate Moon landings and satellite fly-bys; all with exquisite accuracy. One can but imagine the thrill of being able to understand the motion of the heavens in terms of a very few simple rules.

Newton demonstrated for the first time that gravity is a completely democratic force through which all bodies are mutually affected to a greater or lesser degree (depending on their masses and separations); Aristotle's explanation of stellar motion, his fifth and perfect element, the geocentric hypothesis and the whole edifice of cosmology built on them, collapsed never to rise again. From Newton's simple set of rules follow the majestic motions of stars and planets, of galaxies and nebulae.

Gravitational and inertial masses

The expression for the force of gravity on a body of mass m by a body of mass M is proportional to m itself, and this implies that the more massive the body is, the stronger the gravitational force. So the mass m plays two logically different roles:

- On the one hand m in $F = ma$ is a measure of the body's inertia, it determines how strongly a body is accelerated by any given force. In this role m is called the *inertial mass*.
- On the other hand m in F_{grav} is a measure of how strongly a body is affected by the force of gravity and also how strong a gravitational force is generated by it. In this role it is called the *gravitational mass*.

These two quantities refer to different properties of a body and need not be equal. Extremely precise measurements, however, indicate that they *are* equal (at least to one part per trillion). Newton simply assumed this equality as part of his hypothesis for the form of F_{grav}, and in his statement of the Second Law; we will see that this apparently innocuous assumption led Einstein to one of the most profound insights about the structure of the Universe.

Consider now the application of the Second Law to the case of the gravitational force:

$$mMG/r^2 = F_{grav} = ma \Rightarrow MG/r^2 = a$$

so that the factors of m on both sides of the equation *cancel*. The acceleration suffered by the body of mass m is independent of this quantity and, in fact, the final expression for a makes no reference whatsoever to any internal property of the accelerating body: the same acceleration will be generated for an elephant as for a mouse. This implies, for example, that if no other force (such as air resistance) acts on two falling objects and these are released at the same height they will reach the ground simultaneously.

I will illustrate this property of motion under the influence of gravity with the following anecdote (Figure 4.8). One day a hunter, armed with a very powerful rifle, decided to shoot a monkey who was hanging from a tree. The monkey, however, espies the hunter and guesses his nefarious purpose. "I will thwart him," thinks the monkey. "As soon as he fires I will just let myself drop, and his bullet will hit the branch while I am safely falling towards that other branch down there."

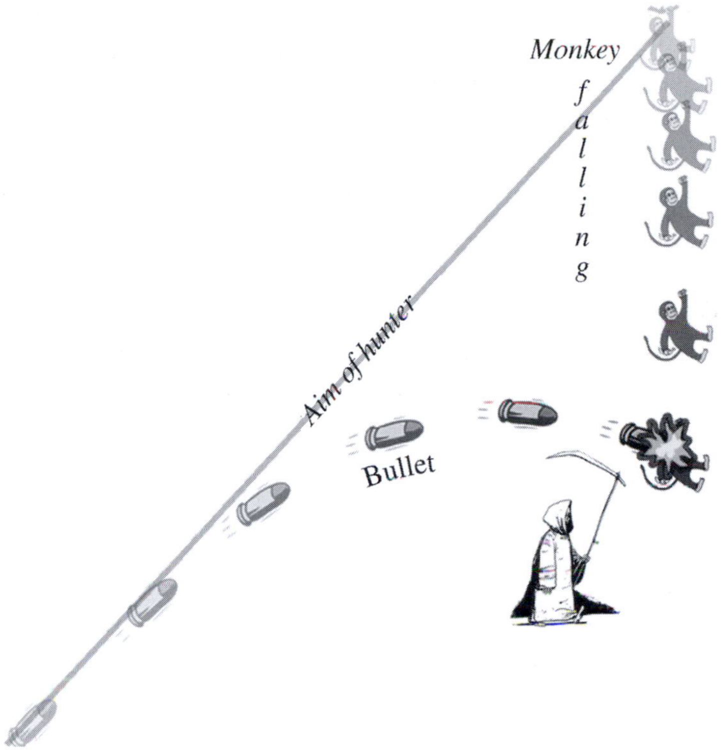

Figure 4.8 Shoot the monkey, an illustration of the equivalence of the inertial and gravitational masses. If a monkey starts falling freely at the same instant a hunter shoots at him the acceleration experienced by the ape and the bullet are identical and produces the same rate of fall for both. As a result the bullet hits the monkey. If the monkey had stayed on the branch he'd have been safe.

And he carries out his plan: as soon as he sees the gun being fired the monkey lets go. But instead of cleverly escaping, the unhappy monkey finds himself dead.[19]

The monkey's argument fails for the following reason: in the absence of gravity there would be no forces acting on the bullet or the monkey, hence the former would travel in a straight line (as indicated in the figure), and the latter would not fall; so the bullet would hit the monkey. Now, in the actual "experiment" there is a force (gravity), and the monkey will not stay at rest but will accelerate downward. But gravity also acts on the bullet and produces precisely the same downward acceleration on it. Hence the bullet will no longer travel in a straight line, but will follow the curved path indicated in the figure. Since the deviation from their force-free motions (state of rest for the monkey, motion along a straight line for the bullet) are produced by a force which generates the same acceleration in both objects, their vertical positions are then perfectly correlated leading to the monkey's demise . . . had he only stayed on the branch!

Now, given a force of *constant strength*, it will affect bodies differently; the more massive the object the smaller the effect: a blow from a hammer will send a small ball flying, the same blow will hardly affect a planet. On the other hand gravity produces the same *acceleration* on the monkey and the bullet; that is why the monkey is hit. Since the mass of the monkey is very different from that of the

Box 4.5

On the nature of gravitation

Another interesting feature of F_{grav} is that it is time independent. This implies that if a body moves its change is perceived instantaneously by all the bodies throughout the Universe. Because of this instantaneous action at a distance Leibnitz (among others) criticized Newton's hypothesis, but these complaints fell by the wayside due to the enormous predictive power of Newtonian gravity. In fact we will see that this criticism *was* justified and that gravitational effects are not transmitted at infinite speeds. The effect of this phenomenon is small in ordinary circumstances, but very significant in various cosmological objects (such as black holes).

Leibnitz also complained that neither the nature of gravitation nor the mechanism by which it is transmitted through apparently empty space, were specified. Newton responded that he was interested in the properties of the observable phenomena and their mathematical elucidation, not in their ultimate nature, offering his often misunderstood statement *"Non fingo hypotheses"* – I do not invent hypotheses – that was not an arrogant claim of infallibility, but more likely, an expression of his refusal to speculate on the nature of gravity. Newton also professed his belief in the existence of some unspecified medium used by gravity to transmit its effects, but would again hazard no guess as to its properties.

bullet but their accelerations are identical, gravity's *force* is different for each of them. One might rightfully ask *why* is it that the accelerations are independent of the mass, Einstein posed this question, and answered it by proposing the General Theory of Relativity.

The large number of successes of Newton's theory led to an almost blind faith in its accuracy. To give an example of this accuracy consider the history of Neptune's discovery. In 1843 a young astronomer at Cambridge, J. C. Adams, discovered an anomaly in the orbit of Uranus, and by the end of 1845 had concluded that this was due, not to a failure of Newton's law of gravity, but could be explained by the presence of a new planet whose orbit could be inferred. Adams submitted his results to G. Airy, his boss, who was unconvinced and dropped the matter. Meanwhile U. Leverrier in France had independently done a similar set of calculations, and published his results in 1846. This spurred Airy into action, but the Cambridge Observatory lacked an up-to-date chart of the region of the sky where the new planet was supposed to reside (at that time). Meanwhile Leverrier had written to J. G. Galle at the Berlin Observatory who did have the appropriate charts, and promptly located the new planet. After much discussion this planet was called Neptune, and its discovery is credited to both Adams and Leverrier (Airy is now best known for having described a special set of functions).

Optics

Newton's first work as Lucasian Professor was on optics. Every scientist since Aristotle had believed light to be a simple entity, but Newton, through his experience when building telescopes, believed otherwise: it is often found that the observed images have colored rings around them (in fact, he devised the reflecting telescope, to minimize this effect). His crucial experiment showing that white light is composite consisted in taking a beam of white light and passing it through a prism; the result is a wide beam displaying a spectrum of colors. If this wide beam is "collected" using a lens and then made to pass through a second prism, the output is again a narrow beam of white light. If, however, by using a screen, only one color is allowed to pass, the final beam has this one color only. Newton concluded that white light is really a mixture of many different types of colored rays, and that these colored rays are not composed of more basic entities (Figure 4.9).

Concerning the nature of light, Newton believed that it consists of a stream of small particles (or corpuscles) rather than waves; for he believed the latter could not explain why light travels in a straight line, while for corpuscles he could use his laws of motion to predict how light would reflect and refract. Perhaps because of Newton's already high reputation this "corpuscular" theory was tentatively accepted; but it soon became clear that it could not account properly for the phenomena of diffraction, interference, and polarization, while the wave theory could. By the end of the nineteenth century light was understood as a wave phenomenon.

Newton's legacy

It is hard to underestimate the influence of Newton in the progress of science. The simplicity of the assumptions coupled with the wide range of phenomena that could be understood provided a sense of power over Nature unheard of before. For the first time natural philosophers could claim to understand (at least a part of) Nature's behavior. Humanity was certainly at the mercy of Nature's forces, but these forces were amenable to knowledge and study, and their effects could be predicted, and sometimes tamed and harnessed for the convenience of the species. Technological applications soon followed these mathematical conclusions, lending credence to these claims. Newtonian mechanics made possible the industrial revolution (though, of course, was not its only catalyst).

At the scientific level the *Principia* provided the standard for research publications given its clarity of hypotheses, the rigor of derivation, and the significance of the results. It has served as a model since its publication more than 300 years ago.

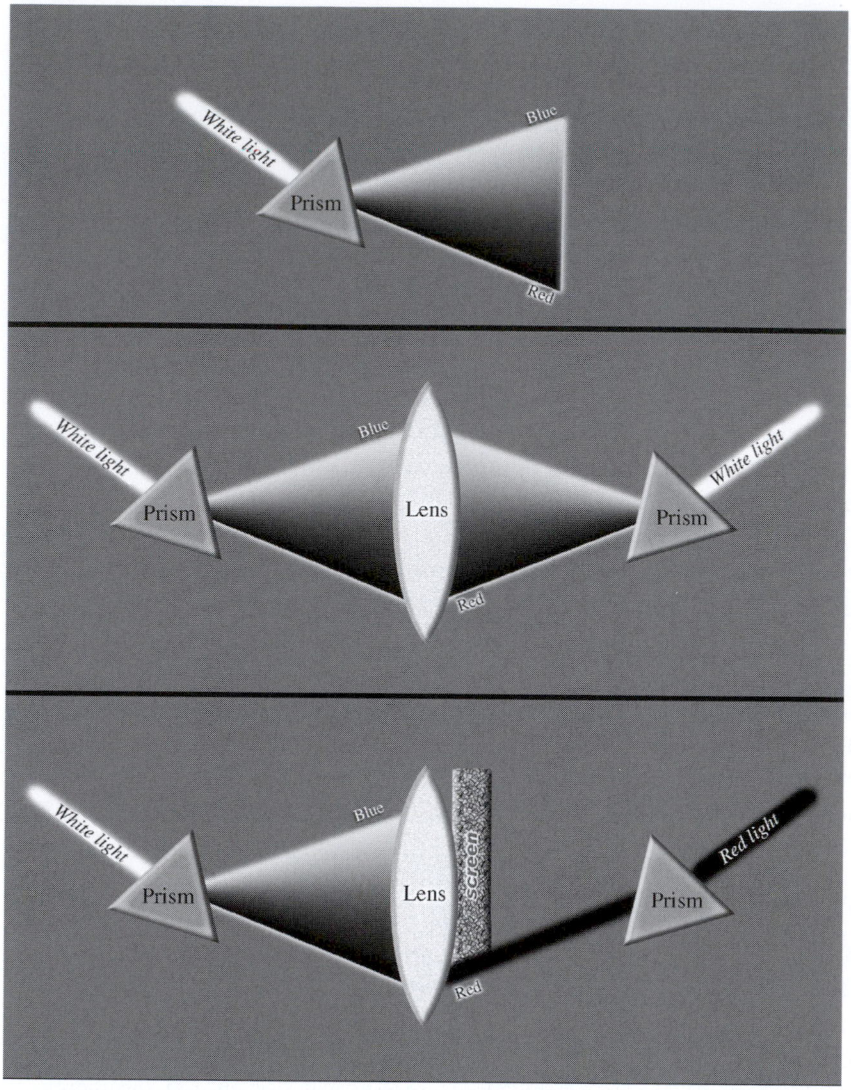

Figure 4.9 Diagram of Newton's experiments on the composition of white light.

Appendix: Galileo and the Inquisition

The Copernican views of Galileo, when combined with his prestige and notoriety, would almost inevitably bring him into conflict with a Catholic Church that felt besieged on all sides (for Luther's schism was in full progress at the time). This was exacerbated when Galileo's friendship with Jesuit astronomers cooled after his dispute with Fabricius (a member of the order) over the primacy of the discovery of sunspots; so that the Society of Jesus refrained from any further defense of Galileo and his views.

By 1613 Galileo began to openly support the Copernican hypothesis and to dismiss the geocentric view presented by the Bible. To Father Castelli he wrote, "... it seems to me that as an authority in mathematical controversy it" (i.e. the Bible) "has very little standing. ... I believe that natural processes which we either perceive by careful observation or deduce by cogent demonstration cannot be refuted by passages from the Bible." This position seriously worried Galileo's allies among the clergy; Cardinal Bellarmine advised him (through a friend) to describe the Copernican ideas only as convenient supposition and not as a factual description of the Solar System.

Galileo ignored all such warnings and so the attack began in 1614. A Dominican preacher, Tommaso Caccini, pointing out the irreconcilable differences between the Scriptures and the Copernican hypothesis, lodged a formal complaint against Galileo before the Congregations of the Holy Office (the Inquisition) on March 20, 1615; and this accusation was supported by other minor complaints. Again Galileo was warned to moderate his views, this time by Monsignor P. Dini, a well-connected Vatican official, but Galileo again refused. That same year he published the *Letter to the Grand Duchess of Tuscany* reaffirming his adherence to the Copernican ideas and asserting that "Nature .... never transgresses the laws imposed upon her, or cares a whit whether her ... methods of operation are understandable to men ... nothing physical ... ought to be called in question ... upon the testimony of Biblical passages." He did temper these statements by claiming to be interested only in providing a serviceable description of the world, and if it be found lacking or wrong by the Church, he would be ready to submit to this judgment, and even to see his book burned.[20] In December 1615 and armed with friendly support letters, Galileo set out for Rome, hoping to convert the ecclesiastical authorities to the heliocentric idea.

It was to be an unsuccessful attempt. On February 26, 1616, the Inquisition directed Bellarmine to summon Galileo and command him to stop defending and publicizing the Copernican hypothesis under threat of imprisonment. That day Galileo appeared before Bellarmine and declared his submission to the ruling. On March 5 the Holy Office published its historic anti-Copernican edict:

> The view that the Sun stands motionless at the center of the Universe is foolish, philosophically false, and utterly heretical, because contrary to Holy Scripture. The view that the Earth is not the center of the Universe and even has a daily rotation is philosophically false, and at least an erroneous belief.

On the same date the Congregation of the Index forbade publication or reading of any book advocating the condemned doctrines; forbade the use of Copernicus' *De Revolutionibus* "until it is corrected"; in 1620 it allowed Catholics to read editions from which nine sentences that represented the theory as a fact had been removed. Subdued, Galileo returned to Florence and lived a retired life outside the heliocentric controversy for several years.

In 1619 a disciple of Galileo published an essay criticizing the views of the Jesuit Orazio Grassi on the nature of comets. Grassi responded with an irate attack on Galileo and his followers, which Galileo answered in his 1622 manuscript *The Assayer* where he again reaffirmed his belief in observation, reason, and experiments as the dictating procedures in scientific endeavors. The text was revised and somewhat softened by members of the *Accademia dei Lincei*, after which the Pope Urban VIII accepted its dedication and sanctioned its publication (October 1623). Encouraged by this Galileo returned to Rome in April 1624 hoping now to convert the Pope to the heliocentric hypothesis, but was again unsuccessful. Urban refused to lift the inquisitorial ban, but they still parted on friendly terms.

Over the next couple of years Galileo worked on what was to become his chief work, the *Dialogue of G. G., . . . Where, in Meetings of Four Days, Are Discussed the Two Chief Systems of the World, Ptolemaic and Copernican, Indeterminately Proposing the Philosophical and Natural Arguments, as Well on One Side as on the Other*, a critical exposition of the Ptolemaic and Copernican systems. In 1626 his pupil B. Castelli was appointed mathematician to the Pope and another disciple, N. Riccardi, became censor of the press. Thus encouraged Galileo hastened to complete a draft of his manuscript; in May he again traveled to Rome and obtained the Pope's permission to publish. He was to work on this book for six more years; finally he issued it in February 1632.

The *Dialogue* might have been less troublesome to Galileo had he not included a comment on the prohibition of the Holy Office to discuss the heliocentric ideas, the tone and phrasing of his statements effectively telling the reader that the dialogue format used in the text was chosen as a device to dodge the constraints imposed by the Inquisition. The *Dialogue* tells of the discussions of three people, Salviati and Sagredo who support the Copernican ideas, and Simplicio, the simpleton, who rejects them, with transparent sophistry; the general conclusion is that the Ptolemaic system is superior, but the arguments presented are unsound and it is quite clear that Galileo intended the book as a support of heliocentrism. Moreover, Galileo dared put into Simplicio's mouth an almost verbatim phrase that Urban insisted be added: "God is all-powerful; all things are therefore possible to him; ergo the tides cannot be adduced as a necessary proof of the double motion of the Earth without limiting God's omniscience" (to which Salviati comments sarcastically: "An admirable and truly angelic argument").

The Jesuits (who did not escape Galileo's sharp tongue) insisted on the dangers of the Copernican ideas, raising them above those of Calvin and Luther; they also pointed out to the Pope that his statement was put into the mouth of a simpleton. Thus pressed Urban appointed a commission that examined the work and reported that it treated the Copernican system as fact, and that it secured the Pope's imprimatur through misrepresentations. Galileo's appearance before a tribunal of the Inquisition was now inevitable, for Urban would not interfere further (though he kept himself informed of the development of the trial).

In August 1632 the Inquisition forbade further sale of the *Dialogue* and confiscated all remaining copies and in September it summoned Galileo despite his advanced age and many infirmities; his devoted and beloved daughter and even the Grand Duke of Tuscany urged Galileo to submit to the Church. Galileo reached Rome on February 13, 1633 and took residence in the house of the Florentine ambassador. He appeared before his judges on April 12 and was charged with having broken his pledge of 1616 not to defend the heliocentric hypothesis; he was urged to confess his guilt but he refused, denying the charges. He was then ordered imprisoned in the Inquisition palace until April 30, and though he was not tortured, he might have been led to fear it. After a second appearance before the tribunal he confessed that he had stated the arguments in favor of heliocentrism too forcefully. He was then allowed to return to the ambassador's house.

On May 10 he was examined a third time, and he offered to do penance for his transgressions, though he asked for consideration for his age and deteriorating health. He was examined a fourth time on June 21, when he admitted his sins, and on June 22 he was pronounced guilty of heresy and disobedience, and was offered absolution on condition of full abjuration. He was to be detained in the prison of the Holy Office for an unspecified period, and was required to recite daily for the next three years the seven penitential psalms as penance. In addition he was made to kneel and in full penitential garb to repudiate the Copernican ideas, declaring:

> I, Galileo Galilei . . . having before my eyes and touching with my hands, the holy Gospels swear that I have always believed, do now believe, and by God's help will for the future believe, all that is held, preached, and taught by the Holy Catholic and Apostolic Roman Church. But whereas – after an injunction had been judicially intimated to me by this Holy Office, to the effect that I must altogether abandon the false opinion that the Sun is the center of the world and immovable, and that the Earth is not the center of the world, and moves, and that I must hold, defend, or teach in any way whatsoever, verbally or in writing, the said doctrine, and after it had been notified to me that the said doctrine was contrary to Holy Scripture – I wrote and printed a book in which I discuss this doctrine already condemned, and adduce arguments of great cogency in its favor, without presenting any solution of these; and for this cause I have been pronounced by the Holy Office to be vehemently suspected of heresy, that is to say, of having held and believed that the Sun is the center of the world and immovable, and that the Earth is not the center and moves . . .

The sentence was signed by seven cardinals, but was not ratified by the Pope.[21] He spent three days in the Inquisitorial prison and then, by order of the Pope, was allowed to move to the Grand Duke's villa in Rome. A week later he moved to Sienna to the comfortable palace of his former pupil, Archbishop Ascanio Piccolomini.

Finally in 1633 he was allowed to return to his own villa in Arceti, near Florence, where he remained until his death. Though technically still a prisoner and forbidden to wander outside his property he was free to pursue his research, to teach and to write, and to receive visitors. Milton came in 1638; his nun daughter came to live with him, and took upon herself the penalty of reciting the psalms.

The long trial and forced abjuration apparently left Galileo a broken man; but his capitulation can only be criticized by those persistent souls that refused to comply under similar circumstances; and then it must be remembered that such men were often tortured and burned for their insistence. Yet Galileo was not defeated: the abjuration did not kill his curiosity and interest in Nature, and he continued his investigations; and, of course, the *Dialogue* spread, and the *Dialogue* did not recant. In 1835 the Church withdrew the works of Galileo from her Index of Prohibited Books, and in 1983 admitted the veracity and accuracy of Galileo's ideas. The broken and defeated man had triumphed over the most powerful institution in history.

Notes

1. Protestant writers condemned sacerdotal exorcism as magic, but the Church of England recognized the value of exorcism as a healing rite.
2. It is a sad commentary on the Salem witch trials of 1692–3 that the peculiar behavior of the accused might have been due to poisoning by ergot (*Claviceps purpurea*), a hallucinogenic fungus that plagues rye, which was abundant in New England at the time and may have contaminated the bread eaten by the victims.
3. Lima (1551), Mexico City (1553), Jena (1558), Geneva (1559), Lille (1562), Strasbourg (1567), Leiden (1575), Helmstedt (1575), Wilno (1578), Würzburg (1582), Edinburgh (1583), Franeker (1585), Graz (1596), Dublin (1591), Lublin (1596), Harderwijk (1600), Giessen (1607), Groningen (1614), Amsterdam (1632), Dorpat (1632), Budapest (1635), Utrecht (1636), Turku (1640), Bamberg (1648).
4. I will have much more to say about the ether when we discuss electricity and magnetism and the Special Theory of Relativity.
5. When Pascal stated that the angels themselves could not produce a vacuum, Descartes remarked that the only existing vacuum was in Pascal's head.
6. This is true only when the pendulum's swing is not too large compared to its length.
7. Of course it is important for the experiment to be fully contained inside the ship; for example, measuring the position of the ship with respect to shore will give a different result when done on shore, but then the experiment is not done fully inside the ship: one must look outside. Galileo reached this conclusion by performing a variety of purely mechanical experiments, but the result is true for experiments of *all* types.
8. We refer here to forces that might make the banana move with respect to the boat.
9. Titus Lucretius Carus (*c*. 99–55 BC). Roman poet and philosopher, author of *De Rerum Natura* (On the Nature of the Universe), a comprehensive exposition of the Epicurean world view.
10. This is true for any motion where the acceleration is constant.

11. This landed him a permanent position as "Chief Mathematician of the University of Pisa and Philosopher and Mathematician to the Grand Duke of Tuscany."
12. Jupiter is now known to have a series of faint rings as well as a large number of satellites, but Galileo's telescope allowed him to see only four of the latter: Io, Europa, Ganymede, and Callisto.
13. But he did not use Kepler's 1609 results describing planetary orbits.
14. cf. the Gibbs lectures delivered by Lord Kelvin in 1899.
15. Similarly $F = ma$ is also not universally valid.
16. Even though the magnitude of the velocity does not change, its *direction* does, and this also represents acceleration.
17. If we further increase the initial speed of the apple it will reach escape velocity and leave the vicinity of the Earth.
18. In obtaining this result Newton had to prove that the interaction between Sun and planets behaves, to a very good approximation, as if the mass of each of these objects was concentrated at its center; for the purposes of this calculation heavenly bodies behave as point-like objects.
19. No real animals need be used in demonstrating this (see http://phyld.ucr.edu/Mechanics%202/M-9D.htm).
20. But he added, "I do not feel obliged to believe that that same God who has endowed us with sense, reason, and intellect has intended us to forgo their use."
21. That on leaving the trial chamber Galileo muttered defiantly, "*Eppur si muove*!" (And yet it does move!), is probably a legend; yet there is little doubt this phrase does reflect his true beliefs.

5
The clouds gather

For more than two centuries after its inception the Newtonian view of the world ruled supreme to the point that scientists developed an almost blind faith in this theory. And for good reason: there were very few problems which could not be accounted for using this approach. Nonetheless, by the end of the nineteenth century new experimental evidence difficult to explain using Newtonian mechanics began to accumulate, and by the early twentieth century this led to a complete revision of the fundamental ideas about Nature. This transition was far from easy, for example, as late as 1884 Lord Kelvin, still immersed in Newtonian bliss, mentioned a few "clouds" that marred the horizon of physics, expressing his confidence (or perhaps his hope) that a natural explanation using this prevailing theory would soon be found; once this was accomplished we would understand the whole of Nature. These clouds, however, were not appeased, and grew into a storm that completely overthrew the classical ideas of Newton and replaced them with those of relativity and quantum mechanics.

Among the problems of the time (not all were mentioned by Kelvin) were:

- Light had been recognized as a wave, but the properties (and the very existence!) of the medium that conveys light appeared inconsistent.
- The equations describing electricity and magnetism were inconsistent with Newton's description of space and time.
- The orbit of Mercury presented a small but disturbing unexplained discrepancy between the observations and the calculations obtained using Newton's equations.
- Materials at very low temperatures did not behave according to the predictions of Newtonian physics.
- Newtonian physics predicts that *any* oven at *any* stable constant temperature has infinite energy (even the oven in your kitchen).

The first quarter of the twentieth century witnessed the creation of the theories that explained these phenomena. The first two problems require the introduction of the Special Theory of Relativity. The third item requires the introduction of the General Theory of Relativity. The last two items can be understood only through the introduction of a completely new mechanics: quantum mechanics.

Box 5.1

William Gilbert (1544–1603)

Born in Colchester, England, into a middle class family of some wealth. Entered St John's College, Cambridge in 1558, and obtained his B.A. (1561), M.A. (1564) and M.D. (1569). He became a senior fellow of the college, held several offices and set up a medical practice in London (1573), becoming a member of the Royal College of Physicians. He never married. In 1601 he was appointed physician to Queen Elizabeth I, and upon her death in 1603 to King James I.

His main work, *De Magnete, Magneticisque Corporibus, et de Magno Magnete Tellure (On the Magnet, Magnetic Bodies, and the Great Magnet of the Earth)* published in 1600, became the standard work throughout Europe on electrical and magnetic phenomena. It is a comprehensive review of what was known about the nature of magnetism, containing also many new experimental results obtained by Gilbert himself. In it he coins the terms electric attraction, electric force, and magnetic pole; Gilbert is often considered the father of electrical studies. He built a philosophy where magnetism was the soul of the Earth; he believed that a perfectly spherical lodestone, when aligned with the Earth's poles, would spin on its axis, just as the Earth does, and used this to argue that the fixed stars were not all at the same distance from the Earth. He also believed that a magnetic force was responsible for maintaining the planets in their orbits (a view adopted by Kepler).

As a result of these developments Newton's formalism lost its fundamental character. It is of course still a perfectly good theory *but* with a very well defined range of applicability. This does not imply that Newton was "wrong," it merely implies that his theories, although accurately describing Nature in an impressive range of phenomena, do not describe *all* of it. The new theories that superseded Newton's have the virtue of explaining everything Newtonian mechanics did (with even greater accuracy) while extending our understanding to an even wider range of phenomena. In this chapter we describe the history and content of the theory that describes electric and magnetic phenomena, which was instrumental in the creation of the Special Theory of Relativity.

It is worth emphasizing that this demotion of Newtonian mechanics was driven by its inability to explain the *data*; and the theories of relativity and quantum mechanics were accepted not just because of their logical consistency, but because they *could* explain the data, and also make predictions that have been subsequently verified with enviable accuracy.[1] In the ripeness of time new data might be obtained that cannot be understood using these theories of ours, and this will signal their demise. Future scientists will then find replacements that will provide a yet deeper understanding of Nature. These new theories will have to explain everything relativity and quantum mechanics do, together with

Box 5.2

The Leyden jar

In 1746 Pieter van Musschenbroek discovered the so-called Leyden jar, a simple and effective contraption for storing static electricity. It consists of a closed container made of an insulating material covered inside and outside by some metal (or other conducting material) and fashioned so that the inside and outside conductors are not connected. If a positive electric charge is deposited in one of the conductors it will induce a negative charge in the other, and this arrangement will persist for a long time. Should an incautious observer touch both conductors, electricity will flow through him in a most unpleasant way.

that new data that will require their creation; in addition they should provide experimentally verifiable predictions. Though there is no evidence suggesting the need of such a revision, one should not forget that our currently accepted theories are but approximations to reality.

Electricity and magnetism

The story of the discovery of the laws that determine the properties of electric and magnetic phenomena, their surprising and exciting interconnections, and their relationship with light and optical effects is not as involved as the one relating our learning to understand mechanical phenomena and the properties of motion. This is perhaps because the most interesting electric and magnetic phenomena require rather sophisticated machinery such as batteries, long metal wires and, for optics, lenses and mirrors of high quality and low aberration.[2] In the presentation below I provide first a very brief history of the various discoveries, and then describe the physical laws that were obtained through these discoveries, and the implications derived from these laws.

Box 5.3

Charles Augustin de Coulomb (1736–1806)

Was a member of a family renowned in the legal and administrative professions on his father's side. He was born and grew up in Angoulême, the family then moved to Paris and he entered the *Collège Mazarin*. At this time his parents lost their money through financial speculations and they separated; his father moved to Montpelier while his mother remained in Paris. She disapproved of his interest in mathematics and astronomy, so Charles went to live with his father. He joined the Society of Sciences in 1757 and read several papers there. He entered the *Ecole du Génie*

Box 5.3 (cont.)

at Mézières in 1760 after returning to Paris (1758) to receive the private tutoring he needed to pass the entrance examinations. He graduated in 1761 as a trained engineer with the rank of lieutenant in the *Corps du Génie*. For the next 20 years he was involved in a series of engineering projects. From 1764 to 1772 he helped with the fortification of Martinique (a French colony in the West Indies), ruining his health. Returning to France he was sent to Bouchain where we studied applied mechanics; in 1773 he presented a paper at the Paris *Académie des Sciences* and became Basut's correspondent in 1774 as a result. Coulomb was next posted to Cherbourg where he wrote a memoir on the magnetic compass that was awarded a share of the *Grand Prix* of the *Académie des Sciences* in 1777. In this paper Coulomb invented the torsion balance, an instrument for measuring very small torques. In 1779 he was sent to Rochefort to construct a wooden fort near the Ile d'Aix in collaboration with the Marquis de Montalembert. He continued his mechanical studies, concentrating on the effects and properties of friction. He won the 1781 *Grand Prix* from the *Académie* where he obtained a permanent position in the mechanics section; he could then return to Paris, and devote himself entirely to physics. Between 1785 and 1791 he wrote seven treatises in electricity and magnetism that contain, among many other results, the well-known relation describing the attraction and repulsion among charges (Coulomb's law). Coulomb was a prolific writer (25 papers between 1781 and 1806), an efficient member of the *Académie* (member of 310 committees), and a consultant on engineering matters (he was put in charge of a large section of Paris' water supply in 1784). He had two sons (born in 1790 and 1797) with his mistress Louise Françoise LeProust Desormeaux whom he married only shortly before his death in 1802. He resigned from the *Corps du Génie* in 1791 dissatisfied with the changes that followed the French Revolution. The *Académie* was abolished in 1793, and he lost his position as overseer of the water supply as well. The *Académie* was replaced by the *Institut de France* to which Coulomb was elected in 1795. In 1802 he was appointed inspector of public instruction and was responsible for setting up the *lycées* across France.

A bit of history

The ancient Greeks were well acquainted with the basic magnetic and electric phenomena. Magnetic materials were mined in Magnesia as early as 800 BC for their property of attracting iron;[3] and Thales (apparently) studied such forces, as well as the peculiar property of amber that will attract light objects such as feathers after being rubbed with wool (the Greek word for amber is *elektron*). We now know that this second phenomenon, which we call *electrostatic attraction*, is due to the fact that the amber becomes electrically charged (rubbing brings many points of the surfaces into good contact, so that, at the atomic level, electrons are ripped from one material and transferred to the other). It is possible to put an electric charge on any solid material by rubbing it with any other material: an automobile becomes charged when it moves through the air, a comb is electrified in passing through dry hair, etc.[4]

For many centuries electrostatic attraction remained a rather mysterious curiosity. In contrast, magnetic materials were soon used in practical applications, especially as navigation devices. Some historians believe that the Chinese used loadstones as compass needles as early as the twenty-sixth century BC, others that this was an Arab–Italian idea that was exported to China in the thirteenth century AD; the first extant European reference dates from this time, as well as the first records of experiments done with magnets.

The modern studies of electricity and magnetism began with William Gilbert who, noting the attraction and repulsion between magnets, argued that the compass needle works through the same effect, and concluded that this requires the Earth itself to be a magnet: *Magnus magnes ipse est globus terrestris*. Before this hypothesis it was believed that this peculiar behavior of compass needles was possibly due to an "affinity" between magnetic materials and the North Star. Gilbert debunked this and many other such "folk-facts," such as the claim that garlic would deprive a magnet of its properties. He also showed that most materials can be charged using friction and attributed this to the removal or addition of an electric "fluid"; it took another 130 years for Stephen Grey to realize that electricity can flow, and another 40 for Du Fay to hypothesize (correctly) that there are two types of electric "flavors" (an idea criticized by B. Franklin as unnecessarily complicated).

By the middle of the eighteenth century a variety of electrostatic apparatuses were available, and this technological advance allowed the determination of the laws that regulate the behavior of magnetic and electric phenomena. Using these instruments Coulomb obtained the expression for the electric attractions between charges[5] (1785–91). Soon thereafter (1813) Poisson demonstrated the conservation of charge; a result known to Faraday, who also demonstrated that magnets moved inside coils generate electric currents. In 1820 Ørsted showed that electric currents generate magnetic effects, an effect that was quantitatively described by Ampère. Finally, around 1845 J. C. Maxwell noted that the results of Ampère were *inconsistent* with the conservation of charge, but that a simple modification of the equations would dispose of this problem. Maxwell then proceeded to study the physical consequences of the modified equations and realized that they described not only electric and magnetic phenomena, but also all effects related to light: in one fell swoop Maxwell unified electricity, magnetism and optics into a coherent theory. The ramifications of this momentous achievement can hardly be understated; at the scientific level it provided one of the prime motivations for the creation of the theories of relativity that led to our modern view of space and time; technologically Maxwell's results provided the foundation for our modern telecommunication industry. This theory even had a philosophical impact by providing a paradigm for the unification of apparently diverse effects.

Box 5.4

Michael Faraday (1791–1867)

Was born near London, the third son of a blacksmith and a farmer's daughter, both hailing from the north of England. When he was four the family moved to London where another child was born. They were members of the Protestant Sandemanians, who believed in the literal truth of the Bible and tried to recreate the sense of love and community of the early Christian Church. Michael had a difficult infancy for his father's poor health prevented him from providing for his family. He first attended a day school where he got a basic education. From 1804 until 1811 he worked for a bookseller, first as an errand boy and later as a bookbinder. This job gave him access to a variety of books and so he had his first brushes with science. He attended a variety of lectures in 1810–12 and even wrote to the president of the Royal Society asking for information on getting involved in scientific pursuits (he received no reply). In 1812 he set out as an independent bookbinder, but still intent on his scientific vocation he wrote to Humphry Davy who advised him to stick to bookbinding. However, in 1813 Davy fired his assistant for improper behavior and offered the position to Faraday. That same year Davy, with Faraday, traveled through Europe; on this trip Faraday met Ampère and Volta. After this broadening and enlightening experience Faraday returned to London to the Royal Society working and lecturing on chemistry, and submitted his first paper in this topic in 1816. In 1821 he was made Superintendent of the House and Laboratory at the Royal Institution which allowed him to marry Sarah Barnard. Late on that year he also began working in electricity and magnetism, and published his results in the *Quarterly Journal of Science* recording for the first time the conversion of electrical into mechanical energy. From 1821 to 1831 he worked on chemistry, in 1823 he succeeded in liquefying chlorine (up to then believed to be a "permanent gas" incapable of condensing), and in 1825 isolated benzene. He was elected a fellow of the Royal Society in 1824 despite being opposed by the then-president Davy (who still thought of Faraday as his assistant). In 1826 he began his yearly children's Christmas lectures on science, a tradition that continues to this day (with other speakers). Faraday returned to his research on electricity and made his most important discovery, that of electromagnetic induction and was then able to convert mechanical into electrical energy. He read his paper before the Royal Society in 1831. In 1832 he received an honorary degree from Oxford, and a year later he became Fullerian Professor of Chemistry at the Royal Institution. He was awarded the Royal and the Copley Medals, both from the Royal Society. In 1836 he was appointed by the Crown a Member of the Senate of the University of London, and, continuing his work on electricity he was able to produce "a coherent theory of electricity." This extremely heavy workload told on his health and in 1839 he suffered a nervous breakdown from which he recovered completely only six years later. In 1845 he succeeded in showing that strong magnetic fields affect the polarization of light. Following this line of experiments he discovered the magnetic property of

Box 5.4 (cont.)

certain materials known as diamagnetism. By 1850, while Maxwell was constructing the final theory of electromagnetic phenomena (never a mathematician, Faraday never mastered Maxwell's approach), Faraday's mental abilities began to decline; he continued lecturing at the Royal Institution but declined its presidency. In 1972 Faraday's laboratory was restored to its form of 1854 and an adjoining museum was opened.

Box 5.5

André-Marie Ampère (1775–1836)

Was born into a prosperous family; he lived in Lyons until he was seven years old and then the family settled in Poleymieux where his father imparted and designed André-Marie's education instilling in his young mind a passion for learning. Legend has it that Ampère had "mastered" mathematics by the age of 12, but this appears unlikely since by his own account he started learning the subject only when 13; at this young age, however, he did teach himself calculus and mechanics. This idyllic environment was not to last: in 1791 André-Marie's father accepted a position as Justice of the Peace in Lyons, his sister died in 1792 and shortly thereafter the city was besieged for refusing to accept order from the revolutionary government in Paris. After the fall of Lyons, Ampère's father was imprisoned and then guillotined. The young André-Marie was devastated, recovering only a year and half after meeting his future wife Julie, whom he married in 1799; his first son was born one year later. They lived on his tutoring fees until 1802 when he was appointed professor of physics and chemistry at the *Ecole Centrale* in Bourg. Julie remained in Poleymieux due to illness. Despite a busy teaching schedule Ampère managed to submit two papers to the Paris Academy in 1803. Later that year he moved to a position in Lyons where he endured a second blow when his wife died in July. Consumed by guilt for what he believed was his having abandoned Julie as she was dying, he quit Lyons for Paris, a decision he was to regret. He missed his friends from Lyons as well as that city's peaceful and friendly scientific environment. Despite his lack of formal education he was appointed *répétiteur* at the *Ecole Politechnique* in 1804. In 1806 he embarked on his second and disastrous marriage. He separated from his wife Jenny only two years later, obtaining the custody of his daughter Albine; she was later to embark in an abusive marriage with one of Napoleon's lieutenants. In 1809 Ampère was appointed professor at the *Ecole Politechnique*, a post he held for 19 years, where he shared teaching duties with Cauchy. In 1811 he suggested the existence of fluorine; three years later he submitted a memoir on differential equations to the *Institut National des Sciences*, and was elected as a member. Around 1820 Ampère heard of Ørsted's experiments in electricity and magnetism and over the next six years he worked on this subject producing various important results including the law that bears his name relating the generation of magnetism by changing electric forces. From 1826 he became a

Box 5.5 (cont.)

professor at the University of Paris where he remained until his death. Ampère received a measure of satisfaction from his son's success, but they were never on easy terms being of the same temperamental character.

It is an interesting sideline that when Faraday announced his discovery that a magnet and a coil can be used to generate electricity he was criticized by an observer for a lack of practical applicability of his results; Faraday responded that one cannot gauge the usefulness of a new-born babe. Within the next hundred years these "impractical" investigations lead to the creation of electric generators and motors, devices now essential for the proper functioning of our society.

Maxwell's equations

The culmination of 150 years of intense studies of the properties of electric and magnetic materials yielded a set of four equations (known as Maxwell's equations) that accurately describe all electric, magnetic and optical phenomena; with the proper re-interpretation they are also valid in the quantum regime. They are applicable even when considering objects moving close to the speed of light (where Newton's equations fail). In the previous section I gave a short summary of the historical developments that led to the discovery of these equations, in this section I will summarize the physical content of the equations themselves.

After suggesting the modification that rendered the equations for electromagnetism consistent with charge conservation, Maxwell proceeded to investigate the mathematical structure of this new set of hypotheses. He found, as expected, that they expressed in mathematical language the results of Coulomb, Ørsted, Faraday, Ampère, and others, pertaining to the motion and interaction of magnets and charged particles; so this new theory did encompass its predecessors. The big surprise came when he looked at the solutions to his equations for situations containing accelerating charges and/or magnets. For in this case he *predicted* that such objects would generate waves spreading at the speed of 299,792 km s^{-1}, which is numerically identical to the speed of light. It is not that Maxwell *chose* this speed to be that of light, this was a *prediction* of the theory: if his equations were a good description of Nature, moving charges ought to generate waves that travel at the speed of light.

By Maxwell's time it was also known that light behaves as a wave (as I discuss below), and the fact that his equations described waves moving at the very same speed seemed to him impossible to be coincidental:

> We can scarcely avoid the conclusion that light consists in the transverse undulations of the same medium which is the cause of electric and magnetic phenomena.

Maxwell's surmise was quite correct: the same equations describing electric and magnetic phenomena also describe the behavior of light. In addition they show that visible light is but a member of a vast family of waves, all moving at the same speed and differing only in their wavelength. Just as Newton created a unified theory of motion that encompassed both Earth-bound and celestial bodies, Maxwell created a theory that described all electric, magnetic, and optical phenomena, all expressed in terms of four simple laws that I now describe.

Charge conservation

By the end of the eighteenth century it was known that electricity comes in two flavors conventionally (and unimaginatively) labeled "positive" and "negative." By the middle 1800s it was known that the net charge of an isolated system does not change irrespective of the internal transformation the system might undergo. In fact, it was through the imposition of this condition that Maxwell arrived at the final form of his theory of electromagnetism.

The persistence of charge can be illustrated as follows. Imagine we have a certain number of objects inside a box, some positively and some negatively charged, and we define the *net charge* of the system as the total positive *minus* the total negative charge. Imagine we now perform all sort of experiments to the objects in it, breaking them, shining lasers on them, heating them, etc., but taking care that no charge sneaks into or leaks out of the box. If after all such torments, we look into the box and calculate the net charge again, the result will be the same as initially, even though the total positive and negative charges will in general change.

Box 5.6

James Clerk Maxwell (1831–79)

Was born in Edinburgh but the family soon moved to a village near Dumfries where he enjoyed a country upbringing and displayed an early scientific curiosity. When eight his mother died and his parents' plan of a home education fell through; the family moved back to Edinburgh to live with James' aunt, and he was sent to the Edinburgh Academy where he was at first regarded as shy and rather dull. When 14 he wrote a paper on geometry that was read to the Royal Society of Edinburgh. In 1850 Maxwell entered Cambridge where his exceptional intelligence was recognized by faculty and students (despite his eccentricities). He graduated from Trinity College in 1854, and remained in Cambridge first tutoring pupils, and then supported by a fellowship from Trinity. In 1855–6 he read a paper (in two parts) to the Cambridge Philosophical Society where he expressed in concise and elegant manner the equations obeyed by electric and magnetic fields. In 1856 his father fell ill and Maxwell left Cambridge for Scotland in order to spend more time with him; his

Box 5.6 (cont.)

father died later that year and he returned to Cambridge only to learn he had been appointed Professor of Natural Philosophy at Marischal College in Aberdeen. In 1857 he obtained the Adams Prize for his paper on Saturn's rings. A year later he became engaged to Katherine Mary Dewar whom he married a year later. In 1860 Marischal and King's Colleges merged and Maxwell lost his position (despite his wife being the daughter to the Principal of Marischal). He unsuccessfully applied for the Chair of Natural Philosophy at Edinburgh, but soon was appointed to the chair of Natural Philosophy at King's College in London where, despite demanding teaching duties, he did his most important work. In 1862 he calculated the speed of propagation of the electromagnetic field and showed it was the same as the speed of light, he then hypothesized that these wave phenomena were in fact identical. He also completed his work, begun at Aberdeen, on the kinetic theory of gases based on a probabilistic interpretation of thermodynamics. Maxwell left London for his estate in Scotland in 1865. He made periodic trips to Cambridge and in 1871 accepted the position of first Cavendish Professor of Physics, at this time he designed the Cavendish laboratory, which opened in 1874. During the years 1865–71 he constructed the set of equations that bear his name and which summarize the whole of electricity, magnetism, and optics; they are one of the great scientific achievements of the nineteenth century. Einstein described the change in the conception of reality in physics that resulted from Maxwell's work as "the most profound and the most fruitful that physics has experienced since the time of Newton." In 1879 Maxwell's health began to fail and he returned with his wife, who was also ill, to Scotland. Despite intense pain he remained cheerful; but when the couple returned to Cambridge later that year he could scarcely work, and he died on November 5.

For example, suppose we fill the box with neutral hydrogen atoms, so the initial net charge is zero. If we then shine a high-intensity ultraviolet laser into the box, some of the atoms will be ionized, that is, the electron of each of these atoms will be stripped from the atomic nucleus. As a result we end up with a box that has some neutral atoms, some electrons and some hydrogen nuclei: the total positive and negative charges, originally zero, are no longer so. And yet, there will always be an equal number of electrons and nuclei, so that the net charge is still zero.

The way charges talk

By the beginning of the eighteenth century it was known that two positive or two negative charges repel, and that a positive charge will attract a negative one. The manner in which this attraction and repulsion occurs was first published by Coulomb in 1785. He found that the force between them is very similar in form to the gravitational one: it is proportional to the charges of each body, directed along the line joining them, and decreases as the distance squared. There is,

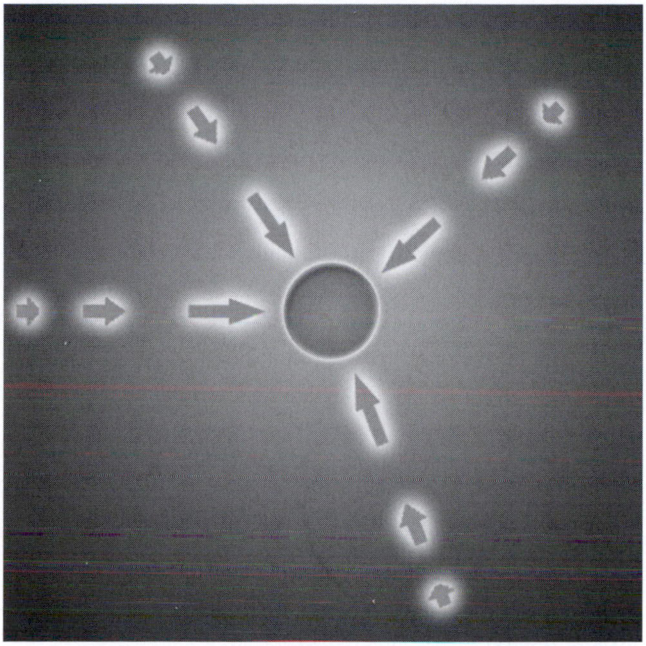

Figure 5.1 Illustration of the field concept: the central charge affects the region surrounding it (simulated by the light shading that fades as one moves away from the charge); the arrows indicate the magnitude of the force a charge of the opposite sign would experience at various locations.

however, an important difference: this electric force can be attractive or repulsive; the gravitational force is *always* attractive. The first equation of Maxwell is a restatement of Coulomb's law, generalized to arbitrary charge distributions, but before stating it I would like to introduce the idea of a field.[6]

The idea of a field

If we fix a charge at one point we find that it influences all charges in its vicinity, either attracting or repelling them, and that this effect wanes with distance. If we look at the behavior of some trial charges in a region of space that does not contain the fixed charge, we can still tell that there is something near by since the trial charges will be affected: a force will act on them.

Because of this one can say that the fixed charge affects the surrounding space, and that the change it induces will affect all charges (though the effect is weaker the larger the separation). A very fruitful description of this situation, due to Faraday, is to represent the forces between charges as a two-step process: (1) the fixed charge affects the space around it and then (2) this "affectation" is experienced by other charges. The change in the space surrounding the fixed charge then acts as a mediator of the force between charges. Faraday called this mediator the *electric field* (Figure 5.1).

All charges generate an electric field around them, and the intensity of this field determines the strength of the force the charge will generate. If we look for the effects of two charges that we call A and B, on a third one we call Z, we can simply calculate the electric fields generated by A and B at the location of Z and add them together, the result determines the force on Z.[7] Using the concept of a field Coulomb's law becomes:

> ***Maxwell's First Law*:** Point-like charges generate electric fields that are proportional to the charge and decrease as the square of the distance. The combined field from various charges is the sum of the individual fields for each charge.

One can also use similar arguments for magnetic phenomena: we can imagine that magnets modify their surroundings by generating a field, naturally called the magnetic field. The interaction between magnets can then be described as the interaction of one magnet with the field generated by the other.

The field concept is applicable to a wide variety of situations quite separate from electricity and magnetism. For example, a given mass will generate a gravitational influence in its vicinity, and so we can associate with it a gravitational field. As another example we can imagine a cold room with a heater in one corner, the heater generates an "influence" that permeates the room, changing the air temperature, and so we can talk about a temperature field, etc.

More importantly, however, is the fact that the introduction of a field provides an alternative to the unsatisfactory idea of "action at a distance." Newton's expression for the gravitational force between two objects, $F_{\text{grav}} = m M G/r^2$, makes no reference to time, and this implies that the effects of this force are felt instantaneously (though weaker the larger the distance between the objects). For example, should a planet suddenly materialize in some location, any observer anywhere in the Universe may immediately learn of this miraculous occurrence, provided he/she (it?) is capable of measuring gravitational forces with sufficient precision. Objects within Newton's theory of gravity are assumed to be able to "act" instantaneously at arbitrarily large distances, a peculiarity that was strongly criticized from its inception (by Leibnitz and others).

In contrast the description in terms of fields as mediators for interactions easily accommodates non-instantaneous effects. In this picture an object generates a field, which then *propagates* at a characteristic speed; another object located some distance away will experience a force only after the propagating field reaches it. The lapse between the time the field is generated and the time at which the effect is felt can be shorter or longer depending on the speed of propagation for the field.

In Newton's theory of gravity there is no delay, and this corresponds to an infinite speed of propagation. We will see later on that infinite speeds do not occur in Nature, for nothing can reach a speed above that of light; in particular no interactions can be transmitted instantaneously (and this is one of the central defects in Newtonian gravity that is corrected by Einstein's General Theory of Relativity). The idea of field-mediated interactions is perfectly suited to this situation: it provides the clearest and most succinct description of electromagnetism and also of the gravitational interactions, where, as it turns out, fields propagate at the speed of light.

If we split a magnet . . .

Since early antiquity it has been known that magnets contain two opposite poles, commonly labeled "north" and "south"; and that equal poles repel and unequal ones attract. The nature of this magnetic force has been studied quantitatively from the time of Gilbert with as much interest as that devoted to the electric interaction. Curiously enough certain parallels were obtained between charge and magnet interactions. For example, if one takes two very long magnets then the repulsive force between the north poles decreases as the square of the separation (though when the separation becomes comparable to the size of the magnets this is no longer true).

The above might suggest that there are some instances in which magnets will behave precisely as charges and that Coulomb's law will apply to both; alas this is not the case. For magnets have one unique peculiarity: if a magnet is split in two we do not obtain separate north and south poles, but instead two smaller magnets are engendered, each with its north and south poles. If the daughter magnets are themselves split, we obtain four tiny pieces, each being the spitting image of their grandfather, with its own north and south poles. This result is observed for magnets of all shapes and sizes; and occurs for any type of splitting (symmetric or not, into 2 or 34 pieces, etc.). The conclusion is that, whereas lone electric charges are common, lone magnetic poles are *never* found in Nature; and this is the essence of the second of the Maxwell equations:

***Maxwell's Second Law*:** There are no lone magnetic poles.

When magnets move

Up to the nineteenth century no clear relationship between electricity and magnetism had been observed. At this time, however, new instruments and, especially, new reliable ways of generating electricity became available, and with these technological developments came the realization of the deep connection between electric and magnetic phenomena.

The simplest example of this occurs when a magnet is moved through a loop of wire, for if the wire is connected to a current meter a non-zero current is observed as long as the magnet moves. If one moves the magnet back and forth through the wire loop an alternating current is induced and it persists as long the motion is continued (Figure 5.2). The first quantitative expression for this effect was provided by Faraday.

Using the concept of fields Faraday expressed this effect as a two-step process: when a magnet is moved back and forth an electric field is generated, and this electric field will affect the electrons in a wire making them move and generating a current. He then concluded that changing magnetic

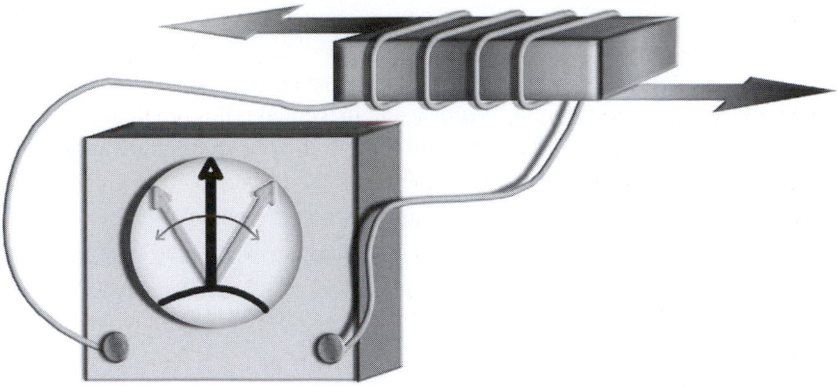

Figure 5.2 When a magnet moves through a wire loop it generates a current; if the direction of motion is reversed so is the current.

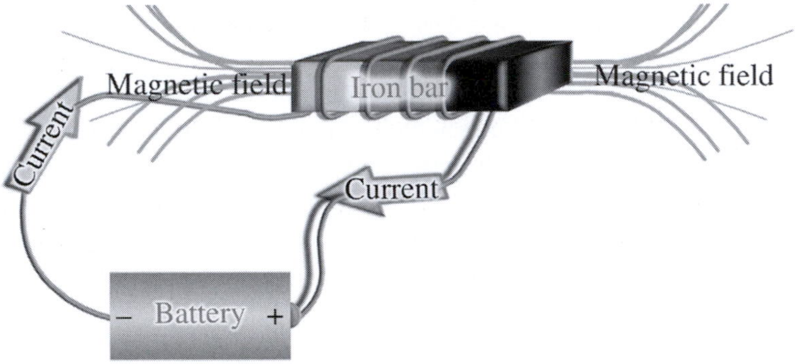

Figure 5.3 When a current passes through a wire a magnetic field is generated (the iron bar amplifies the effect).

fields generate electric fields, and this constitutes the third of Maxwell's equations:

***Maxwell's Third Law*:** Changing magnetic fields generate electric fields.

When charges move

In 1820 the Danish physicist H. C. Ørsted organized a public demonstration of electric phenomena taking advantage of the electric battery recently discovered by Volta. According to legend, during this demonstration he accidentally placed a magnet parallel to a long wire and then proceeded to let current flow along this wire. To his amazement the magnet was affected as if the current going through the wire mimicked another magnet. If the wire was disconnected from the battery the effect disappeared, only to reappear upon reconnection. His conclusion was that currents generate magnetic fields (Figure 5.3), an effect that was afterward described quantitatively by Ampère.

When Maxwell studied the consequences of Ampère's formulas and the three laws described above he found that they were inconsistent with charge conservation. That is, a mathematical consequence of the laws of Coulomb–Poisson, Faraday, Ampère, and the absence of lone magnetic poles, was that under certain conditions the net charge of an object can change. These conditions are far from rare and, in particular, are realized as electrons move inside atoms, and in no case is a change in the net charge observed: the predictions of the theory were not confirmed by the data. One or more of these laws was incorrect or, at best, incomplete.

Maxwell then observed an asymmetry between the laws of Faraday and Ampère: the first predicted that a changing magnetic field would generate an electric field, but the second one just stated that moving charges generate magnetic fields. Perhaps it is possible also to generate magnetic fields through changing electric ones? Maxwell tried this possibility (which entailed adding a term to Ampère's equation) and found that with this rather simple modification the equations were consistent with charge conservation. In addition, as mentioned above, the equations now described not only all electromagnetic phenomena, but also all optical effects. Maxwell's fourth equation summarizes these results:

> ***Maxwell's Fourth Law*:** Currents and changing electric fields generate magnetic fields.

Before combining the above four laws and "producing" the kind of waves we call light, it is worth making a brief detour to describe what a wave *is* and what are the properties that differentiate waves from particles. Then we will return to reap the rich crop sowed by Maxwell.

Waves vs. particles

The word "wave" is colloquially used in a dizzying variety of circumstances (e.g. "New Wave"), but in physics it has a rather precise meaning. The *American Heritage Dictionary* provides the definition:

> A disturbance or oscillation propagated from point to point in a medium or in space.

Thus when a stone is dropped on a calm pond we see a series of circular waves emanating from the spot where the stone hit the water, spreading out at a certain speed. If a bigger stone is used, the resulting waves are more pronounced, the distance from crest to trough becomes larger. If instead of dropping a stone we attach it to a rod and move it up and down we find that the faster we move it the closer together the crests and troughs of the waves, so that if we look at one point on the pond's surface we will see the water swelling and ebbing faster.

These properties of waves have specific names (Figure 5.4, Figure 5.5):

- The number of wave-crests that go through a fixed point on the pond every second is called the *frequency*.

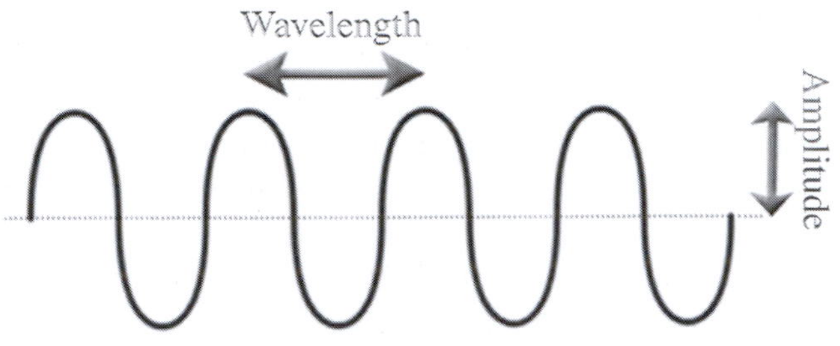

Figure 5.4 Definition of amplitude and wavelength.

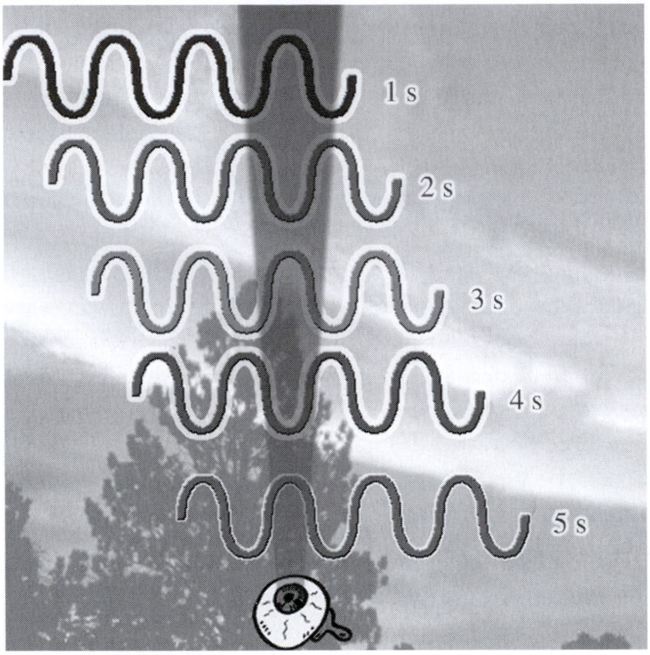

Figure 5.5 Definition of frequency (one crest passes every 2 s, the frequency is 0.5 s^{-1}).

- The distance between two crests is called the *wavelength*.
- The vertical distance between crest and trough is called the *amplitude*.

In general there is a relation between the frequency, wavelength, and speed of a wave, and this relation depends on the medium in which the waves propagate. For example, if we use honey instead of water the "honey waves" will spread slower than the water waves in the pond, and their wavelength will be larger when the plunger is moved at the same rate.

Let us illustrate the above definitions using a familiar situation. Imagine a cork floating on the pond. As the wave goes by the place where the cork is floating it will bob up and down. Suppose that you measure the time it takes for it to go down from its highest point, to its lowest and then back to its highest point again, then the frequency is the *inverse* of this time. So if the cork takes 2 s to go up and down and back up, the frequency would be 1/(2 s) or 0.5 inverse-seconds (s^{-1}). This is just a way of counting the number of oscillations per second: if each oscillation takes two s, there will be half an oscillation per s, and so the frequency is 1/2 s^{-1}; a frequency of 7 s^{-1} indicates that there are seven oscillations each second, etc.

Going back to the pond, note that as the cork bobs up and down it does not change position; even though the wave spreads, it does not carry the cork with it. The same thing can be said of the water itself, the waves spread through it but do not carry the water along with them (if it did one would have a Biblical parting of the pond waters every time a wave goes by). In fact, if you look closely at particles suspended in water (ponds usually have many of those) as the waves pass, they make circular motions about their initial positions but are not carried along. These waves use water as their *propagation medium*, without producing a net motion of it; similarly sound waves use air (or water or other materials) to propagate. Without a medium these waves simply do not propagate: there is no sound in a vacuum. A reasonable question in connection with these observations is whether *all* waves need a medium to propagate in, the answer is (perhaps surprisingly) *no*, and the way this was discovered is the subject of many of the following sections.

A particle is characterized by its mass and other measurable properties (for example, its charge). Ordinary everyday experience shows that waves behave very differently from particles.[8] For example, if you are taking cover behind a wall from a person shooting tomatoes at you, you will not be hit; yet when they scream abuse, you hear them perfectly well. Sound waves (and all waves in general) have the ability to go around obstacles (up to a certain extent: if the wall is very tall and wide the insults will not reach you); particles have no such ability.

These properties of sound waves are well known. But, if light is a wave, should it not behave in the same way? And if it does, how come we do not see a person standing behind a wall (whom we can clearly hear)? We will now consider this (apparent) paradox.

Light

It is now known that under all common circumstances light behaves as a wave propagating at a speed close to 300 000 km s^{-1}. This, however, is a recent realization; in fact, whether light traveled at finite or infinite speed was the subject of much debate and was left unanswered for a long time. Galileo tried to measure the speed of light using the following experiment: he put two men on hills separated by a bit less than a mile, and then told one to open a blind lantern, the other was to raise his hand when he saw the light and the first note any lapse between his opening the lantern and seeing the raised hand. No time delay was observed (which is not unnatural, the lapse is about 10^{-5} s!). So the question remained unanswered.[9]

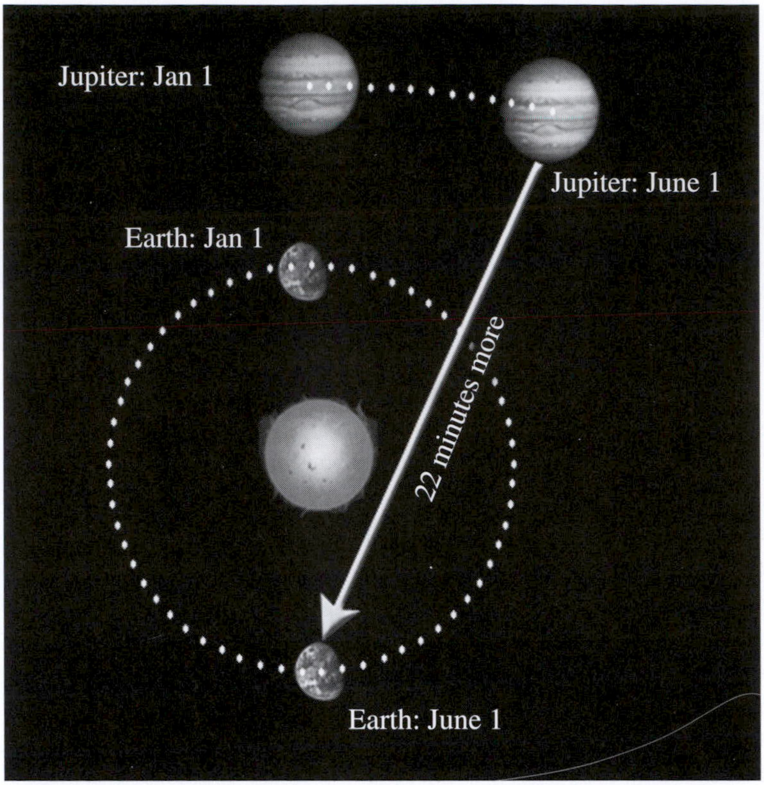

Figure 5.6 Rømer's argument explaining the lag in Jupiter's eclipsing its satellites due to the finite speed of light.

In 1676 the Danish mathematician Olaus Rømer observed that the eclipses of Jupiter's moons were 11 minutes ahead of schedule when the Earth was closer to Jupiter, and they lagged behind (also by 11 minutes) when the Earth was farthest from Jupiter. Assuming that there are no problems with the predictions of Newtonian physics concerning the motion of Jupiter's moons, he concluded that the discrepancy was due to the different times light takes to get to Earth at the two extremes of its orbit (Jupiter moves very little during one year, it takes 12 years for it to circle the sun, Figure 5.6). Rømer then calculated that the speed of light would be 210 000 km s^{-1} (an error of about 30 percent).

This is, of course, not the only possible explanation, Rømer could have argued, for example, that Newton's equations could simply not account for the motion of Jupiter's moons; still the hypothesis that light travels at a finite speed furnished the explanation that required the smallest number of new assumptions and, following Ockham's razor, it is the one which ought to be examined first. It was not until 1849 that H. Fizeau measured the speed of light using purely Earth-bound machinery.

These arguments and experiments showed that light traveled at a finite speed, but provided no information as to its nature. Newton believed it to be made of corpuscles, while his contemporaries Hooke and Huygens proposed an

Table 5.1. *Wavelengths of electromagnetic waves*

Name	Wavelength
Radio	10 cm or larger
Microwave	$\sim$1 cm
Infrared	$\sim 10^{-3}$ cm
Visible	$\sim 10^{-5}$ cm
Ultraviolet	$\sim 10^{-6}$ cm
X rays	$\sim 10^{-8}$ cm
Gamma rays	10^{-9} cm or smaller

alternative hypothesis according to which light behaves as a wave. These competing ideas can be tested by carefully observing the shadow cast by a sharp edge. If light is a stream of particles, the shadow will be as sharp as the edge itself, providing an abrupt transition from light to darkness. If, on the other hand, light is a wave, the transition will not be sudden, and some light will creep into the shadow region. This is so because light waves, like any other kind of waves, would be able to go around obstacles provided these are not too large. Careful measurements showed that shadows are not, in fact, abrupt, but exhibit the precise features predicted by the wave hypothesis. This is not common knowledge because it is a small effect, light dies out almost as it turns the corner, and if the corner is not very sharp, light is scattered in many ways and the effect disappears. By 1817 overwhelming experimental evidence supported the wave hypothesis and the argument was considered settled.

In 1905, however, Einstein provided a successful explanation of the photoelectric effect based on the quantum-mechanical version of the corpuscular hypothesis (for which he was awarded the Nobel prize). So, under certain conditions light behaves as a wave, while under others it behaves as a particle. This ambiguity is not a result of a fickle nature, but a reflection of the limitation of our words in attempting a description of quantum phenomena. The mathematical treatment is quite unambiguous, but when we try to find similes using language, we find that light's dominant characteristics are often similar to those of a wave, though on occasion its behavior is closer to that of a particle.

Still, by the beginning of the nineteenth century, and before the advent of quantum mechanics, the hypothesis that light is a wave traveling at a large (by our standards) but finite speed[10] was universally accepted. As for any wave we can now ask what is its wavelength, amplitude, frequency, etc.? Maxwell's equations predict that visible light occupies but a small region in a continuum of electromagnetic waves, which can have any frequency. For historical reasons waves of different wavelengths have different names (Figure 5.7 and Table 5.1 (the symbol $\sim$ means "about")).

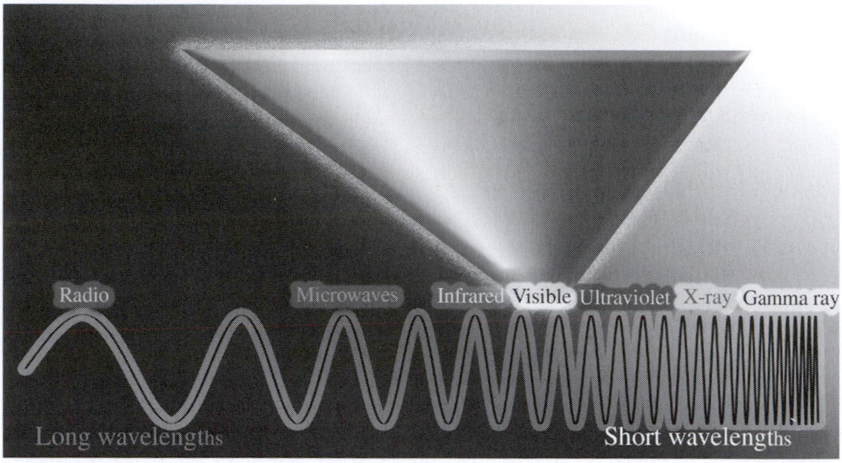

Figure 5.7 The electromagnetic spectrum.

All of these are common names. Every one of these waves travels at the same speed in a vacuum[11] equal to the speed of light (labeled "visible") also in a vacuum; the only difference between them is the wavelength – the distance between two consecutive crests in the corresponding wave trains. The only distinguishing characteristic of visible light has nothing to do with its properties as an electromagnetic wave, but is related to the peculiar conditions of our Earthly atmosphere that is transparent in this range of electromagnetic waves.

If light is indeed a wave, it is natural to speculate that it will have properties similar to those of other waves, such as sound waves, or water waves. In particular, all these waves are produced by the undulations of some medium: water for water waves, air (for example) for sound, etc. Thus it was *hypothesized* that light waves are the undulations of a medium called the *ether* (nothing to do with the anesthetic). Light will then propagate in any region containing ether, but will not whenever this substance is absent (much like sound will not propagate in a vacuum since it has no transmitting medium). The ether cannot be some common substance, for example it cannot be any type of gas, since light propagates in a jar from which all gases have been pumped out. And yet it must be one of the most abundant substances in the Universe, filling space to its deepest recesses. The very fact that we see the stars implies (according to this hypothesis) that the entire region separating us from them should be filled with ether (otherwise light would not be able to propagate toward us, and we would not see them). Because of this one of the initial and most important tasks that followed the ether hypothesis was the determination of the properties and, indeed, the very existence, of this mysterious substance.

Problems

The end of the nineteenth century witnessed the growth of evidence against classical Newtonian physics. I will mention two such problems, the first concerns the ether, the medium in which light was supposed to propagate, and which appeared to have inconsistent properties; the second refers to a contradiction between Galilean relativity and the theory of electromagnetism. The resolution of these conflicts cannot be achieved within Newtonian physics: it requires the theory of relativity.

Ether

Having postulated the existence of the ether, it becomes interesting to determine the properties of this substance. First and foremost, since the light from distant stars does reach us, we must assume that the ether permeates the whole Universe up to its farthest reaches (remember that under this hypothesis light cannot go where no ether is present). We must then imagine that the Earth plunges through this ocean of ether as it circles the Sun. The ether must then be very tenuous, for otherwise loss of energy through friction would have driven the Earth spiraling into the Sun long ago. And yet, the ether would have some rather interesting effects. To see these let us first describe a simple situation here on Earth that serves as an analogy for what will follow.

Imagine a windless day in which you take a ride in your red convertible which, unfortunately, has no windshield. As you speed up you will feel the air blowing in your face, and the faster you go, the stronger this wind is. Visualizing now the Earth moving in a sea of ether (just as the car moves in a "sea" of air), we conclude that observers on Earth's surface should be able to detect an "ether wind" – though this might be very hard, the ether being so tenuous.

But there *are* effects produced by this ether wind that one can measure. To understand how they come about imagine yourself back in your convertible (still with no windshield), now driving your nagging grandmother to the store; she of course sits in the back seat ... it's safer. She talks all the time, but, fortunately, her words get blown back by the wind. In contrast she hears *everything* you say, for your words also get blown back by the wind, and right into her ears. In the same way, as we stand on Earth, the ether wind should affect the direction of the light coming from the stars. At different times of the year, the ether wind blows in different directions since the Earth is also moving in different directions, and so this effect should have a seasonal variation. As a result, careful observations should show an annual shift in the positions of the stars (Figure 5.8), and, in fact, this periodic shift *is* observed. This result supports the ether hypothesis and, in addition, it shows that the Earth must not drag the ether with it (for if the Earth moved surrounded by a cocoon of ether there would be no ether wind near its surface). The ether should then go

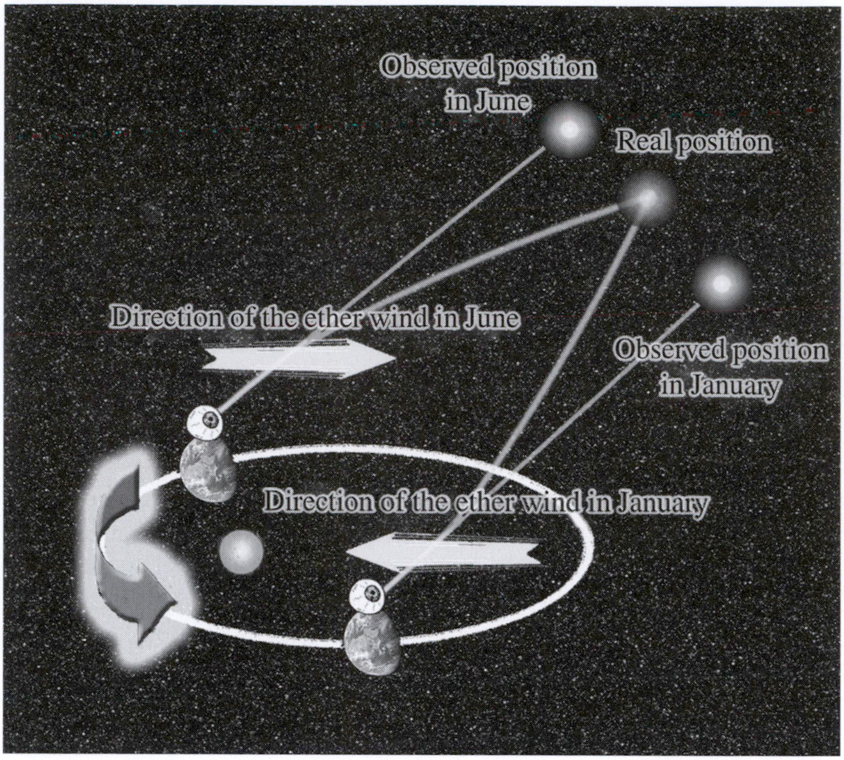

Figure 5.8 Illustration of the effects the ether would have on starlight. If the Earth moves through a sea of static ether the observed positions of the stars should shift periodically.

through the Earth's atmosphere "much as the wind goes through a grove of trees," as poetically described by T. Young.

So we can score one confirmed prediction for the ether hypothesis. However, there are other predictions that can (and were) tested. In order to derive them we need to briefly revisit Newtonian mechanics. Suppose you are on a train that travels at a speed of 1 $\mathrm{m\,s^{-1}}$ with respect to a train station, and that you kick a ball in the direction of the train's motion. If the ball then moves at 2 $\mathrm{m\,s^{-1}}$ *as measured on the train*, then an observer in the station will see the ball move at $1+2=3$ $\mathrm{m\,s^{-1}}$: the velocity of the source (you) adds to the velocity of the object (ball). Thus the two parallel velocities (the train's and the ball's with respect to the train) add up. Consider now the same situation but with a beam of light replacing the ball. If the train moves at speed v then light traveling forward will move at speed $v+c$, where c denotes the speed of light in the train.

Suppose you compare two beams of light, one going in air at a speed c_{air}, and another traveling through glass at a speed c_{glass} (both measured at rest with respect to the ether); these two numbers are not identical because light interacts differently with the two materials, in fact, one approximately has $c_{\text{glass}}=0.75\,c_{\text{air}}$.[12] Now, if there is an ether, and the Earth is moving at a speed v with respect to it, then one can select the orientation for the apparatus such that both beams happen to lie along the velocity v.[13] In this case the speed of light in air and in glass will be altered, becoming $c_{\text{air}} + v$ and $c_{\text{glass}} + v$ respectively.

The experiment I want to discuss (whose details are unimportant here) measures the ratio of these speeds, and should give the result $(c_{\text{air}} + v)/(c_{\text{glass}} + v)$. Now, if the beam is rotated 180°, the direction of the ether wind with respect to the experiment is reversed, and the measurements ought to produce the value $(c_{\text{air}} - v)/(c_{\text{glass}} - v)$. The amazing thing is that this experiment gives the *same* value no matter how it is oriented with respect to the motion of the Earth through the ether. Curiouser and curiouser: the speed of light in glass actually depends on the color of light, and yet the above experiment gives no effect for any color, and the very same effect can be repeated with glasses of all types and, in fact, with any pair of transparent substances, with the same negative results. The only way this can be reconciled with the ether hypothesis is by assuming that the ether is trapped by transparent materials and, in particular, this ought to be also a property of water.

So, on the one hand, the ether is a medium so tenuous that all objects move through it without any appreciable frictional effects, but on the other hand it is dense enough for it to be trapped by transparent substances. In order to test this Fizeau performed a very important experiment. He sent light through tubes where water flowed in different directions. The water was supposed to drag at least some ether, which would then alter the speed of light depending on the velocity of flow, but when the speed of light for moving and stationary water was compared no discrepancy was found.

The most famous of the experiments constructed to detect the motion through the ether was the Michelson–Morley (or M&M) experiment. The basic idea was to compare the speed of light in different directions with respect to the ether wind. To this end, Michelson and Morley constructed an apparatus where an initial narrow light beam was split by an optical device, and the two daughter beams were sent in perpendicular directions. These beams were then reflected back, recombined, and observed through a telescope. Upon recombination there would be a slight mismatch between the crests of the two light wave trains, for they had traveled different distances. Because of this mismatch the light from the two beams would sometimes add up and sometimes cancel out, and so the observer would see a pattern of bright (where the beams reinforce each other) and dark (when they cancel) fringes (Figure 5.9).

Now suppose we rotate the table where the experiment is placed; the speeds of the two beams with respect to the ether will change, and so will the times taken for the beams to recombine. Because of this the mismatch between troughs and crests in the two wave trains should also change, and a *shift* in the pattern of dark and bright lines should to be observed . . . except that it wasn't! No detection of the motion through the ether could be measured.

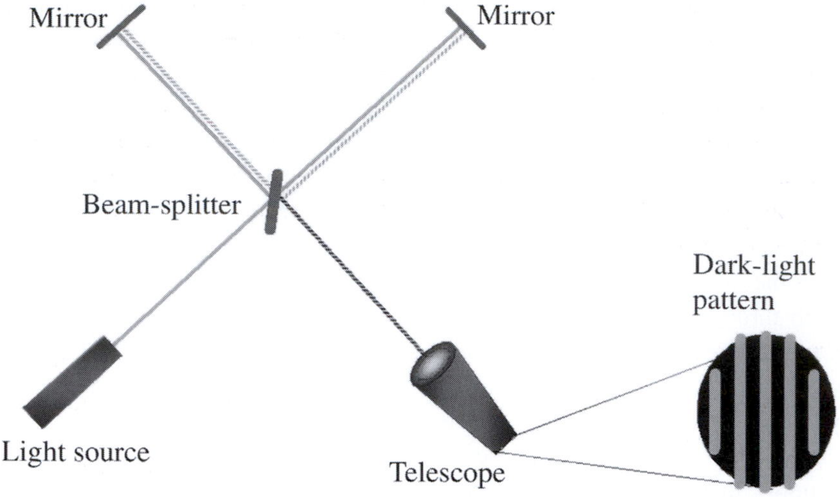

Figure 5.9 An illustration of the Michelson–Morley experiment. When the whole contraption is rotated, the ether wind should induce a shift in the pattern of dark and bright lines; no such shift was observed.

It was then claimed that the only thing proved was that the ether in the basement where the experiment was done was dragged along with the air. But the experiment was repeated a large number of times, in particular it was done on a hilltop, and no effects were ever obtained. This last result was the death blow to the ether hypothesis: M&M's experiment showed that the ether must be dragged along by the air, while stellar observations denied precisely that!

And so we have finally reached a contradiction: the ether is trapped inside the same transparent substances for some experiments while not for others: the ether hypothesis is *inconsistent* and must be discarded . . . and yet, how does one explain the observed shift in stellar positions?

The predicted behavior of light beams as they progress through the ether was a direct result of the manner in which velocities should add if Newtonian mechanics are correct. The simple addition of parallel velocities is an inescapable conclusion that can be drawn from Newton's ideas about space; the fact that it leads to a contradiction when applied to the propagation of light beams signaled the demise of the Newtonian view of Nature.

Galilean relativity

Galileo formulated his principle of relativity by stating that one cannot use any mechanical experiment to determine absolute constant uniform velocity. Now Maxwell's equations contain a velocity c but *they do not specify with respect to what this velocity is to be measured!* We must conclude that either absolute velocities can be determined using experiments involving light, or else, light must move at speed c in *all* reference frames.

The idea that c is absolute is impossible to accept within Newtonian mechanics, for within this theory velocities simply add: if we then have a source of light moving at speed v, the light from it ought to travel at speed $c + v$. This is in direct contradiction to Maxwell's equations, which predict that light travels with speed c, no matter how fast the speed of the source. This contradiction does not result from the application of these theories to certain very specific effects or arcane processes. On the contrary, the disagreement stems from the basic structure of the theories and cannot be corrected by small modifications: neither Newton's mechanics or Maxwell's electromagnetism are able to describe accurately phenomena associated with light.

Prelude to relativity

This was the situation before 1905: the ether was postulated, but its properties were inconsistent. Newton was believed to be right, but the corresponding mechanics were inconsistent with the results of electromagnetism. If Newton's theory is correct, then, are the equations describing electricity, magnetism, and optics inaccurate? If Newton is wrong, how can we understand all its successes? And, if the theory of Maxwell is correct, how can one understand light as a wave if the thing in which it travels cannot be described consistently? All these problems were solved with the advent of the Special Theory of Relativity to which I now turn.

Notes

1. One of the best measured quantities, the anomalous magnetic moment of the muon (a subatomic particle) is accurately *predicted* by quantum theory with an accuracy of one in a trillion.
2. Though these items appear to us now rather trivial they are in fact rather novel inventions. Alessandro Volta, the inventor of the modern battery, announced his discovery in 1800. Other technological developments became readily available only at the end of the eighteenth century.
3. According to Lucretius the term "magnet" was derived from Magnesia; Pliny the Elder, however, claimed that it came from the name of the shepherd Magnes who first discovered its properties.
4. One does not always get an electric shock from cars or combs because they often get discharged; this happens, for example, if there is high humidity.
5. He was anticipated by H. Cavendish and J. Robison, but they did not publish their results.
6. In physics and mathematics, as in other fields, certain words are given precise technical meanings quite distinct from their colloquial ones. In the present context a field does not refer to a broad, level expanse of land, but to the object defined in the following section.
7. This is not a trivial result: there is no a priori reason to suppose that the total effect is obtained by simply adding the contributions from each of the charges, and yet this is what is observed experimentally and is embodied in the mathematical structure of Maxwell's equations.

8. This is not true when phenomena at very short distances are examined; at distances below 10^{-8} cm (atomic size) the difference between waves and particles becomes blurred.
9. One can, however, use this result to get a limit on the speed of light: if the human response time is, say, half a second, then this experiment shows that light travels faster than 2 miles s^{-1}.
10. The speed depends on the medium in which light travels; the value given above corresponds to the speed in space.
11. In a medium there is some interaction between the atoms and the waves and the speed can be different.
12. For some materials the difference can be enormous.
13. In practice the experiment is set on a rotating table and is repeated for a variety of orientations.

6
The Special Theory of Relativity

Introduction

The properties of light and the ether remained a puzzle through the late nineteenth century and up to 1905. In direct contradiction to Newtonian mechanics, the speed of light (as described by the equations of electromagnetism) did not depend on the motion of the observer; stranger still, the ether, the medium in which light was believed to propagate, apparently could not be described consistently, as implied by the negative results of the Michelson–Morley (M&M) experiment.

Just before the advent of Special Relativity, however, a final effort was made to understand the negative result of the M&M experiment from a Newtonian perspective. In 1904 H. A. Lorentz and G. F. FitzGerald postulated (independently) that as objects move thorough the ether, this most subtle of substances would nonetheless exert an irresistible pressure resulting in a contraction of all lengths along the direction of motion. This effect was precisely (and miraculously) arranged to cancel the effects of the ether wind, thus reconciling the negative M&M result with the ether hypothesis.

To see how this might work, recall that the M&M experiment is based on the idea that the time a light beam will take to cover the distance between two mirrors depends on the direction of the light beam with respect to the ether. If there is an ether wind, one should detect a change in this travel time as the beam direction changes, *provided* the distance between the mirrors is fixed. What Lorentz and FitzGerald proposed is that the ether wind will *in addition* compress the apparatus holding the mirrors, an effect so contrived so as to precisely cancel the change in the travel time.

A calculation shows that for this cancellation to occur, an object of length l moving with velocity v with respect to the ether should be contracted to a length l' given by

$$l' = l\sqrt{1 - (v/c)^2}$$

(c denotes the speed of light). This rule should apply to all materials independent of density, weight or rigidity.

But, instead of resolving the ether puzzle, this proposal added wood to the fire, for it required that the ether possess yet other fantastic properties. Despite its being a very tenuous medium that could not be felt or tasted, even the strongest of materials would be squashed when moving through it. And all materials would be squashed by precisely the same amount, and in such a perverse way so as to make it impossible to detect the motion of any object through the ether. Finally, this amazing substance might or might not be carried along by some transparent media, and would opt to do one or the other according to which possibility made it impossible to detect it. Cast in this light, the story of the ether was just like the one about "little green men on the moon" described earlier: the main property of the ether was that no experiment could determine its presence. The ether hypothesis was not falsifiable.

Enter Einstein

In 1905 Einstein published three papers. The first (dealing with the "photoelectric effect") gave a very strong impulse to quantum theory, and got him the Nobel Prize in 1921. The second dealt with the effect of molecular vibrations on the movement of small particles in a fluid (Brownian motion). The third paper was called *On the electrodynamics of moving bodies*; it changed the face of physics and the way we understand Nature.

This last paper starts with a very simple (and well-known) observation: if a magnet is moved inside a coil of wire a current is generated, and the same effect is produced if the magnet is kept fixed and the coil is moved: only the relative motion of the magnet with respect to the coil is relevant (Figure 6.1). This and similar results suggest that in electromagnetism, as in mechanics, the rules describing the behavior of electric and magnetic phenomena will be the same in any two reference frames in constant relative motion:

> *The same laws of electrodynamics and optics will be valid for all frames of reference for which the laws of mechanics hold.*

This postulate, called by Einstein the *Principle of Relativity*, is the fundamental assumption on which he based his Special Theory of Relativity.

The speed of light is absolute

One particular consequence derived from the Maxwell equations is that all electromagnetic waves in vacuum travel at speed c, a value that does not depend on the frame of reference; and this is a very puzzling result. This implies, for example, that, if we travel at 99 percent of the speed of light with respect to a laser, and measure the speed of the beam coming out of it, we will find it to be precisely c (and *not* 1 percent of that value). The same result will be obtained if

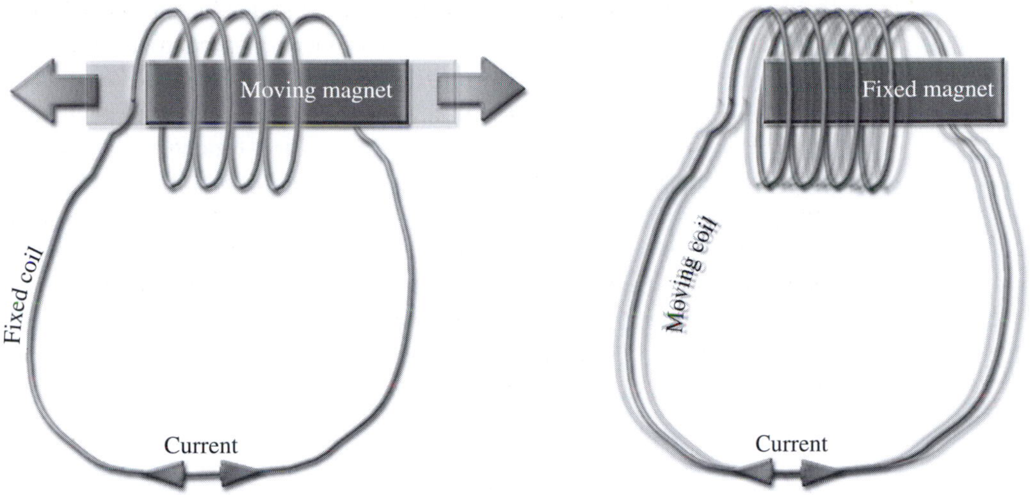

Figure 6.1 Illustration of one of the experimental facts that led Einstein to the Principle of Relativity; the same current is generated by moving the magnet or moving the coil.

we increase our speed to 99.9 percent or 99.999 percent of the speed of light: the speed of the light source with respect to the observer has no effect on the speed of the light produced, c is *absolute*. It is ironic that the first result drawn from the theory of relativity is that a certain quantity is absolute. This property of light is completely at odds with the expectations from Newton's mechanics: within this theory if a light beam has speed c in one reference frame it will have speed $c + v$ in a reference frame moving along the direction of propagation with velocity v.

Once Einstein accepted the Principle of Relativity he was faced with a choice: either Maxwell's electromagnetism or Newtonian mechanics must fail to accurately describe the correspondence between the observations of inertial observers when their relative speed is close to c. Newton's mechanics had survived for about 250 years, it was universally accepted by the physics community, and its predictions agreed with all experiments (done up to 1905). Maxwell's equations, in contrast, were scarcely 50 years old, had not been tested as thoroughly as Newton's, and were not even universally accepted. And yet, Maxwell's equations did describe accurately all known electromagnetic and optics phenomena and, in particular, the behavior of light. Einstein took the daring path of siding with Maxwell, and so challenged the whole edifice of the Newtonian theory. He was right.

Einstein's rejection of Newtonian mechanics implied his belief that this theory is but a good approximation to a more fundamental description of the world around us and cannot be an accurate description of Nature under *all* circumstances. His next (enormously challenging) step was to construct a new (relativistic) mechanics consistent with the absolute nature of c, and whose predictions were at least as accurate as those derived from Newton's equations.

Because of this success the Special Theory or Relativity replaced Newton's mechanics, melding mechanics with electromagnetism in one coherent edifice.

Though I will not describe the relativistic generalization of Newton's laws (this would require a somewhat technical digression), there is one point that is worth clarifying: the relativistic addition of velocities. I mentioned above that objects traveling at speeds much smaller than c behave quite as Newton expected: their velocity in two reference frames differs by the relative velocity of the frames; in contrast light behaves in a most non-Newtonian way, blithely disregarding the speed of the frame. But, one might rightly ask, what happens at intermediate speeds? As we consider an object moving at higher and higher velocities, can we find its speed in any reference frame using Newtonian mechanics, as long as its speed lies below c?

In fact this is not the case: the Newtonian addition of velocities, though extraordinarily accurate at ordinary speeds, it is always an approximation to the relativistic result. The explicit formula obtained by Einstein is the following: if the speed of an object in one reference frame is v_{obj}, and the relative speed between this frame and a second one is u_{fr}, then the speed of the object measured by an observer in the second frame is:

$$\frac{v_{\mathrm{obj}} + u_{\mathrm{fr}}}{1 + v_{\mathrm{obj}} u_{\mathrm{fr}}/c^2}$$

When both v_{obj} and u_{fr} are much smaller than c, then the denominator is almost equal to 1 and the speed of the object measured by the second observer is almost identical to $v_{\mathrm{obj}} + u_{\mathrm{fr}}$, as predicted by Newton. This is the case for ordinary velocities, for the deviation from the Newtonian expression is very small indeed; for example if $v_{\mathrm{obj}} = u_{\mathrm{fr}} = 100\ \mathrm{km\,h}^{-1}$, the speed in the second frame would be $199.999\,999\,999\,9983\ \mathrm{km\,h}^{-1}$, while Newton would have predicted $200\ \mathrm{miles/h}^{-1}$ instead, a deviation of about 0.000 000 0002 percent.

If, however, either v_{obj} or u_{fr} (or both) approach c, the deviations become quite significant. For example, if $v_{\mathrm{obj}} = c/2$, $u_{\mathrm{fr}} = c/2$, the speed of the object measured by the second observer would be 80 percent of the speed of light (and not c as Newton would have expected). In the extreme case where $v_{\mathrm{obj}} = c$, a computation shows that its speed in the second frame also equals c, independently of the relative speed of the frames, in agreement with Maxwell's predictions.

The above expression for combining velocities is more complicated than the one provided by Newton, but it has one advantage: it is verified by experiments. It also reduces to the Newtonian expression at low velocities. Though Newton might have been displeased by this behavior, one cannot deny the experimental results: a light beam coming from a moving source will reach distant Pluto at the same time as the beam coming from a stationary source, with or without Newton's consent.

In this day and age thousands of experimental results attest to the veracity of the Einstein expression for the addition of velocities, but in 1905 no such data

Box 6.1

High energy accelerators

Most of the studies in subatomic physics are done in enormous machines commonly called "colliders" where electrically charged particles such as electrons and protons are accelerated to speeds very close to that of light and then forced to crash into each other. The resulting debris provides important clues as to the fundamental structure of matter. A popular design for a collider consists of one or more concentric rings in to which the colliding particles are piped and accelerated using electric and magnetic fields. Given the enormous speeds of the particles the design must be extremely accurate, even a very small error can send all the particles crashing into the walls of the ring. All calculations are done using Einstein's mechanics, and the behavior of the particles perfectly matches the predictions of the theory; a design of a collider using Newtonian mechanics would lead to a useless machine.

existed. Significant discrepancies between Newton's and Einstein's mechanics become noticeable only at speeds close to c which explains why no problems were detected with Newton's theory before 1905 as at that time all experiments involved speeds small compared to c. The best examples of the accuracy of Einstein's mechanics occur in experiments done since the 1950s where subatomic particles are accelerated to speeds approaching c. The behavior of such particles completely vindicates Einstein's approach, but dramatically disagrees with the Newtonian predictions.

The Principle of Relativity (combined with Maxwell's equations) implies that there is a universal speed whose value is the same to all inertial observers. This required a change in the formulas that describe the motion of objects, which might be regarded as a technical feature, needed to accurately predict the trajectories for all speeds; useful in a boring sort of way. But the unchanging value of c has other more profound consequences, and these required a sea change in our understanding of the nature of space and time. Absolute c is *inconsistent* with an absolute space and an absolute time, and accepting the former then denies the latter: within the theory of relativity the properties of space and time are *not* independent of the observer. These conclusions can be reached without any technical complications, a surprising simplicity that is one of the strengths of the Special Theory of Relativity.

The motion of free bodies

It is important to note that the speed of light is the *only* absolute velocity, all other speeds change when we shift the reference frame. In particular, the speed of any material body can be specified only with respect to a reference frame: for Einstein as for Galileo there is no absolute motion.

This is also embodied in the Principle of Relativity itself: if all the laws of physics are to be the same for all inertial observers, then any two such observers working in enclosed laboratories will derive the same properties of Nature irrespective of their relative state of motion. Not only are they incapable of determining whether they are moving, but the very question "am I moving?" still makes no sense unless a reference frame is specified. Moreover the two observers cannot determine whether they are in relative motion unless they communicate in some manner (for example, by looking at each other's laboratories).

The top speed

The absolute nature of c also implies that this speed represents an absolute upper limit for the speed of motion of any physical entity. I will argue that this is so by contradiction, that is, I will assume the opposite and show that this position contradicts the Principle of Relativity.

Thus I start by imagining that there is a physical object, which I call **R**, which moves faster than c. I will then compare some observations made from a laboratory on Earth to those made from a reference frame in which **R** is at rest. Let us then imagine that **R** is in space moving away from Earth,[1] and that a laser beam is fired from the Earth towards **R**. From the Earth's point of view this beam of light travels at speed c, but it will never reach **R**, since, by assumption, **R** moves faster than c and continuously gains on it. This implies that in the frame of reference in which **R** is at rest the laser beam will be moving in the *opposite* direction, that is, *back towards Earth*. This would then violate the absolute character of c, for not even its direction of motion would be the same in these two reference frames. So, as claimed above, the existence of **R** implies that the speed of light is not absolute, and this would then imply that either Maxwell's equations or the Principle of Relativity (or both) are wrong. We conclude that if we accept Maxwell's equations and the Principle of Relativity we *predict* that no object can be propelled to a speed larger than c.

One might still try to visualize, as Einstein did when a teenager, what would happen if someone were to travel at the speed of light. As that observer moved through a village (for example) they would be moving at the same speed as all the light coming from all objects in that village. So, if he were to look around, he would see the same things with the same relative positions all the time, nothing would ever change since he would be riding along with a single image: the one carried by the light from the village at the time he passed it. Thus in his frame of reference time would stand still! This, however, is somewhat simplistic;[2] for example, at very high speeds a forward-facing observer would see his field of vision reduced to a small circle that shrinks as his speed approaches c (and would become a point at that speed). Within this reduced field he would see some of the objects lying *behind* him, all objects would be deformed in a rather peculiar way, and their color would shift (becoming redder when receding and bluer when approaching). We will see below that it is impossible for anything having mass – such as you – to reach the speed of light (though you can come arbitrarily close), and even if you travel at a speed very close to c all light beams, including the ones carrying the images from the objects in the village, will still travel at speed c in your reference frame.

Goodbye to the ether

When Maxwell wrote his theory of electromagnetism he did it as a man of his time, and so his writings contain a full discussion of the ether and its alleged properties; properties that eventually led to logically inconsistent conclusions. It was impossible for Einstein to simply ignore this problem, and in fact he solved it; but his solution was as unexpected as the way Alexander the Great dealt with the Gordian knot: Einstein simply observed that there is no need for ether to exist at all!

The Principle of Relativity implies that the Maxwell equations are the same for all inertial observers, and that all observers in constant relative motion should derive equivalent consequences from them. The ether, if present, would belie this idea, for in the reference frame in which the ether is at rest, there would be no ether wind, and electromagnetic phenomena would take a markedly simple aspect, not observed in other reference frames. So, accepting the Principle of Relativity requires we discard the ether hypothesis: light requires *no* medium to propagate.

Of course, even for Einstein, saying something did not make it so: he needed to show that electromagnetism without ether was a viable and consistent theory whose predictions (especially those concerning light) were consistent with all the experimental results. In this section I will briefly revisit two such experiments: the Michelson and Morley experiment,[3] and the semiannual shift in the observed stellar positions. As I indicated the first was inconsistent with the existence of an ether wind while the second appeared to require its presence.

The consistency of ether-less electromagnetic theory followed from the observation that none of Maxwell's equations contain any reference to the ether per se, so that the whole idea could be shed while retaining all the experimental confirmation of the theory. It is true that two parameters in the equations were identified with specific properties of the ether, but this identification is not mandated by the theory itself, it is rooted in the prejudice that all waves needed a medium in which to undulate, it is only after one asserts that such a medium *must* exist that the connection between the ether and the electromagnetic equations follows. Einstein observed that one could simply state that the few parameters contained in Maxwell's equations can be experimentally determined without making reference to the ether at all. This approach provides a clear illustration of the use of Ockam's razor: the basic theory is contained in the mathematical structure we call Maxwell's equations, but this theory had been top-loaded by adding the ether postulate which is irrelevant to the experimental predictions derived from the theory and was included to satisfy certain preconceived ideas (for example, that *all* waves require a medium to propagate). Einstein, noting this fact, simply removed this additional unnecessary postulate.

The M&M experiment was designed to measure the changes in the speed of light produced by the ether wind, but obtained a null result, interpreted first as the absence of an ether wind. Within the framework of Special Relativity, however, this negative result is to be expected: the speed of light is absolute and is completely unaffected by the motion of the source (in this case the Earth); no ether wind was detected because there is no ether.

More interesting is the manner in which Special Relativity explains the periodic shift in the position of the stars. Imagine observing a light beam coming to Earth from a distant star; let's study this beam first from the point of view of a reference frame where the star is at rest, and then change to the Earth's reference frame. In the star's frame we can decompose the speed of the light beam into a component parallel to Earth's velocity (with respect to the star) and another perpendicular to it. Now, when we change to the Earth's reference frame the horizontal component is modified according to the addition of velocities formula given above, while the change in the vertical component is fixed by the requirement that the total speed of the beam remains equal to c; since the horizontal and vertical velocity components change when we change reference frames, the *direction* of the light beam also changes. This change in direction depends on the relative velocity, and, carrying out the computation, Einstein found that when Earth's velocity is reversed (as happens after six months) the change in direction is also reversed. These results are mathematically identical to the ones obtained using the ether hypothesis, but Einstein did not need to assume the existence of the ether at all, the effect is but one more consequence of the absolute nature of c. With the coming of Special Relativity the ether went the way of the epicycles.

Simultaneity is relative

Everyday experience indicates that the statement "two events happened at the same time" (i.e. they were simultaneous) is universal; for example, the claim "the skunk sprayed the dog across the yard just as I rang the bell" would be verified by anyone looking into the matter, and if verified by one observer it would be confirmed by all others. Thus I can say, "I got home at the same time you got to work," and nobody (usually) wonders about the consistency of such a statement. Everyday experience then suggests that simultaneity is an *absolute* concept.

And yet, it is not. The combination of the Principle of Relativity with the absolute value of the speed of light implies that simultaneity is, in fact, *relative*: events simultaneous for one observer need not be simultaneous for others. The reason we are misled by everyday events is that (as in the expression for adding velocities) the deviations from the "standard" expectation are very small when all speeds involved are much smaller than c. If one observer agrees that I got home at the same time as you got to work, another observer moving with respect to the first would not see these two events occurring at the same time, but in order to verify this surprising discrepancy extremely accurate clocks are required (unless the relative speed of the observers is close to c).

Below I present an argument leading to the conclusion that simultaneity is relative. I cast the arguments in the form of a murder mystery in the hope that this will make it easier to swallow the bitter pill that so "natural" a concept as simultaneity depends on the observer.

A futuristic murder mystery

In this futuristic tabloid story a murder is committed in the cargo bay of the starship *Orbit*, where the victim is found with two head wounds caused by high-power laser beams. The crime was observed from three places: a space station, the cargo bay itself, and from another ship, the *Conquest*. At the time of the crime the *Orbit* was moving at a speed $c/2$ with respect to the space station; and the *Conquest* was moving in the same direction as the *Orbit*, but at a speed $3c/4$ with respect to the space station (and was ahead of the *Orbit*), as illustrated in Figure 6.2.

To simplify the language I will say that both ships as seen from the space station were moving to the right (Figure 6.2). Note, however, that the *Conquest* continuously gains on the *Orbit* (since it moves faster), so from the point of view of an observer inside the *Conquest*, both the space station and the *Orbit* are moving to the *left*. Finally an observer inside the *Orbit* would see the space station moving to the left and the *Conquest* moving to the right.

Everyone agrees that the dead man was hit in the head by two laser beams simultaneously, as confirmed by sensors in all three spaceships; this cannot be doubted for there is available a photograph of the laser bolts as they hit the victim's head. The (alleged) assassins are also in custody, they call themselves Alma and Omar; Alma entered the cargo bay through the rear door (the one closer to the engines) and Omar through an opposite door; in fact, they confessed that they *did* fire the lasers, specifically they stated that each entered the cargo bay and immediately fired. The crime investigators in the *Orbit* found that the victim was sitting precisely at the same distance from both doors.

During the trial Omar's lawyer made the following very interesting observation: as seen from the space station the *Orbit* was moving to the right, carrying with it the victim (Figure 6.3). Now, Alma entered the rear door and fired, and her laser bolt came out of the gun moving at speed c, but as it traveled the victim was moving *away* from that laser bolt. In contrast the victim was moving *towards* the bolt fired by Omar; and yet both bolts hit the victim at the same time. The only way to reconcile these facts is for Alma to have shot *before* Omar. He then argued that since she fired first the victim was

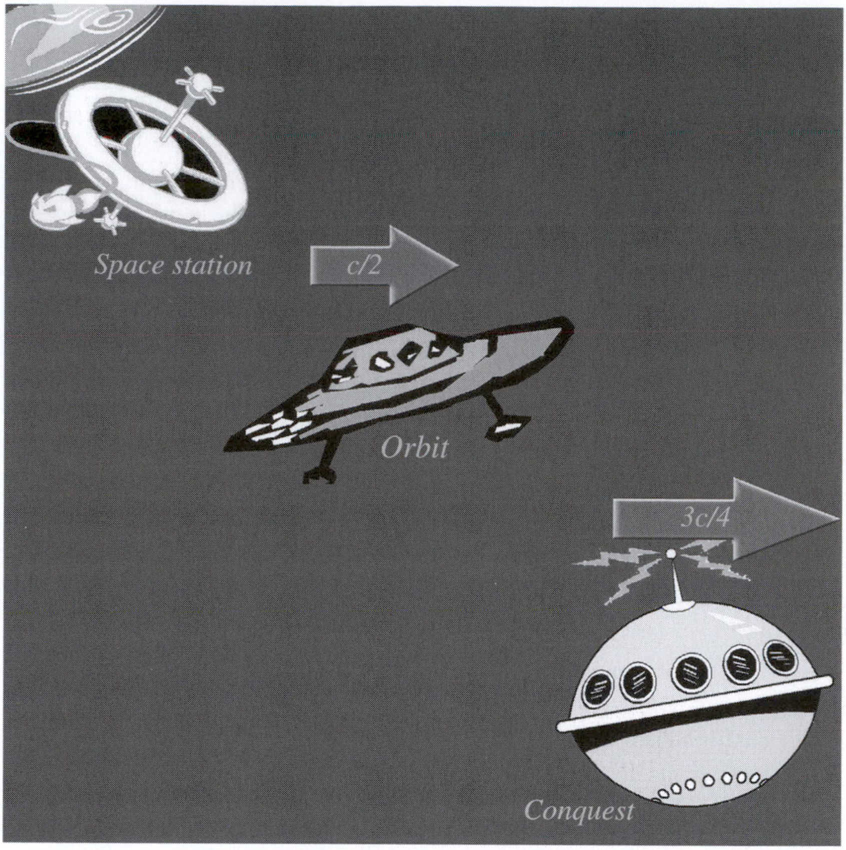

Figure 6.2 Setup for the murder mystery (velocities are measured with respect to the space station).

effectively dead by the time Omar decided to fire and that his guilt is correspondingly smaller.

Alma's lawyer paid very close attention to this statement and immediately provided the following counter-argument (Figure 6.4). As seen from the spaceship *Conquest* the *Orbit* was moving to the *left*, so that in that frame of reference the victim was also moving *left* and away from the beam fired by Omar. Then the same reasoning tells us that Omar must have fired before Alma.

At this point the judge intervened and pointed out that as observed from the cargo bay of the *Orbit* the situation is perfectly transparent (Figure 6.5): both bolts hit at the same time, both bolts traveled the same distance, and so it must be that both assassins fired simultaneously, Omar and Alma are equally guilty.

The point of this story is to show that the absolute nature of c requires that simultaneity be relative, at least for the events described above. The answer to the question: "who fired first?" depends on the observer; just as all speeds except c must be specified with respect to a reference frame, so

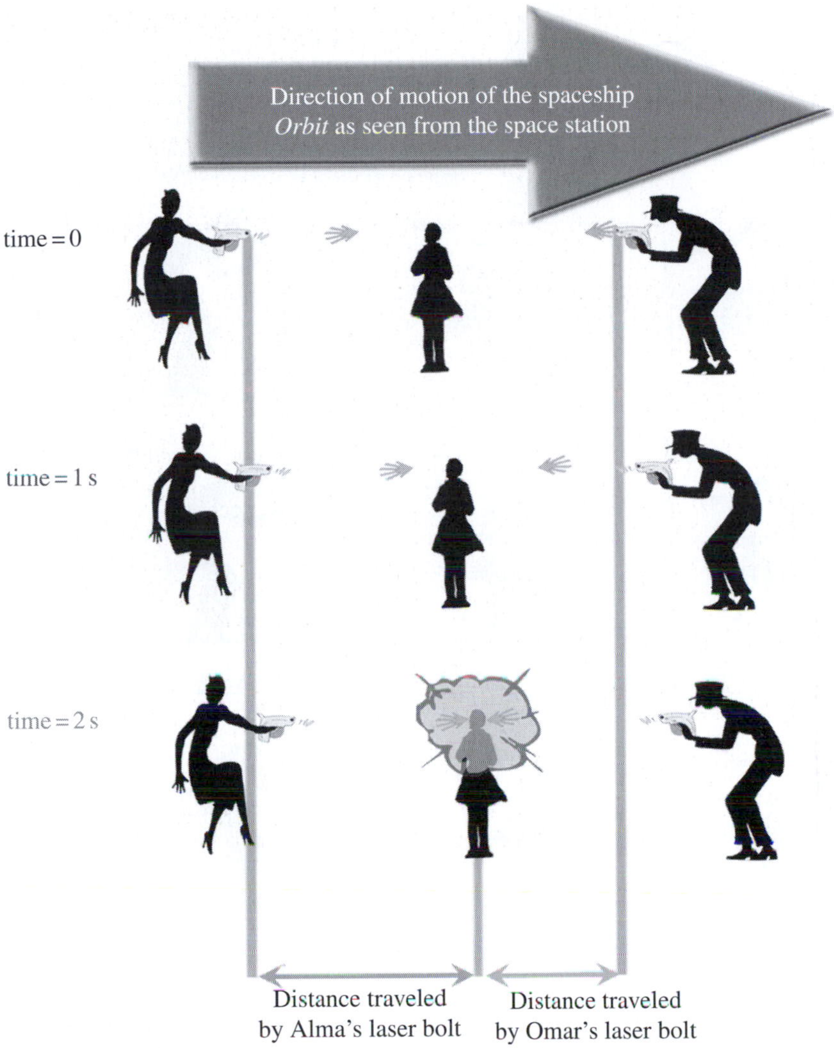

Figure 6.3 The murder as seen from the space station.

must the time ordering of the events connected with this crime. In any one reference frame one can determine who shot first, but the conclusion will change with the observer's state of motion. But, is this true for *all* possible sets of events?

Time orderings

In fact it cannot be true that the time ordering of *all* events depends on the frame of reference. We already saw one example: all interested parties could *see* both bolts entering the victim's head at the same time, and there cannot be any observer dependence about that! Since the victim had a very small head we

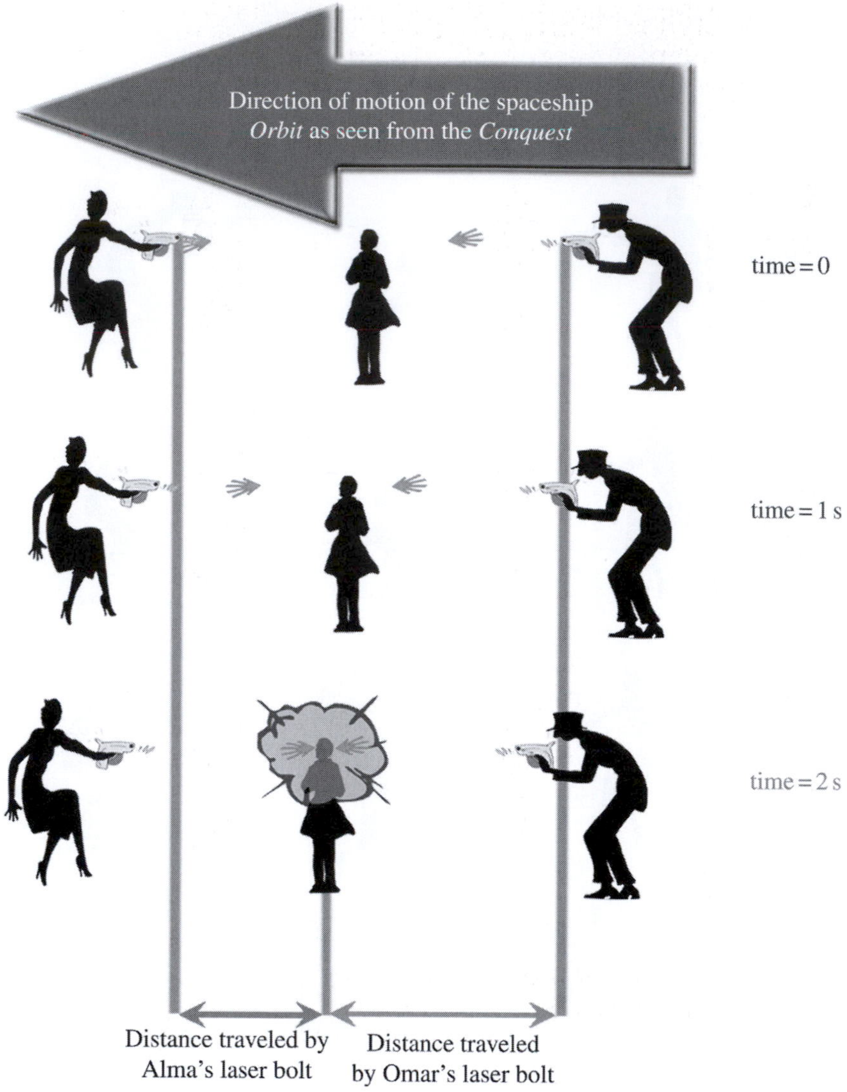

Figure 6.4 The murder as seen from the spaceship *Conquest*.

can say that events that occur at the same time in the same place must be simultaneous to all observers.[4]

We saw in the murder mystery that for one observer Alma shoots before Omar, while for another observer the opposite is true. In contrast to this situation there are events that happen at different places but whose ordering in time cannot depend on the observer. For example, let us imagine that Sigfrid met Brunhild in Hamburg, and that their son Thor is later born in Munich. Then it must be the case that for *all* observers Sigfrid and Brunhild met before Thor was born. Otherwise Thor would be born before his parents' meeting (which makes no sense) and might even

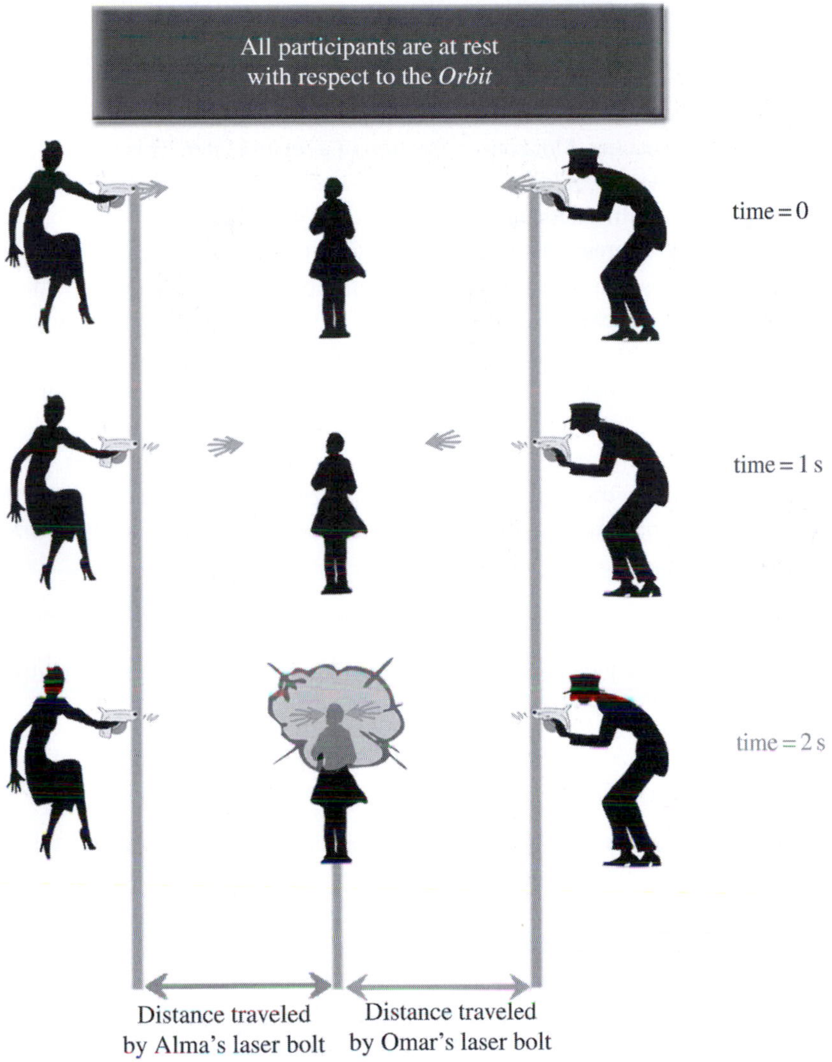

Figure 6.5 The murder as seen in the spaceship *Orbit*.

prevent this meeting, thereby *insuring* that he will not be born at all![5] In the same way, all observers will agree that Thor's birth happened before his death.

So there are events such as the birth and death of a person whose time orderings are the same for all observers, while others, such as the ones described in the above murder story where the sequence of events is relative. What is the difference?

The one clue is the following: in the murder story the assassins came in from opposite sides of a cargo bay and shot the victim, but though Omar and Alma may have agreed beforehand to do this dastardly deed, at the moment they pulled the trigger neither could be sure that their accomplice was faithfully following their plan. The reason is that, for example, the image of Omar shooting travels towards Alma at the speed of light, the very same speed of the laser beam, thus both image and laser bolt pass by the victim; the bolt stays in the victim's head while the image reaches Alma only a fraction of a second later. This implies that when she fired, Alma could not have been aware of Omar' actions, and could confirm Omar's faithfulness only *after* the victim was dead.

In contrast all (interested) observers are able to follow the romance of Sigfrid and Brunhild: they can watch as they meet and follow their . . . ahem . . . until Thor is born. We can witness the influence of the first event (their meeting) on the second (the birth). So in this case there are physical entities that sequentially relate both events making their time ordering absolute (the same for all observers).

The Special Theory of Relativity automatically insures that if an event **A** can influence event **B** by transmitting a signal of some type, then **A** occurs before **B** for *all* observers, otherwise **A** and **B** are independent of each other and their time ordering is relative:[6]

Simultaneity is relative for independent events.

All this is a consequence of the absolute nature of c: simultaneity is relative because the speed of light is absolute.

The relativity of time

We saw above that the Principle of Relativity and the absolute nature of c imply that the time ordering of a certain class of events is relative. This suggests that Newton's basic assumption that time flows in the same way for all observers is inconsistent with the Theory of Relativity; this is in fact the case, and provides one of the most dramatic conclusions derived by Einstein. As in the section above I first provide a simple situation (originally described by Einstein) that illustrates the fact that time is relative (meaning that time intervals measured by observers in relative motion will differ); I then generalize the argument and briefly discuss some of the consequences of this discovery.

Consider first a "light-clock" consisting of a light source and detector. The source emits a light pulse that goes up to the mirror, is reflected, and is then detected; immediately after detection a new pulse is sent upwards. Since the speed of light is not infinite, the round trip for each pulse takes some time, and this interval is the same provided the distance between the mirror and the detector/emitter is kept fixed (Figure 6.7); hence I can use this contraption as a clock. I will assume that this distance is such that the round trip measured by an observer at rest with respect to this "light-clock" is precisely 1 s.[7]

Now let us imagine two such clocks, that we label **A** and **B**, in relative motion at constant speed and direction. Each clock travels with its own observer: Alan

Box 6.2

Time dilation and Pythagoras' theorem

The distance the light has to travel in Figure 6.8 can be determined by using Pythagoras' theorem. In this reference frame light travels along the long sides of the triangles, each has a length which I call *l/2*; let's call *T* the time it takes to complete the trip, then, by Pythagoras' theorem:

$$l/2 = \sqrt{h^2 + (vT/2)^2}$$

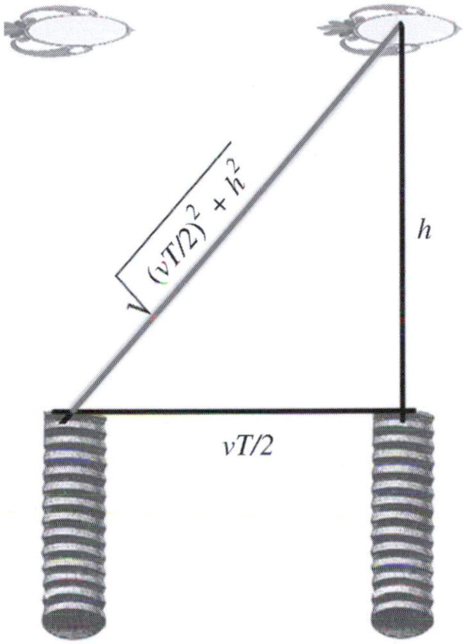

On the other hand $l = c\,T$ since light moves at speed *c* for any observer and it takes a time *T* (according to the moving observer!) for it to get back to the detector. Solving for *T* gives:

$$T = \frac{2h/c}{\sqrt{1-(v/c)^2}} = \frac{T_0}{\sqrt{1-(v/c)^2}}$$

(T_0 is the time it takes the beam to make the up-down trip in the rest frame of the clock). An observer moving with respect to the clock will measure a time T greater than T_0, with the precise expression being given by this formula.

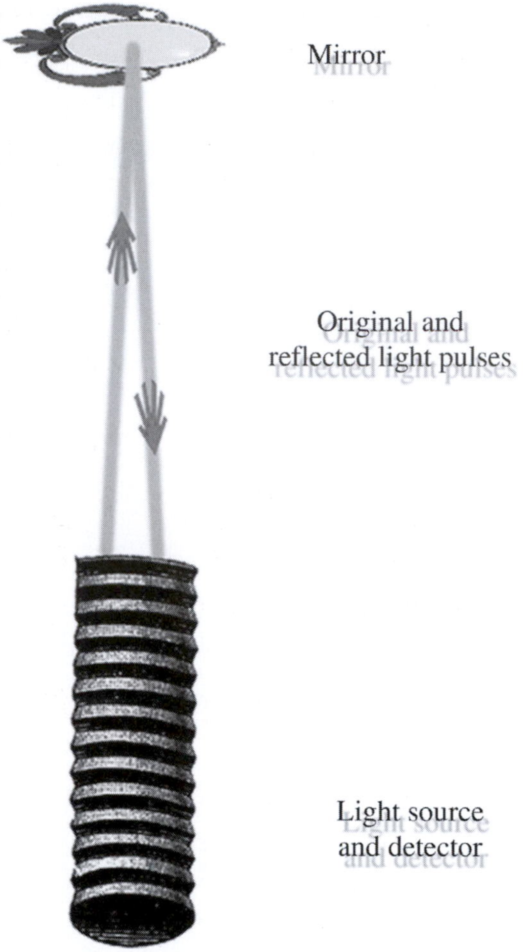

Figure 6.7 A light-clock at rest with respect to the observer.

travels with **A**, and Boris with **B**. Since each observer is at rest with respect to his clock both can attest to the fact that their clock measures 1 s per up-down trip of the light beam.

However, from Alan's point of view the light beam of Boris' clock does not move simply up and down but follows a triangular path (Figure 6.8). This is the case since in Alan's reference frame the mirrors will have moved some distance during the time the light travels between them; only if the light beam of **B** is tilted can we account for its being reflected by the mirrors at their changing position. The triangular path is clearly longer than the simple up-down path followed by the light of **A** (again in Alan's reference frame), and since light always travels at a fixed speed in *all* reference frames, the trip of **B**'s light beam will take a longer time: Alan finds that **B**'s light beam takes *longer* than 1 s to make its trip; Boris' clock is slow with respect to his. This effect is commonly known as *time dilation*.

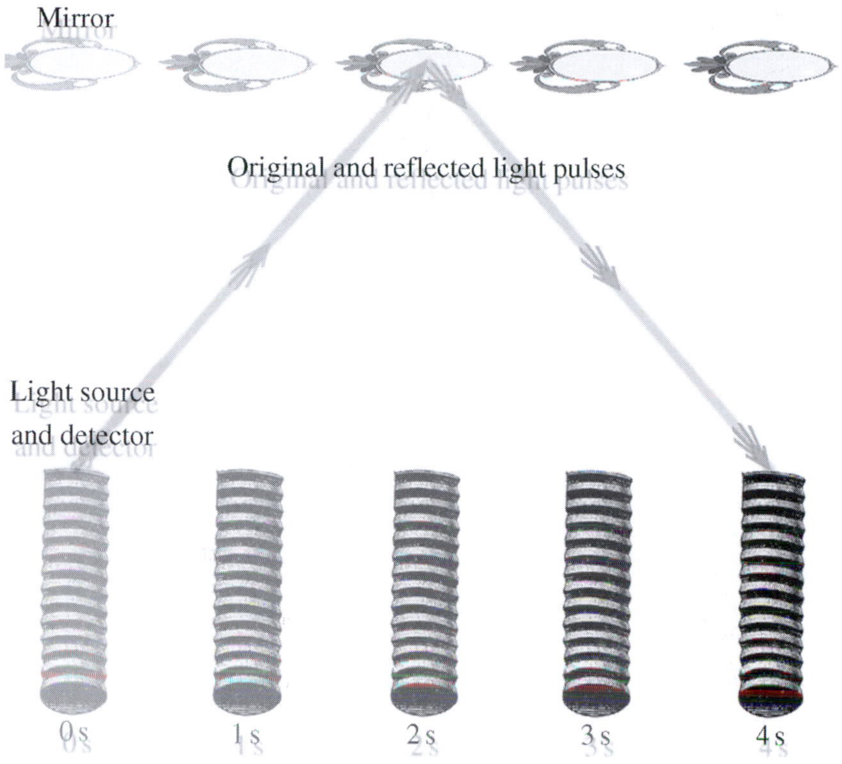

Figure 6.8 Light-clock in motion with respect to the observer (dimmer images correspond to earlier times).

But nothing in the argument is specific to Alan, so we can repeat the same reasoning, but now in Boris' reference frame, and find that in that frame **A** is slow compared to **B**. We conclude that in any reference frame a moving "light-clock" will slow down; a consequence of the absolute value of c. The discrepancy between the time intervals measured by Alan and Boris becomes more pronounced the larger the relative velocity between **A** and **B**; for ordinary speeds it is almost unnoticeable. But, tiny or not, the effect should be present if the Principle of Relativity and Maxwell's equations are correct: time intervals measured by the "light-clocks" are relative and depend on the state of motion of the observer.

Striking as this is, one might question the importance of this result as it was obtained using rather contrived time-keeping devices; could this conclusion be reached using regular clocks? To examine this possibility I will be the devil's advocate and assume that the discrepancy between **A** and **B** is peculiar to the light-clocks used,[8] and that there is at least one special type of clock that always ticks at the same rate no matter how fast it moves with respect to the observer; I will call this miraculous clock **M**.

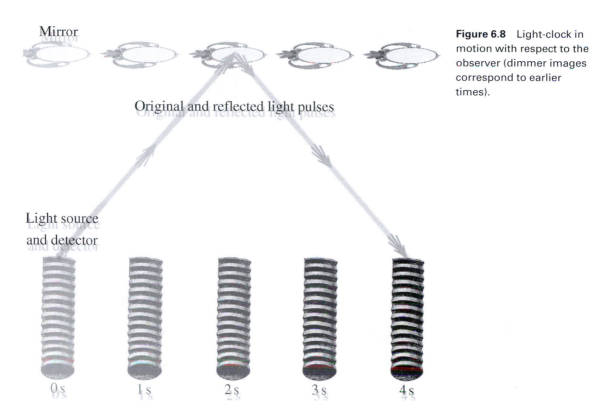

Imagine then that we first synchronize all clocks (**A**, **B** and **M**), then we put Boris, his light-clock **B**, and clock **M** inside a spaceship and send them all to a near-by star. While they are en route we will compare the rates of **M** and **B** with that of **A**, which stays behind on Earth with Alan. By assumption **M** is not affected by its state of motion while we saw that **B** is; then **M** will stay synchronized with **A**, but **B** will not. So Boris will find that **M** continually gains on **B**, and the faster his rocket moves with respect to **A**, the larger the gain. So, even though Boris is in an inertial reference frame, he can use the discrepancy between **M** and **B** to tell he is moving, *and this without reference to anything external to his rocket.*

The existence of **M** implies there are physical effects that can be used to distinguish one inertial frame from another, which is in direct contradiction to the Principle of Relativity. Thus accepting the Principle of Relativity precludes the existence of **M**; equivalently it implies that *all* clocks carried by Boris will slow down with respect to **A**, *precisely* as **B**. This behavior is common to all time-keeping devices, and so it can be associated with time itself: the time measured between any two events depends on the inertial observer, that is,

Time is relative to the observer.

The absence of a universal time is a surprising result that is completely at odds with Newton's concept of time (which he assumed to flow evenly under all circumstances).

Time dilation is a *prediction* of the theory which must not be accepted as dogma, but should be verified experimentally. As for the addition of velocities and the relativity of simultaneity, time dilation is a minuscule effect in our everyday experiences. For example, imagine an ordinary man driving at, say, 100 miles h^{-1} trying to get his wife to the hospital before she delivers. His clock tells him he made the trip from home to the hospital in three minutes flat; in contrast a clock at the hospital would tell him he in fact took about three minutes *plus* two trillionths of a second.

There are some instances, however, in which the effects are clearly observable, and these usually involve subatomic particles. Physicists use such particles to illustrate relativistic effects not out of a perverse proclivity for esoteric objects, but for the very practical reason that these particles are relatively light

Box 6.3

Another view of time dilation

In Alan's reference frame the light pulses in clock **B** must have the same horizontal velocity as **B**, for if they did not they would not be able to keep up with the mirrors. But if the light pulses have a horizontal velocity and the *total* velocity is *always* equal to c, then the light's *vertical* velocity must be smaller. So in Alan's reference frame the light beam in **B** will take longer to make the up-down trip than in **A**.

and it is correspondingly easier to have them move at high speeds. Even in Nature these particles are often found moving at speeds close to c. For example, our Sun spews out a considerable amount of matter through the solar explosions that grace its surface, this material is composed mainly of atomic nuclei (the atomic electrons having been stripped off in the very hot environment of the solar surface), though the average speed of these particles ranges between 300 and 800 $km\,s^{-1}$, some nuclei reach speeds of up to 99.9 percent of c.

When one of these nuclei hits the Earth's atmosphere it generates a shower of daughter particles through a series of nuclear processes (whose details are immaterial). Some of the daughter particles are unstable and will soon decay into others; their average *lifetime* turns out to be a few fractions of a second. And yet, to the initial surprise of the experimenters, these particles survive the trip down to surface of the Earth, which takes longer, *as measured on Earth*, than their lifetime. The surprise evaporated when it was noted that the lifetime was measured with the particles at rest, while in the Earth's reference frame the particles move at speeds close to c, so that all time intervals, including particle lifetimes will be dilated, long enough, in fact, for them to live to see the Earth's surface.

In all cases where the time dilation effect is measurable (that is when the discrepancy between moving and stationary clocks lies above the experimental accuracy) the predictions of Special Relativity have been verified, qualitatively *and* quantitatively. This is the main reason this theory has been accepted, not because of its beauty or simplicity, but because it describes Nature.

Time dilation has an additional consequence that is not often stressed and will certainly affect the future of space travel. One of the inescapable realities about our Universe is the enormous distances separating us from the rest of the objects in it; for example, the nearest galaxy to our Milky Way is 10^6 light-years away. This would seem to insure that no human will ever agree to travel that far: for even if the spaceship was to travel at almost light-speed, who would be willing to embark in a voyage of discovery that will take 20 000 generations to complete? This conclusion, however, is *wrong*; in fact, such trips may be viable thanks to time dilation.

It is true that the ship travels at a speed very close to c and so it will take close to a million years to travel one way to Andromeda, but this refers to time measured by *Earth* clocks. In contrast, clocks inside the ship will tick slower, and this involves *all* clocks, including the biological ones; hence the crew will also age slower. If the spaceship cruising speed is 99.999999999 percent of c, the clocks on board would measure about 4½ years for the trip from Earth to Andromeda. So the crew could go there and back in under a decade . . . *their* time. Of course when they return the Earth would be two million years older, and all their friends and acquaintances would be long dead, as even perhaps the civilization that managed to construct their ship; but one cannot have everything, at least they went, and saw!

Length contraction

Unexpected as the above results may be, they give rise to one particularly puzzling question. Let us go back to the brave crew that decided to go to Andromeda and back. From the Earth we would see their spaceship moving away towards Andromeda at a speed close to c, and it is certainly the case that it will take them close to a million years (*Earth time*) to get there, for we know the distance they will cover is about a million light-years. But in the ship's reference frame the astronauts see Andromeda approaching them at a speed close to c, *and* they arrive in less than five years, how could they cover one million light-years in half a decade without moving faster than light?

The puzzle is rooted in the tacit assumption that the distances measured from the Earth and spaceship reference frames are the same. This assumption is based in Newton's assertion that space, just as time, is absolute, so that the distance separating two points would be unaffected by the state of motion of the observer that decides to measure it. This assumption, however, is *inconsistent* with the Principle of Relativity and Maxwell's equations: if the spaceship crew sees Andromeda approaching at a speed close to c, and they reach their destination in about five years, it follows that *in the ship's reference frame* the distance from Earth to Andromeda is about five light-years (Figure 6.9); the Theory of Relativity then predicts that the distance between any two events depends on the state of motion of the observer. In other words:

> *Length is relative to the observer.*

Not all lengths are affected in the same way: the astronauts in the spaceship find that the distance between Earth and Andromeda is contracted, and this is a length *along the direction of motion*. In contrast, lengths *perpendicular* to the direction of motion are unaffected. To see this imagine two observers, Yanni and Zoe inside a very large ice rink; they both have laser guns which they hold at the same height (measured when both are at rest with respect to each other). Zoe is standing in the middle of the rink, while Yanni zips by her, almost touching, at constant speed. Just when he goes by her they both fire their laser guns at a far wall.

If Yanni were not moving both bolts would hit the wall next to each other; suppose now that vertical lengths are contracted when measured by a moving observer. Then Yanni would find that Zoe's bolt would hit the wall at a lower point than his, while Zoe would find precisely the opposite (since in her frame it is Yannie who is moving). Now, since each laser bolt leaves a big hole, there cannot be any ambiguity about this: either the holes coincide, or one is above the other, and so our assumption that the vertical distance is affected by the observer's state of motion is untenable.[9] We conclude that distances *perpendicular* to the direction of motion are unaffected, while those along the direction of motion are "contracted." This implies that a moving object will be seen thinner

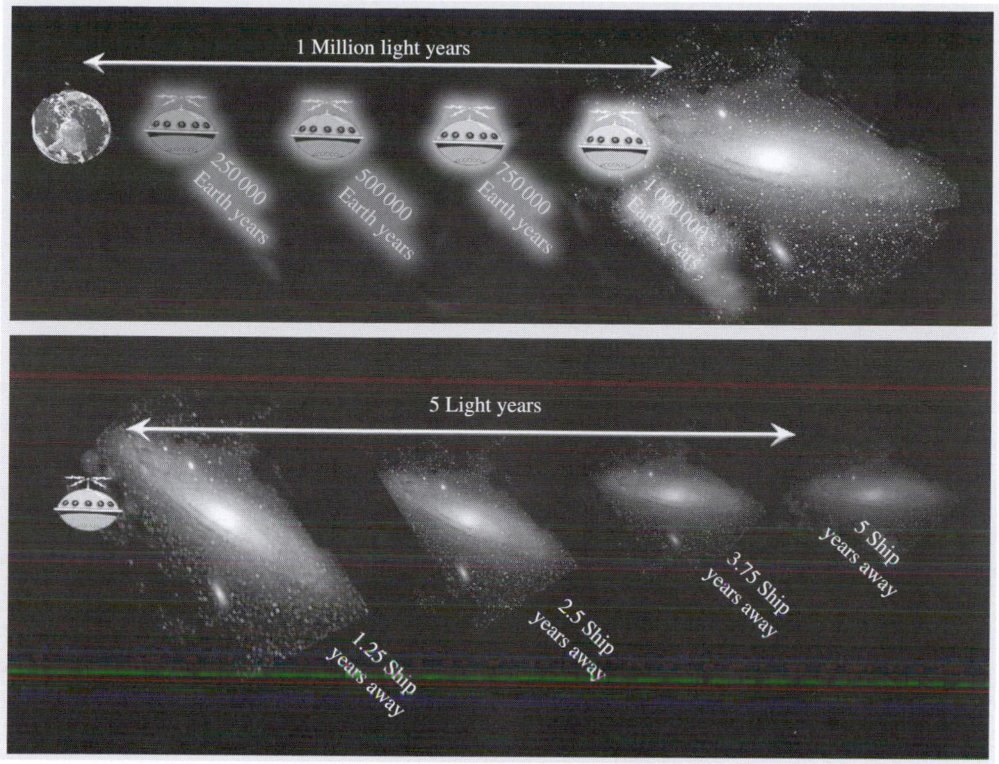

Figure 6.9 An interstellar trip from Earth's (top) and the ship's (bottom) view points (dimmer images represent earlier times).

(due to length contraction) but not shorter. Thin fellows will look positively gaunt at speeds close to that of light.

Length contraction is a consequence of the relativity of space. Just as for time intervals, it is pointless to ask what the length of an object is[10] unless we specify the reference frame where the length is to be determined. The very same object will have two different lengths in two different frames when these are in motion with respect to each other,[11] and it is not that the object is squashed or that its molecular structure is affected in any way, it is only that length is not an absolute quantity. Even though lengths change with the reference frame, this change is not arbitrary: once the length is measured by one (inertial) observer, the mathematical apparatus of Special Relativity determines unambiguously what any other observer would measure.

Paradoxes

Time dilation, length contraction and the relativity of simultaneous events are quite foreign to our ordinary experience, and because of this they can be manipulated into producing apparent logical contradictions. Though such paradoxes are sometimes presented (innocently or willfully) as arguments against the Theory of Relativity, a careful consideration of their details, however, invariably shows that the "inconsistency" arises from a mistaken application of the theory; when the situation is studied carefully all such problems disappear. Still, the way in which these paradoxes are resolved is sometimes interesting, and for this reason I briefly discuss three of them that I believe illuminating.

The one about the castle

Imagine that William the Idle, a rather infamous errant knight, has decided to storm a castle. His idea is to put a battering ram on a cart and send that cart moving very fast toward the castle; it will burst the front door and his men can then finish the job. But his plan has leaked out and the inhabitants have concocted a defense. They know that Bill's battering ram is but slightly shorter than the full length of the castle, reaching from the front door to the back exit; with this in mind, they agreed to open the front gate just as the tip of the ram is about to hit it, then, just as the back of the ram passes through that entrance they will *simultaneously* close the front gate (blocking the incoming horde) and open the back door to let the ram out (they dare not open the back door before in case William has posted some people there). As soon as the rear of the ram goes through the back door they will close that also, the cart will then have simply passed through, and the castle will remain secure. This idea appeared reasonable only because the ram fitted inside the castle, otherwise it would either smash the rear door or block the closing of the front one.

Now, in the reference frame of the castle the ram will be moving at a high speed and so it will be shorter due to length contraction, because of this it will fit even more loosely inside the castle making it more likely that the plan will work. In the cart's reference frame, however, the situation is different: for in that frame the castle is moving towards the ram at a high speed, and so it is the castle's width that will be shortened, so it appears that in this frame the ram would not fit inside, and will necessarily hit either the front or the back door (or both), smashing them and leaving the castle vulnerable. What *does* happen?

To answer this we note that there is one crucial piece of information that I underemphasized: the closing of the front gate and the opening of the back door are to happen simultaneously. But these events, being independent of each other and occurring at different places, will in fact be simultaneous *only* in the castle's reference frame. In the cart's reference frame the back door is opened before the front gate is closed (remember that the cart first passes through the front gate and then attempts to go through the back door).[12] Because of this the ram will be seen to go through the castle in both reference frames, though the circumstances surrounding this passage will be different.

The one about Tweedledum and Tweedledee

Imagine that Tweedledum is sent on a fast trip to a near-by star, while his identical twin Tweedledee stays back in the Looking-glass Land. From Tweedledee's point of view Tweedledum's clock is moving fast and so his biological clock will tick slower, as a result Tweedledum ought to be *younger* than his brother on his return. However, from Tweedledum's point of view it is Tweedledee who moves (in the opposite direction) so in *that* reference frame it is Tweedledee that should be younger when they meet again. What *does* happen?

The solution to this paradox comes from the observation that time dilation is an effect observed in inertial reference frames. Tweedledee's frame is inertial, but Tweedledum's is not, for the latter must have accelerated in order to take off, decelerated when he got to his destination, and then repeated the process in reverse order. So, Tweedledum (had he been interested) could have known that he had been moving, or more precisely, that his speed had changed, and therefore that he was not in an inertial reference frame. His conclusions regarding Tweedledee are therefore unwarranted.

Later on we will see that in a small region of space the effects of acceleration are indistinguishable from those produced by a gravitational force, and that clocks inside a gravitational field slow down. This is *not* a consequence of Special Relativity for it involves non-inertial reference frames, but it can be understood using the General Theory of Relativity. When this effect is taken into account a careful calculation yields the same result in either reference frame: Tweedledum will be younger.

The one about the astronaut

The cosmonaut Leonid Alexandrov has a close call during a space walk. While he is moving with respect to his spaceship he sees a big slab of metal traveling in a direction perpendicular to his own; to his distress he realizes they will crash. He is precisely 6 ft tall and this piece of metal has a square hole 6 ft to the side (when measured at rest with respect to the slab). But from Leonid's perspective the length of the hole will shrink to something shorter than 6 ft; so he believes he will not fit through the hole, will hit the corners, his helmet will break and he will probably die! However, from the point of view of an observer on the metal slab it is Leonid who is shorter and will in fact fit snugly through the hole. What *does* happen?

The paradox is resolved in a way similar to the previous ones. For Leonid to be hit a simultaneous coincidence of his head and legs with the two extremes of the slab's hole should occur. An observer in the slab's reference frame will see Leonid's head and feet simultaneously brush the edges of the hole in the slab. Leonid, however, sees his feet barely brushing by one edge of the hole, and a fraction of a second later, his head brush past the opposite edge. This is because these two unrelated events, occurring at separate locations, might be simultaneous in one frame (the slab's), but not in others (such as Leonid's). What is more peculiar is what Leonid sees: since distances along the direction of motion are contracted but those perpendiculars to it are not, he will see the slab tilt in such a way that he goes through the hole harmlessly (as suggested by the fact that he sees his feet go through the hole before his head does).

Space and time

All events we witness are labeled by a series of numbers: three to determine where it happened, and one to tell *when* it happened. These four numbers, usually called the *coordinates* of the event, are measured using some devices such as measuring rods and clocks. According to Newton the properties of measuring rods and clocks can be made completely independent of the frame of reference in which they are located, but Einstein showed this is *not* the case: the distances and times separating two events depend on the state of motion of the observer with respect to these events.

Though the measurements of length and time intervals obtained by two observers in motion relative to each other are not equal, they *are* related; for example, the times measured by two clocks are related by the time-dilation formula given earlier. Suppose now that observer **A** measures the location and time at which an event occurs: Flash ran the 100 yd dash in 3 s flat. If observer **B**, moving with respect to **A**, wants a description of this feat in *his* coordinates, **B** needs to know his velocity with respect to **A**, the distance run as measured by **A** (100 yd) *and* the time it took as measured by **A**'s clock (3 s); it is *not* enough to know the distance and relative velocity, the *time* it took is also needed.

So, in order to translate the measurements from one reference frame to another *all four coordinates are needed*. Special Relativity tells us that space and time intervals are relative *and* that space and time are interlinked. In fact, the mathematical description of the Special Theory of Relativity is most naturally expressed by combining space and time into one object: *space-time*. A point in space-time is specified with four numbers: the location and time of occurrence of an event.

Within Special Relativity the nature of space-time is unaltered by whatever is in it. There are rules that state how the measurements of two observers are related, but these rules are not affected by the objects (and beings) that populate space-time, they are the same whether we look at a pea, an elephant or a star millions of times more massive than the Sun; space-time is still the stage where Nature unfolds. In addition, all conclusions were reached using special kinds of observers and reference frames (the inertial ones),[13] but, why should Nature be prejudiced against non-inertial observers?

We will see that the Special Theory of Relativity can indeed be generalized to include accelerating observers; the resulting theory is called the General Theory of Relativity (thus labeled because it encompasses any type of observer). Once this is accomplished, space-time itself becomes dynamic: its properties determine the way in which objects move, and these properties are determined by the objects in it. The evolution of our ideas about space and time, from absolute and unrelated entities, to their being meshed into an imperturbable space-time, to their final description as a dynamical object which is part of Nature, not merely the scenery, is one of the most profound and exciting developments in modern science.

Mass and energy

The principle of relativity leads to the conclusion that measurements of space and time depend on the state of motion of the observer. On the other hand, the manner in which an object reacts to external influences is determined by its mass, and so it is pertinent to ask whether the mass of an object remains an absolute quantity or not. In Newtonian mechanics the ratio of a force to the acceleration it causes on a body, the quantity F/a, is identified with the object's mass, which is a constant. In this section I will argue that relativistic effects require that this quantity depends on the velocity of the object, but identifying it with the mass of the object is not strictly correct (for example, such a "mass" would depend on the direction of the force); the mass is defined as the ratio when the velocity is zero. In the following m should be understood as a quantity that measures the response of an object to a given external force.

The mass of moving objects

Imagine the following situation: Tweedledee and Tweedledum are now at war; in this variation of their saga they attack each other with sticky clay balls fired from identical guns in a region of space where there is no gravity. When they are at rest with respect to each other and one of Tweedledee's clay balls happens to hit one of Tweedledum's in mid-flight they just stick to each other and stop (since they have the same mass and the same velocity).

Now let us put Tweedledee in a fast moving spaceship and have him zoom by Tweedledum so that their distance of closest approach is d; we will say that the rocket moves vertically, and that the distance d is along the horizontal direction. No truce has been agreed upon and the firing continues: Tweedledum decides to fire one clay ball in such a way that it will hit Tweedledee at closest approach (Figure 6.10).

The spaceship moves at speed u and the clay balls travel at speed v (with u much larger than v); in Tweedledum's reference frame his clay ball will travel the distance d in a time $t = d/v$. In Tweedledee's reference frame this time will correspond to a *longer* time due to time dilation (Figure 6.11). The distance d, in contrast, is the *same* for both reference frames since it is *perpendicular* to the direction of motion. It follows that in the rocket's reference frame Tweedledum's clay ball has a smaller horizontal velocity, since it takes longer to cover the same distance. Since everything is symmetrical, in Tweedledum's reference frame it is Tweedledee's clay ball that is slower.

Figure 6.10 The war as seen in Tweedledum's reference frame (only Tweedledum's clay ball is pictured).

To complete the setup let us allow Tweedledee to fire back at Tweedledum and arrange the timing so that the clay balls will collide in mid-air. Assume first that the mass is unaffected by the motion, then since in Tweedledum's reference his ball is moving faster, when the balls collide and stick, they will keep moving towards Tweedledee. In Tweedledee's reference precisely the same argument indicates that the balls will move towards Tweedledum! In fact, by symmetry the balls should stick to each other and the final mess should have no horizontal velocity at all. But this can happen only if our assumption that the mass is unaffected by motion is incorrect; in the rocket's reference frame Tweedledum's clay ball will stop Tweedledee's only if it is heavier (since it moves slower), while the opposite will hold in Tweedledum's frame of reference.

We conclude that *mass increases with the speed of the object*. As with time dilation and length contraction, this is a real physical effect: if two identical objects are weighed while one is moving at high speed with respect to the balance, the fast one will be heavier:

Mass is relative to the observer.

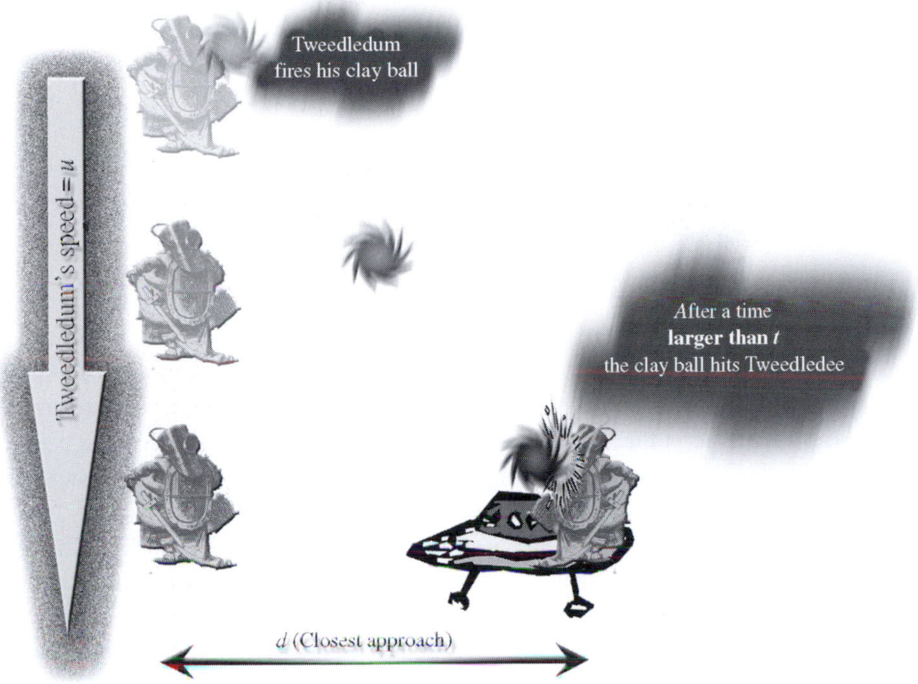

Figure 6.11 The war as seen in Tweedledee's reference frame (only Tweedledum's clay ball is pictured).

Because of the relativity of mass any fixed force will produce a smaller acceleration the faster the object is moving, and the corresponding change in velocity will be also smaller. So the idea that one could kick a ball again and again and again, steadily increasing its speed until it reaches and then exceeds c cannot be realized. The specific expression for this mass increase shows that the mass would become infinite if the speed of an object could reach c, and so one would require an infinite force to reach this speed. Because of this no material object can ever be accelerated to the speed of light, let alone any speed beyond that value.

Mass and energy

The conclusion that mass is relative is drawn from the same physical laws as the other relativistic effects, and like them it becomes noticeable only at very large speeds. Yet, its practical consequences are much more dramatic, having shaped military politics (and everything inside its sphere of influence) since the 1940s. The reason why such an apparently esoteric result produced such societal changes is that the relativity of mass also implies a close connection between the mass of an object and the amount of energy it can produce, and energy and its sources are paramount to our society.

To be precise we will define "energy" as the capacity to do work,[14] and since a moving object can be lassoed and made to do some work, like pulling a car (of course in so doing it donates its energy and slows down), we conclude that moving objects have a certain amount of energy, and that the faster the object moves the more energetic it is.

So the energy of an object, just like its mass, increases with its speed, and this suggests a connection between these two quantities. In fact they *are* connected, and by a simple and by now very famous relation:

$$E = mc^2$$

where E denotes the energy of the object and m its mass. Any object with mass has a corresponding amount of energy; an increase in the amount of energy contained in an object is equivalent to an increase in its mass and it is because of this interrelation that mass and energy are considered equivalent. This correspondence suggests the possibility of transforming mass into energy and vice versa and, in fact, there *are* processes where this expectation is realized.

In particular, even when an object is at rest the energy it contains is not zero: the mass–energy equivalence leads to the remarkable conclusion that, in principle, energy can be extracted from any object as long as it has a non-zero mass, that is, by *virtue of its very existence*. The resulting amount of energy can be enormous. For example the energy in 100 g of garbage will light a 100 watt bulb for three million years; for a more dramatic figure, the energy content of a 2 ton load of trash is almost twice the US yearly energy consumption.[15]

To temper these tantalizing comments I hasten to note that there is no known physical process that converts 100 percent of mass into energy; the most efficient and technologically viable possibility occurs in nuclear reactions. In these processes an atomic nucleus of initial mass M decays (either because the environment is tailored to insure this or because it is unstable and disintegrates spontaneously) into one or more objects of a smaller combined mass m. The difference in mass is released as energy in the amount $(M - m)c^2$; the efficiency

Box 6.4

Robert Oppenheimer on the atom bomb

If the radiance of a thousand suns
Were to burst at once into the sky,
That would be like the splendor of the Mighty One . . .
I am become Death,
The shatterer of worlds

Robert J. Oppenheimer
(on the first nuclear explosion; quoting from Bhagavad-Gita)

for this process is a few percent. The energy released takes the form of electromagnetic radiation that can be used (or misused) for a variety of purposes.

The nuclear problem

It is impossible to provide a thorough discussion of all issues connected with nuclear power and its military and economic complications in this book. The development and use of nuclear weapons, the associated controversies, and the deep moral problems they engendered have filled thousands of pages and I do not presume to provide new insights in this troublesome and murky business. And yet it is also impossible to ignore this problem, it being so inextricably connected with the Special Theory of Relativity. I have then opted for a description of some of the main problems associated with the application of nuclear power to peaceful and non-peaceful situations.

The simple arguments relating mass and energy of the previous paragraphs belie the complications that grew out of them. Unexpectedly, the realization that "in nuclear reactions mass differences are converted into energy" came to us with a demon host, which we have been unable to exorcize. The devil here is hidden in the word "energy" which in these processes appears mostly as radiation, and can be extremely damaging to all organisms. Not only will it hurt tissue in a visible and immediate way, but it can also alter the cell's DNA. This does not generate the gigantic monsters familiar in science fiction movies, but produces something more terrifying: a defective DNA can lead to birth defects, increased incidence of cancer, and to other diseases of genetic origin, and since the DNA is inherited, these health problems recur, sometimes for generations. Radiation hurts the living and the unborn, and these hurts we cannot cure.

It is true that a certain amount of radiation occurs naturally through the decay of radioactive elements present in the Earth's crust, and (presumably) life on the planet has evolved to adapt to this continuous environmental factor. But the *intensity* of naturally occurring radiation is very small compared to that present in an economically viable nuclear reactor, and puny compared to that generated even in the smallest of nuclear weapons. Any decent and humane use of the energy released through nuclear processes requires extreme care in insuring that radiation leaks do not occur, and the complications resulting from this constraint lie at the root of all serious technological and economic obstacles to the development of this type of energy source.

Firstly, reactors must be shielded in order to insure that the radiation released in the main reaction is contained while being transformed into a safer form (e.g. into heat, that is then used to drive steam turbines); the absence of leaks should be verified through frequent and thorough inspections. Secondly, and more troubling, one must dispose of the remainders of the nuclear reactions, the "nuclear ashes." Most of this material remains radioactive, that is, continues to decay while emitting radiation at a harmful level for hundreds of years, and yet its energy output is too low to be technologically useful. So this undesirable byproduct has to be stored in a seismically stable location with a minimum of living creatures, and in containers that for centuries will resist corrosion and other environmental effects. Should a substantial leak occur the radioactive material can be dispersed by wind and water; its effects on any living creature can be (and, unfortunately, have been) disastrous.

There are also, of course, additional "applications" of the mass–energy equivalence that are specifically designed to destroy life: nuclear weapons. It was the very same physical law connecting mass and energy that during the Second World War suggested to both Nazi and Allied physicists the possibility of using the energy released in nuclear reactions in armed conflict. The Nazi effort in this direction was pre-empted by the end of the war, which was partly brought about by the successful development of the first atomic weapons in the Manhattan Project. Shortly after this, bombs were dropped in two Japanese cities, Hiroshima and Nagasaki. In Hiroshima 66 000 people died and 69 000 were injured from the blast; in a few weeks the number of dead had risen to 130 000 and then to 200 000 in five years; in Nagasaki between 60 000 and 70 000 people died from the blast; and 50 000 more died within five years. The fate of more than 300 000 people was sealed in the fraction of a second it took these bombs to explode (Figure 6.12); for comparison, there were about 620 000 casualties in the American Civil War that lasted five years. Current atomic weapons are roughly 100 times more potent than those dropped on Japan in 1945.

The creation of nuclear weapons was one of the watersheds of the twentieth century, and it marks one of the most dramatic instances in which physics has affected the social structure of the planet. Yet the very same formulas also suggest the possibility of obtaining vast amounts of energy which can be used for constructive purposes. The 1905 new-born babe of an idea, the equivalence of mass and energy, has grown into a two-headed monster; it is the burden and duty of post-Second World War physicists to deal with it.

A note on the method of derivation

It will not have escaped the reader that the relativistic effects discussed in this chapter were all derived by considering first some rather special situations (e.g. the "light-clocks") and then arguing that the statement must generalize provided the hypotheses proposed are valid (in the present case, the Principle of Relativity and Maxwell's equations). The first special cases considered are often referred to as "thought experiments" since they are hard to realize in practice (though there are no physical laws that prevent their being performed); for example, the light-clocks used above would have to be 150 000 km high!

Thought experiments are constructed to present the issues as clearly as possible, to provide tests of the logical consistency of the theory, and to provide

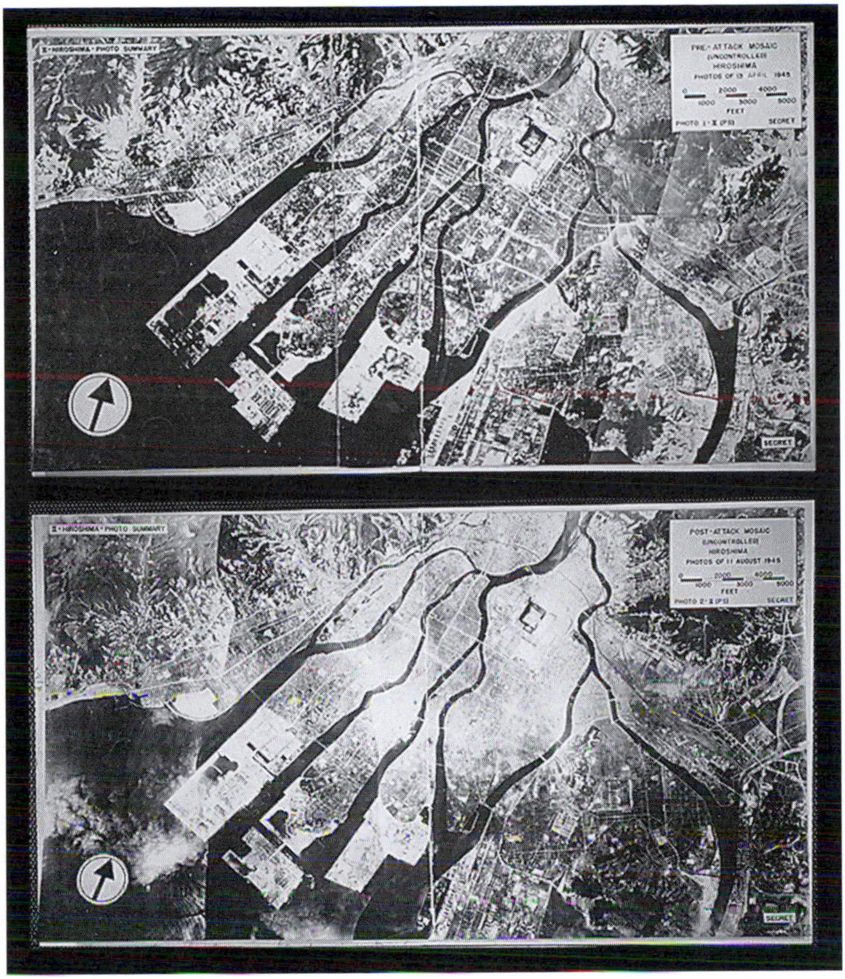

Figure 6.12 Aerial views of the city of Hiroshima before (top) and after (bottom) the dropping of the first atomic bomb. (Courtesy of the US National Archives and Records Administration (http://arcweb.archives.gov/))

some guidance in building the actual experiments that test these hypotheses, and which provide the ultimate justification for the Special Theory of Relativity.

Notes

1. If it happens to move towards Earth, we just wait a bit until it zips by.
2. I present the following statements without proof.
3. Though it is unlikely that the Michelson and Morley experiment was important in the development of the theory of relativity, it does provide the clearest evidence against the existence of an ether wind.
4. As another example: Lincoln's head coincided with his assassin's bullet at one time and place, and this should be true for all observers, for the fact that he died in consequence cannot change from one frame to another.

5. To give a more practical application, you could learn the winning lottery number before the drawing; but that would occur only in some reference frames. So for some observers you'd be a pauper and for others a millionaire . . . but which one *is* you?
6. In this statement, "independent" means that such events cannot influence one another.
7. I am, of course, ignoring the practical difficulties associated with this requirement, for the mirror and the source/detector combination must then be separated by 150 000 km.
8. This approach is often called an "argument by contradiction": we assume the opposite of what we expect and derive a contradiction.
9. The same argument holds if, for example, we assume that the vertical distances increase with speed.
10. One can, of course, say that *the* length of a ruler is the one measured while at rest with respect to it, but this is only a convention.
11. And the direction of motion is not perpendicular to the length being measured.
12. If we replace "closing and opening the doors" by "firing the lasers" the situation is identical to the one presented by the murder mystery discussed above.
13. Hence the adjective "special" in Special Relativity.
14. Of course, this is an empty definition unless we define what work is: so if we imagine a wooden box on a rough surface, then work is the capacity to move this box over some distance despite being opposed by friction (this definition generalizes).
15. The US produces almost 200 million tons of garbage every day.

7
The General Theory of Relativity

The General Theory of Relativity, as the name indicates, is a generalization of the Special Theory of Relativity. It was developed by Einstein with little or no experimental motivation but driven instead by philosophical questions: Why are inertial frames of reference so special? Why is it we do not feel gravity's pull when we are freely falling? Why should absolute velocities be forbidden but absolute accelerations accepted?

The General Theory is certainly one of the most remarkable achievements of science to date, and physicists have exhausted all superlatives praising it. This theory has provided deep insights into the structure of the Universe, altering the very perception of what we call space and time. Rarely has a single person given birth to a field of study, but this is precisely what Einstein did: gravity changed from a tired description of bodies attracting one another with an inverse-square law to the study of space-time and matter as dynamical objects whose interactions are determined by the geometry of the former and the motion of the latter.

Despite its beauty, it might seem inappropriate for the theory, accurate and interesting as it has proved, to receive such praise because its practical applications (those not already available using Newton's theory) are practically non-existent, at least at present. But the popularity of this generalization of the Special Theory stem, not from its applications to daily life, but from the sea change it required in our understanding of space and time, and the deepening of our knowledge that came from it. Once a certain measure of experimental verification became available, this theory that describes the Universe as a whole naturally left the confines of academia and became the subject of news releases, popular articles and, soon thereafter, the standard background of almost all science-fiction literature dealing with space travel.

Wormholes, black-hole horizons, space-time singularities are all engendered by Einstein's gravitational equations, which also describe the more pedestrian planetary orbits. It is because of the way the General Theory has stretched our minds that we respect it and its creator so much.

The happiest thought of my life

Soon after proposing the Special Theory, Einstein realized that the Principle of Relativity clashed with one of the most basic features of Newtonian gravity. The expression mMG/r^2 proposed by Newton for the gravitational force between masses m and M, separated by a distance r, makes no reference to time; this implies that any change in M (for example) will induce a change in m instantaneously, and *any* observer will see cause and effect occurring simultaneously. This represents a physical effect that travels faster than light (though detecting it might be awkward and difficult) and is, of course, inconsistent with the Special Theory of Relativity.

Einstein was convinced of the impossibility of infinite propagation speeds and so set out to find a relativistic generalization of Newton's equations, that is, a theory that would reproduce Newtonian gravity when all the speeds involved are small compared to c, and yet incorporate a finite speed of propagation for the gravitational field. Note also that this implies that there would be a modification of Newton's equations when the objects involved are very massive and compact since, for example, planets moving in a fixed circular orbit round a star of mass M will move faster the larger M is. Newton's equations would then be accurate for small velocities and densities, and Einstein's goal was to extend their reach into the relativistic regime.

Einstein's first paper on this subject was published in 1907; 13 years later he commented that while writing it a thought came into his mind, which he called "the happiest thought of my life":

> *The gravitational field has only a relative existence* . . . because for an observer freely falling from the roof of a house – *at least in his immediate surroundings* – there exists no gravitational field.

To see how this comes about, let's imagine the unfortunate Wile E. Coyote falling from an immense cliff above the desert.[1] As he starts falling he lets go of the bomb he was about to drop on the Roadrunner down below; but the bomb does not drop below Wile, it faithfully stays with him, neither gaining nor lagging, it just floats there. If he were to push the bomb away he would see it move with constant speed in a fixed direction.

Mr. Coyote is fated to repeat this experience with many other objects: on various occasions he finds himself falling in the company of rocks, magnets, harpoons, and anvils. And in all cases the same results are observed: as long as Wile and his accoutrements are freely falling, all such objects *with respect to him* behave as if there was no gravitational force present, irrespective of composition, mass, and any other distinguishing characteristic. Were he to fall freely inside a closed box, he would not be able to tell whether he was plunging to his death (or, at least, severe discomfort), or whether he was in outer space. He could find out, but only by looking outside the box.

This realization is interesting because this is exactly the way Newton's first law described the behavior of force-free objects: if we imagine placing Wile, bomb and box in outer space, a push from Wile would send the bomb moving at constant speed in a straight line, just as when they were freely falling. Any object inside the box, whether freely falling or in a region where there are no forces, will remain at rest or in a state of uniform motion *with respect to the box*, and this behavior is independent of its chemical composition, physical appearance, etc. (as above, air resistance is ignored). A freely falling observer will find his frame of reference to be inertial, as long as he confines his observations to his immediate vicinity. Just as Galileo argued that experiments in a closed box cannot determine the state of uniform motion of the box, Einstein argued that experiments in a freely falling small[2] closed box cannot be used to determine whether the box is dropping in the grip of a gravitational force or simply floating in space far from all gravitational influences.

Why would this be true? The answer can be traced back to the way in which gravity affects bodies. Remember that the quantity we called m (the mass of a body) plays two different roles in Newton's equations. One, as "inertial" mass, determines the acceleration of the body for *any* applied force: $F = ma$. The other "gravitational" mass determines the intensity of the gravitational force between the body and another of mass M: $F_{grav} = mMG/r^2$. As mentioned before these two quantities need not be equal, but, in fact, both numbers *are* equal (to a precision of ten parts per billion).[3]

The most direct implications of the equality between inertial and gravitational mass can be derived using Newton's Second Law, $F = ma$, for a gravitational force:

$$\frac{mMG}{r^2} = ma \text{ so that } a = \frac{MG}{r^2}$$

This is the equation that determines the motion of the body of mass m in the presence of gravity, *and is independent of m*. Any two bodies will follow the *same* trajectory provided their initial positions and velocities are the same: when in gravity's grip, beans, bats, and boulders will move in the same way.

This property of gravity is not shared by any other force. For example, under the influence of electric forces, two objects, one charged and the other neutral, will not behave the same way: the neutral one will not accelerate at all. In contrast, under the influence of gravity these same bodies will receive identical accelerations, charge or no charge.

The equality of gravitational and inertial masses was a tacit assumption made by Newton. Einstein, however, came to realize that this assumption played a central role in understanding the properties of gravitation and of space and time. He also realized that the precise identity of these quantities cannot be experimentally verified to infinite precision, and so he *postulated* their equality. This postulate is in fact the cornerstone of the General Theory of Relativity, and since it asserts that gravitational and inertial masses are equal he called it the *Principle of Equivalence*:[4]

The inertial and gravitational masses are identical.

The conclusions we obtained using Mr. Coyote as a guinea pig are quite general. Consider, for example, the following experiment: Albert is put inside a room-sized box on the Moon (chosen because there is no air and hence no air resistance) with a bunch of measuring devices. This box is placed high above the lunar surface and then let go, so that it freely falls towards the surface. Then, just as Mr. Coyote, Albert would not be able to determine whether he is in outer space or in free-fall (that is, until the box hits the lunar surface!). This follows from the equality of the inertial and gravitational masses, for only then are all objects accelerated in the same way by the Moon's gravitational pull. It is because of the Principle of Equivalence that Albert can be deluded into believing that he is an inertial frame of reference . . . until disabused of the notion by the crash on the surface.

The conclusion is that the Principle of Equivalence implies that for any gravitational force we can always choose a frame of reference in which an observer will not experience any gravitational effects in his/her immediate vicinity (the reason for this last qualification will become clear below). Such a frame of reference is, as stated above, *freely falling*.

Does this mean then that gravitational forces are a chimera, an illusion? Of course not! Consider for example (Figure 7.1) the case where Albert and his box are falling toward the Moon, but assuming now that the box is very large, say about 3500 km long (equal to the Moon's diameter). In Albert's immediate vicinity everything appears the same as before, but this will no longer be the case when he makes observations over large distances. For example, imagine that Albert arranges (perhaps using some machinery) for two apples to drop from a certain height, one at each end of the box. If the box is in outer space the apples will stay put, floating. If, however, the apples, box and Albert are all falling towards the Moon then the trajectory of the apples is not straight downwards, but tilted towards the center of the Moon[5] so, as time passes Albert will see that the apples are drawn together and, since this cannot occur in the absence of forces, he concludes that he is in a freely falling frame. If he were to repeat the experiment with different objects he would verify that they all exhibit the same motion he observed for the apples, as required by the Equivalence Principle.

The whole of the General Theory of Relativity rests on the Principle of Equivalence: this theory would collapse if one were to find a material for which the inertial and gravitational masses have different values. As with the Special Theory, the General Theory of Relativity is falsifiable. One might think that this crucial dependence on a single assumption represents a defect of General Relativity, a blatant Achilles' heel, since a single experiment has the potential of demolishing the whole of the theory. On the other hand one can argue that a theory which is based on a minimum of assumptions is more elegant and profound; from this point of view the General Theory of Relativity is a gem.[6] Still, the Principle of Equivalence is of interest neither because of its simplicity, nor because it leads to philosophically satisfying conclusions, but because of the enormous experimental evidence that confirms the conclusions drawn from it.

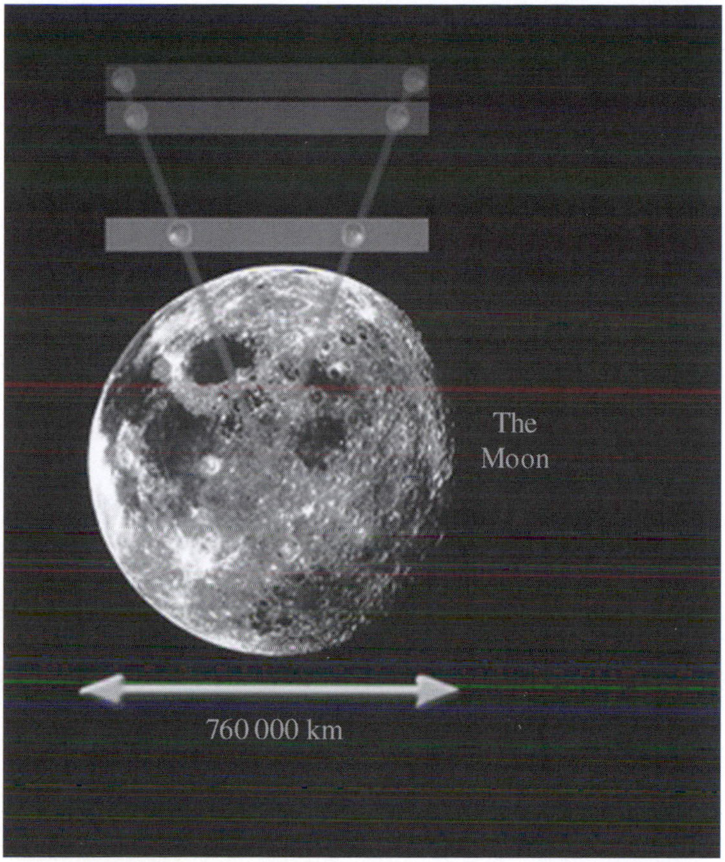

Figure 7.1 Over large distances the gravitational effects can be differentiated from those produced by acceleration (in an accelerated reference frame the apples would not move closer to one another). (Moon image courtesy of Dr. C. Allen, NASA)

Newton vs. Einstein

One might reasonably complain that these arguments supporting the Equivalence Principle rely on the purely Newtonian relations $F = ma$ and $F_{\text{grav}} = mMG/r^2$, but aren't these relations suspect? Leaving "doublethink" aside, how can one justify the most fundamental idea of General Relativity using results from Newtonian gravitation if the former theory claims the latter is inaccurate?

In fact both $F = ma$ and $F_{\text{grav}} = mMG/r^2$ *are* suspect, *except* when all speeds are small compared to c and when all gravitational forces are also small.[7] In this regime the accuracy of the predictions obtained using Newton's equations does support the Equivalence Principle. Motivated by this, Einstein *assumed* that the Principle holds under all circumstances, even when Newtonian gravity fails. So Einstein did not "derive" the Equivalence Principle from Newton's equations, and though he did use Newtonian physics to support his assumptions the ultimate justification was obtained from the verifiable experimental consequences that he derived from his relativistic equations for the gravitational field.

Gravitation vs. acceleration

A person riding a fast elevator will feel pulled downward at the beginning of the ride, when the elevator begins accelerating upward. This perception is similar to the one the person would experience when carrying a heavy load on his head or if he were to suddenly (and distressingly) gain a few pounds in weight. These simple everyday observations suggest a connection between gravity (which is responsible for pulling us down towards the Earth and generates our perception of weight) and acceleration.

In fact, gravity and acceleration are intimately connected. In order to understand their interrelationship let us return to Albert and his box, but let us now place them in outer space, far from any gravitational influences, and attach the box to a rocket that accelerates it (Figure 7.2). Albert's reference frame is then no longer inertial since he can tell without looking outside that the box is in motion. For example, he feels pressed against the bottom of the box, and if he drops an

Figure 7.2 An observer in an accelerated reference frame experiences effects identical to those generated by gravity.

object it will accelerate downward; in fact, all objects accelerate equally.[8] An inertial observer looking at Albert's antics (imagine, for example, that the box is made of one-way mirrors) would explain these results by pointing out that, as seen in an inertial reference frame, any object, once released by Albert, will move at constant speed in a straight line, since there are no longer any forces acting on it; in contrast, the box continues to accelerate, and will soon catch up with the object and hit it. From this point of view it is also clear why Albert sees all objects accelerate at the same rate, for this is a result of having an accelerated reference frame (the box) whose properties bear no connection with any of the objects being considered.

These observations of Albert are identical to the ones made by Galileo long ago, but in completely different circumstances: in the presence of gravity we can tell directly that we are not in an inertial reference frame for we feel pulled towards the center of the Earth, and if we drop any two objects they will reach the ground simultaneously (ignoring air resistance); *this* is the basis of the Principle of Equivalence. Therefore if the Principle of Equivalence is valid, Albert cannot tell whether he is in fact in outer space in a non-inertial reference frame, or whether his box is peacefully resting atop some grassy mound here on the home planet. The Principle of Equivalence also implies that no experiment made by Albert in his immediate vicinity can discriminate between these two possibilities: in small regions gravity mimics, and is mimicked by, non-inertial reference frames.

In order to distinguish between a box under the influence of a gravitational force and one being pulled by a machine we again need a planet-sized box. As in the example above let us imagine that Albert arranges for two apples to be released from the top of the box, one at each opposite end. If the box is being accelerated by a machine, he will see the apples fall straight down to the bottom of the box, while if gravity is present they will follow a slanted trajectory ending up closer than when they started. In the immediate neighborhood of the observer the gravitational force is almost constant and its effects are equivalent to those obtained in a uniformly accelerated reference frame. Over large distances one probes changes in the gravitational field that cannot be simulated by a non-inertial reference frame.

Summarizing: in small regions inertial and freely falling reference frames are indistinguishable, and accelerated (non-inertial) frames perfectly mimic gravity. These various situations can be differentiated by making measurements over distances large enough to allow probing for possible changes in any gravitational effects that might be present.

Light

A very surprising corollary of the above is that light paths are bent by gravitational forces. To show this I will first argue that this effect occurs in accelerated

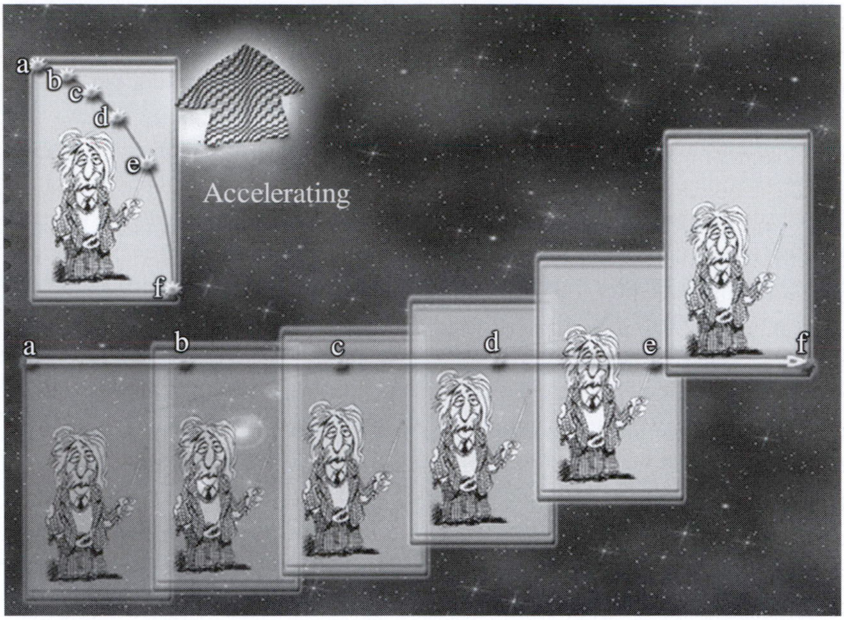

Figure 7.3 In an inertial reference frame (below) a beam of light follows a straight path. In an accelerated reference frame (above) it follows a curved path.

reference frames and then extend it to the cases where gravitational forces are present.

Consider then an elevator being pulled by a crane so that it moves with constant acceleration (that is its velocity increases uniformly with time). Imagine now that a laser beam propagating perpendicular to the elevator's direction of motion enters the elevator through a hole on the left wall and eventually strikes the right wall. The crane operator, who is in an inertial frame of reference, will see the laser propagating in a straight line (Figure 7.3) at speed c. In contrast an elevator passenger finds that the light-path is curved (Figure 7.3). To see why this happens, let us break the trip of the beam into pieces that are 1 cm long (any other choice will do as well); while the beam travels the first cm the elevator will move up some distance d, but by the time the beam covers this first part the speed of the elevator will have increased (since it is accelerating), therefore during the time the beam covers the next cm the elevator will rise a distance *larger* than d, and while it covers the next it will move a distance larger still. As seen from inside the elevator the beam first enters at a height h above the floor, by the time the beam covers 1 cm horizontally its height has dropped to $h - d$, by the time it covers the second cm it has dropped *below* a height $h - 2d$ and so on; the resulting trajectory is then curved. Thus for this simple thought experiment light paths will be curved for observers inside the elevator.

Now, since the Equivalence Principle implies that we cannot distinguish between an elevator accelerated by a machine and an elevator experiencing a constant gravitational force, it follows that the *same* light-bending effect should be observed if we place the elevator in the presence of a gravitational force: *light paths are curved by gravity*.

That gravity affects the paths of material objects such as planets and satellites is not something strange; but we tend to think of light as being different somehow. The above argument shows that light is not so different from other things and is indeed affected by gravity in a very mundane manner (the same elevator experiment could be done by looking at a ball instead of a beam of light and a similar picture would result).

A natural question is then, why do we not see light fall when we ride an elevator? The answer is that the effect in ordinary situations is very small. Suppose that the size of the elevator in the figure is 8 ft high, and 5 ft wide; if the upward acceleration is 25 percent that of gravity on Earth then the distance light falls is less than a millionth of the radius of a hydrogen atom (the smallest of the atoms). For the dramatic effect shown in the figure the acceleration must be enormous, more than 10^{16} times the acceleration of gravity on Earth and this implies that a passenger who weighs 70 kg on Earth will weigh more than 1000 trillion tons in the elevator (in such an environment a human observer would do very little observing as he'd be reduced to a bloody pancake at the bottom of the elevator).

This does not mean, however, that this effect is completely unobservable (it is small for the case of the elevator because elevators are designed for very small accelerations, but one can imagine other situations). Consider, for example, a beam of light coming from a distant star towards Earth (Figure 7.4). Along the way it comes close to a very massive dark object. The arguments above require the light beam to bend; and the same thing will happen for any other beam

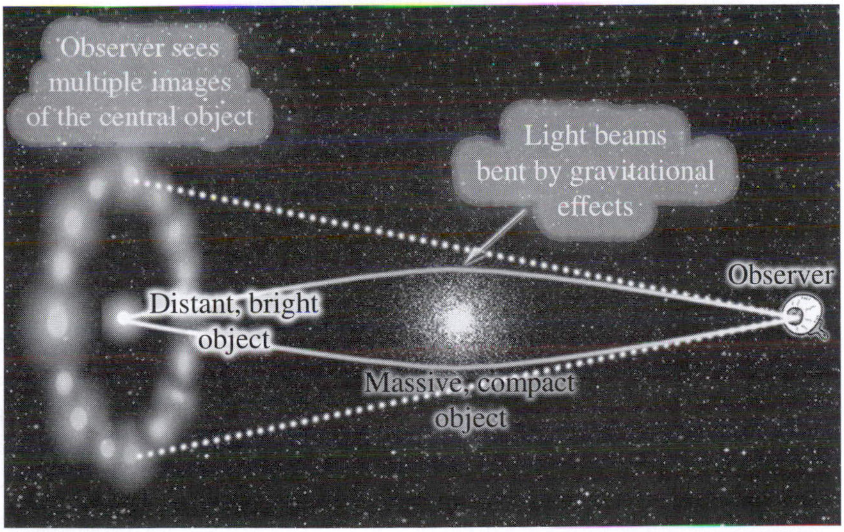

Figure 7.4 Illustration of the lensing effect of a strong gravitational force (produced by a massive compact object).

originating in the distant star. Supposing that the star and the opaque object are both perfect spheres, then an astronomer on Earth will see, not the original star, but a *ring* of stars (often called an "Einstein ring"). If the massive dark object is not a perfect sphere, then an astronomer would see several images instead of a ring. This effect has been christened *gravitational lensing* since gravity acts here as a lens, making light beams converge.

How do we know that the multiple images which are often seen are a result of the bending of light and not those of several stars that happen to be identical twins which ended up widely separated?[9] We know this because these identical twins or triplets, etc. would not only have the same physical properties, but also exhibit simultaneous identical motions, and this despite their being in regions of space separated by many light-years where the various forces affecting the stars would be different. The one explanation that does not need to resort to an ever-lengthening string of coincidences is that these are the images of a single star multiplied by the trick-mirror effect of a gravitational lens.

The bending of light was one of the most dramatic predictions of the General Theory of Relativity, and the first one to be verified by experiment.

Gravitation and energy

When sunlight falls on your skin it will warm it, and eventually burn it. This is because light carries energy which is absorbed by your skin, thus increasing its temperature. Recall also that a body with mass m, by its very existence, has energy in the amount mc^2. There is no way, however, in which we can associate a mass with light. For example, we can always change the speed of an object with mass (even if only a little bit), but this cannot be done with light, which travels at speed c with monomaniacal persistence. Since gravity affects both light and all material bodies, and since both carry energy, but only objects carry mass, it follows that *gravity will affect anything carrying energy*. This conclusion lies at the root of the construction of Einstein's equations, the relativistic generalization of Newton's $ma = mMG/r^2$.

This conclusion leads to some rather strange consequences. Consider for example a satellite in orbit around the Earth: when the Sun shines on it, its total energy will increase (since it warms up), and gravity's pull will increase accordingly. When the satellite is in darkness it will radiate heat, lose energy, and the force of gravity on it will decrease.[10]

Gravity and space

Two of the most dramatic consequences of the Principle of Relativity are that both space and time are relative: distances and intervals separating events have different (albeit related) values for different inertial observers, even simultaneity is relative. The equivalence principle also provides striking conclusions about the nature and properties of space and time, conclusions that will now be valid for *any* observer.

Imagine a region of space near some massive stars but in itself empty. Since there are gravitational forces present inside this region objects moving inside it will not follow straight paths. Assuming that there are no other external influences but gravity, the Principle of Equivalence guarantees that any two objects starting from the same spot with the same speed (that is, if they all have the same initial conditions) will follow the same path.[11]

This situation is similar to the one in the following murder mystery. Imagine a closed room and a line of people waiting to go in. The first person goes in and precisely two minutes afterward, is expelled through a back door, dead; it is determined that he died of a blow to the head. Soon thereafter a second victim enters the room with precisely the same result, she also dies of an identical blow to the head. This goes on for many hours, each time the victim dies of the same thing irrespective of his/her age, occupation, habits, color, political persuasion, or taste in wine or beer. Animals suffer the same fate, be it insects or whales.[12] If a rock is sent in, it comes out with a dent of the same characteristics as the ones suffered by the people and animals. The police finally shrewdly conclude that there is something in the room that is killing people, they go in and ... But the result is not important, what *is* important for this argument is the conclusion that if *everything* is affected in the same way, this effect is a property of the room, not of the things, people, or animals being sent in.

Going back to our region of empty space we can follow a similar argument and conclude that since all objects are affected in precisely the same way, whatever it is that makes the paths curve is a property of the region itself. But it is gravity that makes the paths curve, and inside this region we have nothing ... except a bit of space. And so we conclude that gravity is a property of space; a change in the gravitational effects corresponds to a change in the properties of space itself. Our previous experience with the Special Theory of Relativity indicates a close connection between space and time, and this suggests that gravitational influences will also affect the flow of time. We see below that this is indeed the case, and conclude that *gravity alters the properties of space and time*. We have finally arrived at the modern view concerning space-time: it is not a static absolute stage where Nature evolves, but a dynamic object whose properties depend on the state of motion of the observer *and* on the gravitational forces generated by anything carrying energy; and, in its turn, the properties of space-time determine how objects move. In the words of J. A. Wheeler: matter and energy tell space how to curve and space tells them how to move (Figure 7.5).

So, how *does* gravity affect space?

The natural question is then, how does gravity alter space and time? What kind of deformation does it produce? In this section I will use a standard argument

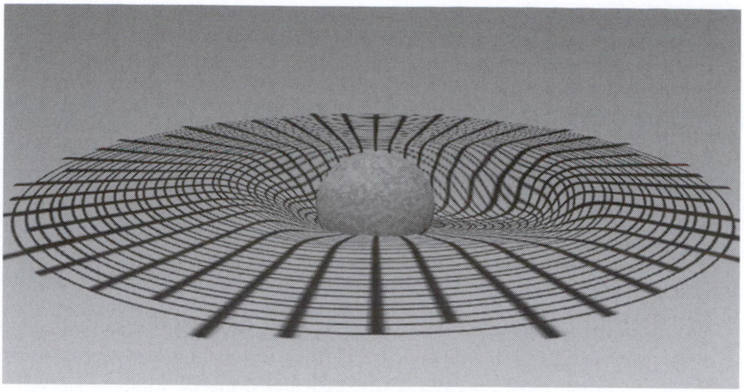

Figure 7.5 An illustration of the bending of space produced by a massive object.

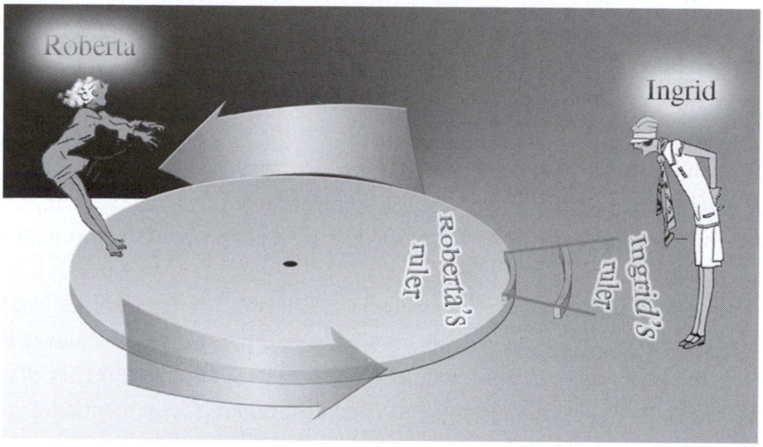

Figure 7.6 Comparing distances measured by inertial and non-inertial observers.

based on what we know about length contraction together with the Equivalence Principle to answer this question.

Imagine two identical observers, Ingrid who is in an inertial reference frame, and Roberta who is in a rotating frame; in Ingrid's frame Roberta will go around in a circle (Figure 7.6). Let us now imagine that Ingrid traces this circle and that she measures its circumference by marking a starting point and then laying down a short ruler along the rim of the circle again and again until she completes the circuit; she then does a similar measurement of the radius. Since she is in an inertial reference frame the usual rules of geometry will hold and she will find that these two lengths are related:

$$\textit{Circumference} = 2\pi \; \textit{radius} \quad \text{(inertial observer)}$$

Now let us imagine that Ingrid, who is rather mistrustful, carefully keeps an eye on Roberta as she, in her turn, does the measurements. When measuring the

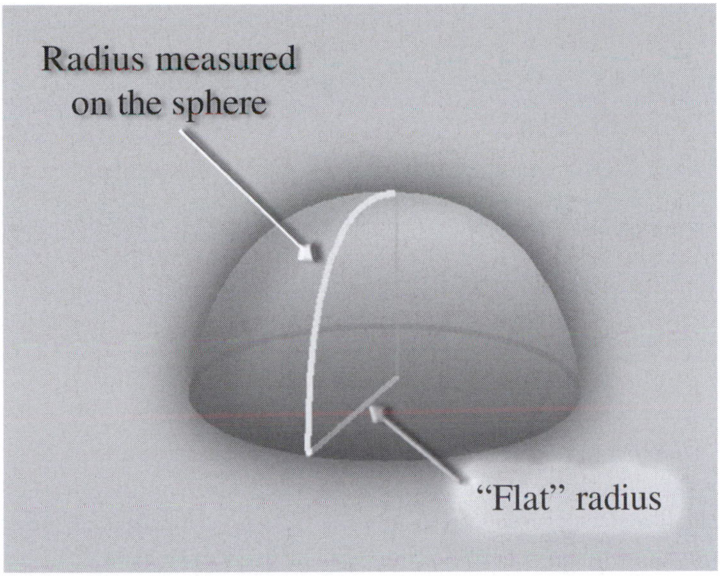

Figure 7.7 The distance from the equator to the pole on the sphere is larger than its "flat" radius.

circumference, Roberta will set the same short ruler along the rim of the circle (to ease her task she might trace a copy of the circle in her reference frame) as she does so it is important to note that her ruler, as seen by Ingrid, will always be pointing in the direction of motion, and so it will be contracted (Figure 7.6): Roberta will then find that the circumference will fit more lengths of her ruler than Ingrid did (the circumference is the same but Roberta's ruler is contracted). No such thing occurs when Roberta measures the radius, however, since in that case the ruler is always perpendicular to the direction of motion. So Roberta will measure the same radius, but a *longer* circumference:

$$\textit{Circumference} > 2\pi \textit{ radius} \quad \text{(rotating observer)}$$

How can this be? Isn't it true that the perimeter always equals $2\pi \times$ radius? The answer to the last question is yes . . . provided you draw the circle on a flat sheet of paper. On curved surfaces this need not hold, and the circumference will be larger or smaller than 2π *radius* depending on the shape of the surface. For example, suppose that you are constrained to draw circles on a sphere, and that you are forced to measure distances only on the surface of this sphere. Then you find that when both perimeter and radius are measured (*on the sphere*) the circumference is smaller than $2\pi \times$ radius (Figure 7.7).

If we repeat the same exercise on a surface shaped like a horse saddle (Figure 7.8) we would find that the circumference is *larger* than $2\pi \times$ radius. In fact, the result of measuring any length, area or angle in the rotating reference frame perfectly matches those obtained for the horse saddle. We then say that the *geometry of space* in a (uniformly) rotating reference frame is that of this surface.

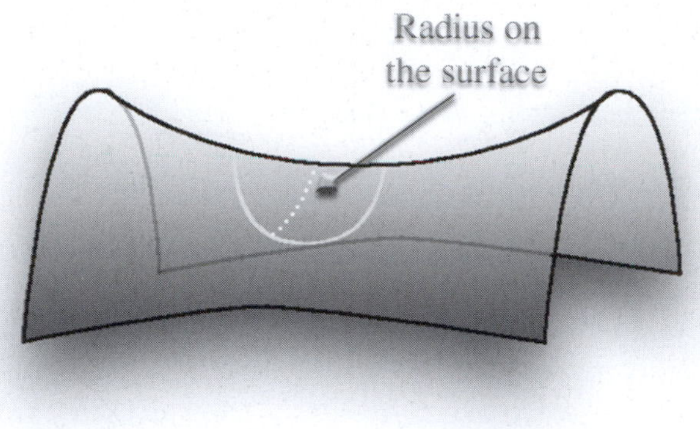

Figure 7.8 Circle drawn on a saddle-shaped surface has a circumference larger than 2π its radius.

In order to relate these observations to the property effects of gravity I shall use the Equivalence Principle, which implies that no experiment contained in a small region of space can distinguish between the effects of gravity and those produced by an accelerated frame of reference. Roberta *is* in an accelerated reference frame (remember that she is rotating, so that her velocity is changing – in direction only, but changing it is! – and is therefore accelerating!).[13] So, if we introduce a small laboratory into the rotating reference frame, no experiment done by Roberta inside her new lab would be able to determine whether the laboratory is accelerating or in the grip of a gravitational force. So, on the one hand any person fated to live in the rotating reference frame would find that space is curved, on the other hand I have also argued that they would not be able to distinguish this situation from what would happen in the presence of gravity. Combining these two results together we conclude that:

Gravitation curves space.

Motion in a curved space

I argued above that gravity affects space by curving it; here I show that the converse is also true: motion in a curved space has the same properties as motion in the presence of gravity. In order to visualize this imagine a world where all things can only move on the surface of a sphere.[14] Consider two beings labeled **A** and **B** (Figure 7.9), which are fated to live on the surface of this sphere, but aside from this, no force acts on them. On a bright morning they both start from the equator moving in a direction perpendicular to it, without any meandering. Our privileged station as three-dimensional beings allows us to see this space from outside, and, should we care to look at them, we would find **A** and **B** moving along two meridians towards the sphere's North Pole. But **A** and **B**, living as they are on the surface of the sphere cannot see this, according to each of them they are just moving in a straight line.

And yet, as time goes on, **A** and **B** will come closer and closer as they themselves can verify (for this one does not need to look from the outside as

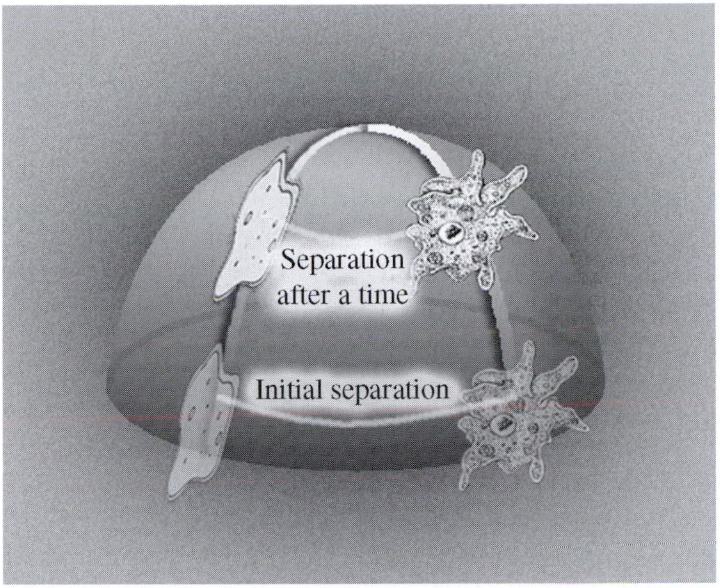

Figure 7.9 Two beings moving on a sphere are bound to come closer just as they would under the effects of gravity. (Courtesy of microscopeworld.com)

we are doing), and this they find most peculiar for there are no forces acting on them. This effect is similar to the experiment done with two apples in a large box that is freely falling towards the Moon (Figure 7.1): an observer inside the box finds that they come closer together despite the fact that no force can be detected in the vicinity of either.

For us it is obvious that **A** and **B** come closer because the space they are in is curved, and because of this the same effect will recur for any two objects or beings. In the absence of forces (excepting the ones that maintain the shape of their space), the motion paths are determined by the geometry of space and will be the same for any object starting at one spot with a given initial velocity. It will not have escaped the reader that these are *precisely* the properties of motion in the presence of gravity. We conclude that associating gravity with the curvature of space naturally leads to one of the main conclusions drawn from the Equivalence Principle, that motion in a gravitating environment is independent of the object.[15]

Now the big step is to accept that the same thing that happens to **A** and **B** on the sphere is happening to us in our four-dimensional space-time. We cannot prove this by looking at our space "from the outside" (because there is no outside, or at least none that we can go to), but we *can* test it in a different way.

Measuring curvature

Imagine that we communicate with **A** and **B** and inform them they are the proud inhabitants of a curved space, and that we are met with point-blank disbelief. We can convince them of our honesty as follows: we first ask them to review the

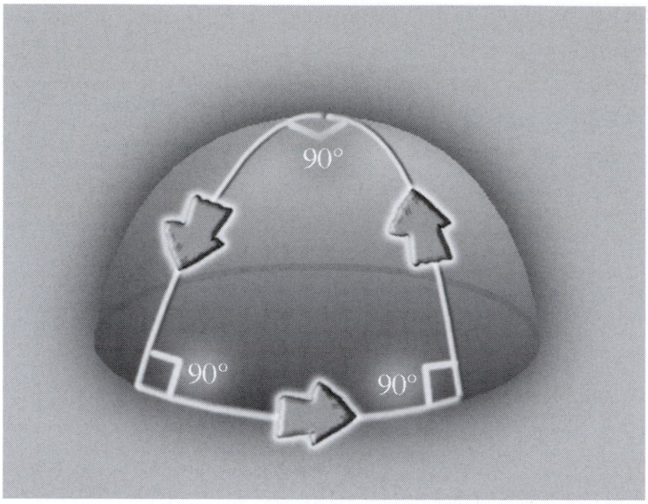

Figure 7.10 The angles of triangles drawn on a sphere add up to more than 180°.

geometry of Euclid (should they be unaware of it), and, in particular convince themselves that the sum of the internal angles of a triangle is 180°. Then we ask **A** to make the following trip (we tell him when to stop and which way to turn): starting from the Equator we ask him to move as before until he reaches the North Pole, then make a 90° turn and travel the same distance (seen from outside he is moving along a meridian back towards the Equator); finally he should make another 90° turn in the same direction as the first and travel again the same distance (Figure 7.10). After all this trekking **A** will find himself back at the starting point.

What **A** finds is then that he covered the sides of a triangle whose internal angles are all 90°; but these add up to 270°, not 180°! The irrefutable conclusion is that **A** lives on a curved space, for in flat space Euclid's results are valid. Note also that we convinced **A** of this fact without taking him "outside" his space; he only had to make careful distance measurements. We can do a similar thing: by measuring very carefully angles and distances we can determine whether a certain region of *our* space is curved or not. In general the curvature is very slight and so the distances we need to cover to observe it are quite large (several light-years). Still there are some favorable situations where these measurements are viable: if space were flat, light would travel in straight lines, but in regions where the gravitational forces are large the trajectories of lightbeams deviate from a straight path, and this provides a measurable effect that quantifies the curvature of space. I will return to this when I describe the classical tests of the General Theory of Relativity.

The warping of time

We have just seen that gravity, from the point of view of the General Theory, is a distortion of space, we also know that the *Special* Theory mingles space and time into a composite object, space-time. We therefore expect that gravity will also affect time, as indeed it does.

Let us go back to Ingrid, the inertial observer, and Roberta in her rotating (non-inertial) reference frame (Figure 7.6), and compare the rate at which their clocks tick. Initially Ingrid sees Roberta move in a circle around her, and, since Roberta has a velocity relative to Ingrid, the clock in the rotating frame will be *slow* compared to the one in the inertial frame. Using the Principle of Equivalence we then conclude that clocks in a gravitational field will slow down compared to inertial clocks, and the slowing down will be more marked the stronger the gravitational effects are.

If Roberta moves closer to the center of the rotating disk her speed increases and her clock will slow down more. So we find that two identical clocks, both in Roberta's frame but placed at different distances from the center will not stay synchronized, one will slow down with respect to the other. From the Principle of Equivalence, it follows that the same will be true in the presence of gravity: clocks will tick at different rates at different locations. Time, far from being an absolute and unchanging, is only *local*, slowing down and speeding up as we move from places where gravitational effects are respectively larger or smaller. The effect of gravity is then to distort the properties of both space and time: space-time is curved by gravity.

One can reach the same conclusions considering the behavior of light in a gravitational field. Since this approach also provides some insight into the nature of light I will re-derive the gravitational time dilation from this point of view.

Color and energy

When a rod of metal is heated up in a flame that is burning low it will shine with a dull red luster, if we stoke the flame the metal soon shines bright red, then orange and then yellow. Thus there is a connection between the color of light and the energy it carries: as we provide more energy the emitted light changes from red to yellow to blue; in other words, as we provide more energy the emitted light has a *shorter* wavelength. If we add energy to a source of light we cannot make the beam it generates go faster, for c is absolute; we can make the beam more intense, which can be accomplished by using the added energy in producing another beam of light of the same color; but we could *also* use the energy to shift the color to a smaller wavelength (in the case of visible light, towards the blue) without increasing the intensity of the beam.

Now imagine we place a laser on the surface of some massive and very dense object with the beam pointing straight up. As the beam moves farther and farther away from the surface it will lose energy;[16] being unable to slow down this loss will be reflected in an increase of wavelength: in the visible the light beam will redden. Conversely, if we shine the beam downward from a height it will gain energy as it falls and its wavelength will decrease; visible light will become bluer. The reddening of light as it moves away from a region where gravity is present is naturally (and unimaginatively) called the *gravitational red-shift*; the observation of this effect constitutes one of the standard tests of the theory and is discussed briefly below.

Clocks, red-shifts, and gravity

A wave of any type provides us with a ready-made clock, for we can sit down and, as the wave passes, watch the procession of crests and troughs, then, for example, use the lapse between two crests as a measure of a basic time interval (like the second). If we use an electromagnetic wave with a wavelength of 3 cm (a type of radio wave) the time interval between crests is about 10^{-10} s; so if we count ten billion crests we know that 1 s has elapsed. Let us construct two identical clocks based on this principle and place one on the surface of some massive planet, while the second one is sent off far into space where gravitational influences are negligible.

Near each clock we place an observer: Ares near the one on the planet, Aphrodite far out in space. Both observers (being deities) have a perfectly regular heartbeat of once per second, that is, once every time ten billion light crests pass by them. Now let us compare the two heartrates; when Aphrodite looks down at the planet she will see that the light beam from Ares' clock is red-shifted, so that compared to her clock, ten billion crests of Ares' clock will take *more* than 1 s to go through (as measured by her clock) since their wavelength is longer. Since Ares' heartbeat is perfectly synchronized with his clock, Aphrodite will find that his heartrate is also slow compared to hers. Similarly, *any* action of Ares is slowed down when seen from Aphrodite's vantage point: clocks slow down in the presence of gravity.

The slowing down of clocks in the presence of gravitational forces affects *all* clocks, including biological ones: a twin traveling to a region where gravity is very strong will come back younger than the twin left in a rocket in the depth of space. The gravitational forces required for a sizable effect, however, are enormous, so the twin will return younger . . . should she return at all.[17] This gravitational effect is distinct from time dilation in Special Relativity; in many situations both these effects will affect clocks sometimes adding up and sometimes partially canceling one another.

The Einstein equations

According to Newton, gravity is generated by any object with mass. Within the Theories of Relativity this statement is modified in two significant ways. First, the equivalence of mass and energy leads us to conclude that any object with energy should also generate a gravitational effect, not only those with mass; second, that gravitational effects correspond to a modification in the curvature of space-time. The Einstein equations are the mathematical formulation of these ideas and replace Newton's expression for the gravitational force:

Mass-energy determines the curvature of space-time.

In addition, we have seen that within General Relativity the motion of objects in the presence of gravity is equivalent to force-free motion *but in a curved space*. When there are no forces and space is flat the resulting trajectory is a straight line, which represents the shortest travel-distance between two points. When the space is curved and there are no forces,[18] objects will follow a curved trajectory, and we note, without proof (since it requires rather sophisticated mathematical machinery), that this trajectory will again be such that it corresponds to the shortest distance between the initial and final points; in this sense the trajectory is "as straight as possible." Such curves that minimize the distance between two points are called *geodesics*; we can then simply state that:

In the presence of gravity all objects (including light) move along geodesics.

This statement replaces Newton's Second Law for motion in the presence of gravitational forces, $F_{\text{grav}} = ma$.[19]

When the curvature of space is small the Einstein gravitational equations and the law of motion he proposed are completely equivalent to Newton's. But for large gravitational fields or at large speeds they predict significant deviations that have been verified in all cases where data from such extreme environments is available. But the General Theory of Relativity is more than an extension of Newtonian gravity: it requires a complete revision of our understanding of space and time. The instantaneous action at a distance represented by Newton's force of gravity is replaced by a description of space-time as a dynamical object that can be perturbed by the matter–energy it contains. These perturbations propagate at the speed of light and correspond to changes in the curvature of space and time; when an object encounters such distortions its motion is modified, not by the action of a force, but because the very space in which the object resides changes.

Within General Relativity gravity is a geometric property of space (curvature) and motion under its influence is a geodesic determined by the *local* curvature of space-time. If there are no other interactions, all objects will be freely falling: freely falling objects follow geodesics. In addition we can imagine looking at a very small region of this curved space, and note that inside it the curvature of space is small, space looks essentially flat.[20] But flat space corresponds to the absence of gravity: in a small region freely falling observers will experience no gravitational effects. We have come full circle and recovered the initial motivations for the General Theory from the perspective of the geometric description of gravitation; in fact, the geometric picture can be used to derive qualitatively *and* quantitatively all properties of the gravitational field. Paraphrasing Pythagoras, gravity and geometry are one.

Black holes

Gravity affects light just as it affects rocks. We also know that we can put rocks in orbit; does it then follow that can we put light in orbit? The answer to this question is yes, but only under unusual circumstances. Ordinary gravitational effects on light are very weak, so in order to bend a beam of light into a closed path enormous gravitational forces (or, equivalently, enormous space-time curvatures) are required. These can be generated by a very massive object of very small size; an example would be an object with the same mass as our Sun (current radius: 7×10^5 km) but squashed down to a radius of less than about 3 km. A beam of light moving near such an object and in the right direction will land in an orbit around it. If you place yourself at any point in this trajectory with a flashlight on your back, some of the light from the flashlight will fall into orbit and will travel around it until it comes back to you: you would see the flashlight (and your back) right in front of your eyes! Light orbits are unstable and a slight error in position or direction will send the beam spiraling off into space or down into the compact object, but if the error is small a light beam will circle the compact object several times before falling in or flying off.[21]

We can go farther and imagine an object so massive and compact that it not only allows light orbits, but whose escape velocity is larger than the speed of light (the escape velocity is the minimum initial velocity an object needs in order to take off from the surface of a planet, star, etc. and not be pulled back again; for the Earth it is about 10 km s^{-1}); such an object is called a *black hole*. *Nothing*, not even light, can leave the neighborhood of a black hole, such a perfectly omnivorous thing will pull onto it anything careless enough to approach too closely, even its own image cannot escape.

Though the fate of any object in the immediate vicinity of a black hole is sealed (it will be devoured), the effect of a black hole on its surroundings does decrease with distance, like all gravitational effects. This means that surrounding a black hole there will be a "boundary," called the black hole *horizon*, such that objects outside the horizon can fly away (which might require a very determined effort), but anything inside the horizon is permanently trapped. For example, if we place a laser close to a black hole and point it towards the horizon, the beam never manages to cross this boundary: as it approaches the horizon the light will become redder and redder as the beam uses more and more energy in its doomed escape attempt; eventually the wavelength tends to infinity and the beam disappears (for this reason the horizon is also called the surface of infinite red-shift). No information can leak out from inside the horizon, and the ultimate fate of anything that crosses it cannot be known on the outside. In this sense the region inside a black hole's horizon is separated from our Universe, it can be affected by stuff from our side (since we can drop things into the black hole), but it cannot react back.[22]

The distance from the center of the black hole to the horizon is determined by the mass of the black hole: the larger the mass the larger the distance. For a black hole with the same mass as our Sun the horizon is about 3 km from the center; for black holes with a billion solar masses (yes, there are such things) this is increased to 3×10^9 km, about the distance from the Sun to Uranus. For such very massive black holes the horizon is so far away from the center that an observer crossing it might not realize what a life-changing step he has taken, only later, when all efforts to leave the area prove futile, will the dreadful realization of what has happened set in.

Let us now try to watch (from a safe distance) as a brave (dumb?) astronaut I will call Thor, carries out his decision to go through the horizon of the nearest black hole. As he approaches the horizon the curvature of space increases and all the effects predicted by the General Theory will come into play. Thus his clocks will tick slower compared to ours, and the light we receive from him will be red-shifted. These effects become more marked the closer he comes to the horizon. According to *our* clocks, it will take Thor's ship a longer and longer time to cover a given distance, and even his movements will appear in slow motion; eventually even the tiniest change will take, in our reference frame, years to complete. As he approaches the horizon we can imagine observing his almost frozen batting of an eyelid, and as we wait to see his eyes again we will grow older and older; we, our children and our children's children will die out, and our very civilization will perish before his eyelid opens again. Stars will be born and die out, and Thor will *still* be approaching the horizon, ever closer, ever slower, ever redder.

In our reference frame Thor takes an infinite time to reach the horizon, but in his frame this life-changing event (for him) takes a finite time. This is the most dramatic example of the relativity of time. After crossing the horizon (infinitely far into our future) Thor stays inside. But even if crossing the horizon might not be a traumatic experience for Thor, the same cannot be said for his ultimate fate. As he approaches the center of the black hole the curvature of space increases dramatically, eventually these distortions of space will stretch Thor's body beyond the resistance of his sinews and he will be dismembered by the gravitational forces.

In reality, the environment surrounding the horizon will be peaceful only for a completely isolated black hole. In realistic situations the black hole will continuously attract and absorb material from its neighborhood, and the horizon will be a region of enormous temperatures produced by violent collisions of this material as it spirals down into oblivion.

Leaving these considerations aside, a careful review of the above original description of a black hole might be construed to represent an unfalsifiable prediction of the General Theory of Relativity: how can we observe something that cannot be seen? There are, in fact, several possibilities: we can imagine observing a black hole in a field of stars that are father away. In this case, due to a gravitational lensing effect, the light of the stars will be concentrated at the rim of the horizon; this, however, is an unreliable characterization for it can be mimicked by many ordinary effects (e.g. a dust cloud between us and a region where stars are more abundant).

A practical black-hole detection technique uses the effects of a black hole on its neighbors. If the black hole is near a star or any other source of matter it will continually siphon material towards its horizon. In this process the material will reach extremely high temperatures, to the point it will start emitting X-rays, and these are expected to vary widely and randomly due to the chaotic nature of this environment. Given the timescale of these variations and the fact that nothing travels faster than light we can estimate the size of the object producing the X-rays. For example, if the fluctuations occur about every 10 ms (a realistic situation), the size of the region generating the X-rays will be about $c \times 10\ \text{ms} = 3000\ \text{km}$ (smaller than Earth).

On the other hand astronomers can see the gravitational effects on near-by stars of whatever is producing the X-ray radiation, and in favorable situations they can use these observations to estimate the mass of the beast. Knowing then the mass and size, one can decide whether this is a black hole or not by comparing the data with the predictions of General Relativity. Following this procedure several black-hole candidates have been identified; they are called "candidates" because the evidence is circumstantial, yet, short of direct observation, it is the soundest way available of identifying black holes.

The best candidate for a black hole found in this way is called Cygnus X1 (the first observed X-ray source in the constellation Cygnus, the swan). This object is a member of a double stellar system (two stars orbiting each other), its companion star is a blue super-giant whose mass can be reliably estimated. From this and the frequency by which the super-giant eclipses the X-ray source, it is possible to determine that the latter has a mass of seven times that of the Sun, which corresponds to a horizon radius of about 21 km. In addition the X-ray flux exhibits variation on a scale of 0.01 seconds, which indicates a compact object of size below 3000 km. A black hole is the only known object consistent with these properties.

The most striking examples of black-hole candidates are furnished by a class of very distant (sometimes billions of light-years away) X-ray sources which, using the variation in this flux, are known to be relatively compact (galaxy size). Then the very fact that we can see them despite the distance that separates us implies that they are extremely bright objects, so bright that we know of only one source that can fuel them: the radiation given off by matter while being swallowed by a black hole.[23] Our current picture of such an object, generically called an active galactic nucleus or AGN, is that of super-massive (a billion solar masses or so) black hole, whose horizon is roughly the size of Uranus' orbit, and that voraciously engulfs matter equivalent to several Sun-sized stars per year. The X-rays we receive are the death cries given off by victims as they cross the AGN horizon and vanish from our world, never to be heard of again.

Waves

One of Einstein's original motivations for the development of the General Theory of Relativity was to find a theory of gravity consistent with the results of the Special Theory and, in particular, with the constraint that no physical entity can travel faster than light. Within Newtonian gravity a redistribution of a group of masses results in a change in the force of gravity they produce, and this change is instantaneously propagated throughout the Universe. Within the General Theory of Relativity the same redistribution of masses results in a change in the curvature of space, and this will also propagate, but now at a *finite* speed equal to the speed of light c (I state this without proof). This effect can be pictured as a traveling ripple of space: a gravitational wave. Gravitational waves are peculiar to General Relativity without any Newtonian counterpart.

A passing gravitational wave corresponds to an itinerant change in the curvature of space. Any object lying in its path will be stretched and strained by it, since the very space it occupies is being deformed; if sufficiently strong, these effects will rip apart the object in question. Fortunately this happens only in the most extreme environments; firstly because such intense gravitational waves are produced only during the most dramatic of cosmic events, such as the collision of two black holes, and, secondly, because the intensity of the gravitational waves decreases rapidly as they move away from their source. We have not yet observed directly a gravitational wave, mainly because (fortunately) our local environment has no significant sources of gravitational waves and the ripples in space that do reach us produce extremely small effects (about one part in 10^{22}).

Nonetheless several experiment designs can reach the sensitivity required to test this unique prediction of the relativistic theory of gravitation. The basic principle is the same as the one used by Michelson and Morley to test the ether hypothesis: the idea is to send two light beams along perpendicular directions, to reflect them and then recombine them, which creates an interference pattern that shifts when in the presence of a gravitational wave. Some of these experiments[24] are currently operational, and though no gravitational waves have been observed to date, this negative result is consistent with the sensitivity of the detectors and the expected intensity of the effect (even for relatively near-by sources). Future experimental upgrades will increase the reach of the detectors to the point that even moderate sources of gravitational waves in our galaxy will become visible. The ultimate goal is to build gravitational wave telescopes with which to study the Universe; these instruments will provide an inside view into the most violent and dramatic events in the Cosmos.

We do have indirect evidence of the presence of gravitational waves through the behavior of systems composed of a pair of massive compact objects. The General Theory predicts that such double systems should lose energy by emitting gravitational waves and, as they do so, they should slowly spiral towards one another. Eventually the objects should coalesce emitting a final burst of gravitational radiation. In all double systems where sufficient data has been gathered the observations are in perfect agreement with the predictions of General Relativity, *provided* the emission of gravitational waves is included.

The first clear evidence of these time-dependent effects was obtained in the early 1980s by Taylor and Hulse (1993 Nobel-Prize winners for this work). They discovered a system where one pulsar[25] circles another compact object. Using the regular pulses received from the system, together with other physical effects (e.g. the companion object regularly eclipses the pulsar), Taylor and Hulse showed that these objects are slowly spiraling into each other, precisely as predicted. Aside from supporting the gravitational wave predictions, the analysis of Taylor and Hulse also provided the first extra-solar test of the General Theory of Relativity.

Tests of General Relativity

After Einstein first published the General Theory of Relativity there was a very strong drive to test its consequences. Einstein himself used his equations to explain a tiny discrepancy in the motion of Mercury. Soon thereafter a dramatic verification of the bending of light was obtained (despite the complications of the First World War). Much later a way of measuring wavelength changes with exquisite precision was used to verify the gravitational red-shift. Since 1916 there have been many other measurements which agree with the General Theory of Relativity to the available accuracy. Here I will describe only the "classical" tests of the theory obtained using data from Earth-bound experiments or from astronomical objects in the Solar System.

Precession of the perihelion of Mercury

Newtonian gravity provides all the necessary tools for calculating planetary orbits with great precision. For a single perfectly spherical planet moving around a perfectly spherical sun the result is, as we have seen, an ellipse. In the case of the Solar System, however, the situation is more complicated for though the main effect is still the solar attraction, all planets also attract one other, and this perturbs the perfect elliptical shape of the orbit of any one planet.[26] One of the main consequences of these perturbations is that after one orbit the planet does not quite return to the same place; as a result the orbit takes the form of an ellipse that is slowly rotating (Figure 7.11), an effect known as the *precession* of the orbit. For all planets *except* Mercury the Newtonian predictions for the rate of precession match the observations; for Mercury the observations indicate that it takes about 64 years, 104 days for its orbit to precess by a full 360°. Newton's equations predict that this should take 182 days more to complete.

Though this represents a discrepancy of less than 1 percent, it could not be accounted for. Many ad hoc fixes were devised (such as assuming there was a

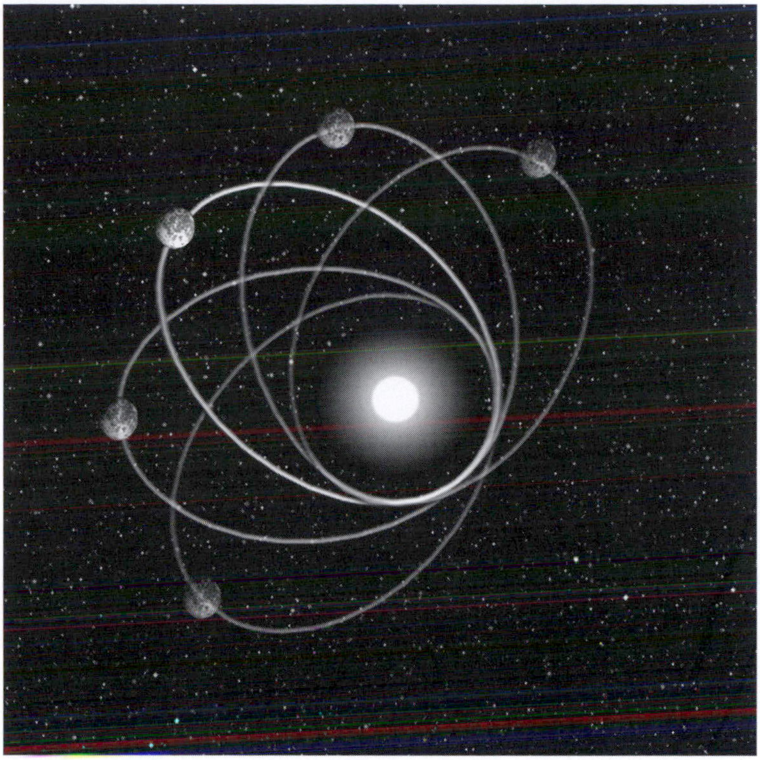

Figure 7.11 The precession of Mercury's orbit; most of the effect is due to the pull from the other planets but there is a measurable effect due to the corrections to Newton's theory predicted by the General Theory of Relativity.

certain amount of dust between the Sun and Mercury or that the Sun is not quite spherical) but none were consistent with other observations (for example, no evidence of dust was found in the region between Mercury and the Sun).

When Einstein applied the equations of General Relativity to planetary motion he found that they predicted a slight deviation from the Newtonian predictions: the effect of the Sun on the curvature of space is not quite the same as a force that decreases as the square of the distance, the main effect is still equivalent to the usual gravitational force mMG/r^2, but there is a small correction in the form of a force that decreases as $1/r^3$; its effects, small as they are, will be more noticeable for small r, that is, for the planet closest to the Sun. This relativistic correction results in an additional contribution to the rate of precession, which Einstein could evaluate as it depends only on quantities already known. The result was the shortening of the precession period by precisely 182 days: the predictions of the General Theory perfectly match the observations.

Gravitational red-shift

We saw that as light leaves a region where the gravity is strong its wavelength increases; for visible light its color will be shifted towards the red. Similarly, visible light falling into a region where the gravitational pull is larger will be shifted towards the blue. This prediction was tested at Harvard by Pound and Rebka in 1960, by looking at the shift in the color of light after it dropped from a tower 22.6 m high. This experiment requires an enormous amount of precision; the change in the gravitational force from the top to the bottom of a tower is minute, leading to a change in the wavelength of barely 0.246 trillionths of a percent; despite this, the experiment did achieve the desired sensitivity and confirmed the predictions of the General Theory.

The gravitational red-shift has also been measured by observing the light emitted from the surface of various stars, mainly the Sun; again the observations match perfectly the theoretical predictions.

Light bending

We saw above that General Relativity predicts that light paths will be bent in the presence of heavy bodies due to the deformations of space they produce. One can test this prediction by the following clever procedure.

When observing a field of stars the General Theory predicts that their paths within the Solar System will be quite straight *unless* they happen to pass close by the Sun (chosen because it produces the largest effect, being the most massive object close to us). The idea is then to compare the position of some stars at two different times chosen so that during one observation the starlight comes close to the Sun, but not during the other. Since the paths are bent in the first case and not in the second, we expect a shift in the observed positions of these stars, and the shift should be more marked the closer the observed position of the star is to the solar rim (Figure 7.12).[27]

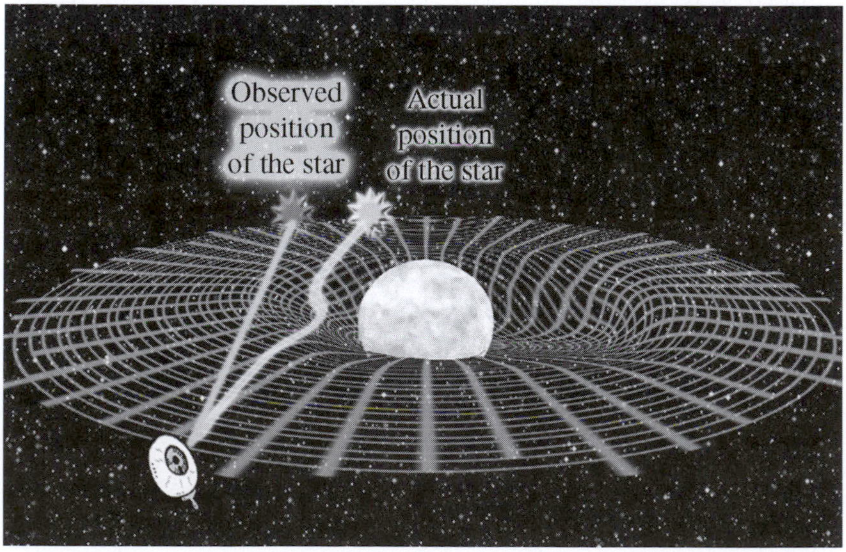

Figure 7.12 The bending of light rays due to the deformation of space produced by the Sun.

Observing stars whose light never approaches the Sun is easy: one merely observes them at night. The observation of stars whose light does travel close to the Sun is much more difficult for they are lost in the solar glare. In order to avoid this we must somehow block the sunlight, and the simplest (and most economical) way of achieving this is to wait for a solar eclipse. Since solar eclipses can be predicted many years in advance one can first determine the positions of the stars without solar interference and then make careful observations of the star field near the Sun during an eclipse.

The first such measurement was carried out by Eddington in 1919. He measured a change in the position of stars when observed near the solar rim; the observed deviation (of barely 1/2000 of a degree) again matched the predictions of the General Theory. His telegram announcing this observation for the first time marked the first dramatic verification of the General Theory of Relativity; the publicity surrounding this event caught the imagination of the public and propelled Einstein from his quiet professorial position into the limelight of public fame (and public scrutiny). Many other observations have since confirmed the presence of a gravitational shift in the stellar positions, all of them agree with the predictions of Einstein's equations (down to the experimental accuracy).

Notes

1. I ignore air resistance.
2. The reasons behind the requirement that the box be small will become clear soon.
3. As mentioned in Chapter 4, the precise statement is that these two numbers are proportional and that the constant of proportionality is the same for all bodies to this precision. Though I do not provide a proof, it is always possible to choose the gravitational constant G so that the constant of proportionality between the inertial and gravitational masses equals one; because of this I will continue to state that the inertial and gravitational masses are identical.
4. *Equivalent*: equal in value, quantity, force, power, effect, excellence, meaning, etc.; interchangeable. (*Webster's New Twentieth Century Dictionary*, 2nd edition.)
5. Remember, for example, that the plumb-line points towards the center of the Earth; a similar situation occurs here.
6. The Special Theory of Relativity is equally powerful; it is based on the one postulate that all inertial frames of reference are equivalent.
7. A gravitational force on a mass m at some location r is small if the energy gained by m when we move it from a very large distance down to r is small compared to mc^2.
8. Note that if Albert throws an object towards the bottom of the box, it will hit the floor sooner than if he had just dropped it; still the *acceleration* is the same in both cases (assuming the machine-produced acceleration of the box does not change with time). The difference in times is due to a difference in initial velocities.
9. The distance between these hypothetical stars can be estimated using standard astronomical tools.
10. Needless to say this is a very small effect, of the order of ten parts per trillion.

11. I assume that the objects coming into this region are not too heavy, so that the gravitational effects they themselves produce can be ignored.
12. The saga does not explain what drives all of them to enter the room disregarding the apparent danger, nor what power lured the various animals to participate (or how whales managed to survive on dry land).
13. To further support the claim that Roberta experiences an effect indistinguishable from gravity note that all spacecraft designed to include "artificial gravity" do so by having the ship rotate about some axis.
14. This example is chosen for convenience in visualizing the various effects described, similar results hold for surfaces of any shape.
15. I am assuming here that the moving things are not massive enough to noticeably curve space.
16. This energy does not, or course, disappear: if we reflect the beam at a certain height it will regain every bit of it as it approaches the surface. This is the same situation as when we throw a ball upwards: as the ball climbs it slows down and eventually stops, but its energy has not disappeared, and the ball will recover it as it falls.
17. For clocks to slow down by a factor of ten the gravitational pull must be 15 million times larger than the one on Earth, a person weighing 70 kg on Earth would weigh ten million tons there.
18. When space is not flat there *is* a gravitational effect, but this is a property of space itself, in this case "forces" refers to any non-gravitational interaction (electric, magnetic, etc.).
19. In the presence of other forces objects deviate from geodesic motion.
20. A similar effect is observed here on Earth, whose inhabitants do not perceive its curvature.
21. In contrast, usual planetary orbits are stable: if the planet is perturbed from its orbit (by, for example, a small meteorite crashing onto it), it will change its orbit, but it will continue to circle the central star; only a stupendous and very well-aimed collision can send a planet crashing down unto the sun or off into space.
22. This is no longer true once quantum effects are taken into account, as described in the final chapter.
23. This is much more efficient than nuclear power which would be incapable of driving such bright sources given their size.
24. The rather colorful acronyms are: LIGO, GEO, VIRGO, EGO, AMA, and TAIGO.
25. Pulsars are the compact remnants of massive stars after they undergo a supernova explosion. They have diameters of about 10 km and are several times more massive than the Sun. These objects rotate rapidly and very regularly while emitting a narrow beam of electromagnetic radiation. Every time the beam shines on the Earth appropriate receivers will register a pulse (hence the name for these objects).
26. In addition, planets are not perfectly spherical, but this affects mainly their rotation.
27. The star itself is, of course, not close to the Sun; this merely implies that the starlight passes close by the edge of the Sun.

8
The relativistic universe

Introduction

Our everyday life is strongly affected by a variety of forces: friction, electric, magnetic, etc. But outside Earth's protective cocoon, in the Universe at large, there is only one predominant force: gravity. It is gravity that determines the structure of stars, and the gathering of galaxies, and the shape of space. It was then natural for Einstein to apply his newly discovered General Theory of Relativity to the study of the Universe at large; scarcely one year after publishing his seminal paper on gravity, he published the first of such investigations, and gave birth to modern relativistic cosmology.

The aim of cosmology is to understand the structure and evolution of the Universe. This is one of the most ambitious projects ever envisaged: to infer the past history of the Universe back to its very inception, to predict its future 'til the end of time, and to understand its behavior and properties throughout its history. This, however, does not mean cosmologists aim at a complete understanding of the whole of the Universe in all its details (which is preposterous and quite outside our capabilities: for example, it is certainly impossible to infer from the properties of the early Universe that you, the reader, will scoff at the idea precisely at the time you did). Modern cosmology aims instead at an understanding of the evolution of the *average* properties of the Universe. A possible parallel is the following: suppose one is interested in the evolution of forests, then one does not consider the specific growth of every single tree in every forest, but one chooses instead certain indicators of growth and change, health, etc. (such as the number of live trees of a certain species and age), and then one measures how these indicators change under varying circumstances (earthquakes, draughts, monsoons, global warming, etc.). As a result one hopes to generate a theory that will allow us to predict how forests are born, grow, and eventually die, without hoping (or intending) to predict how any one individual tree will develop. These predictions will be reliable only provided the indicators are chosen wisely; and this is as true in cosmology as it is in forestry.

The goal of modern cosmology is then to determine the evolution of an average region of space; to extract from present data information about its past history, and to predict its future behavior. Some of these goals have been admirably completed (for example, we have a solid understanding of the average behavior of the Universe's last 13.5 billion years or so). Others, such as being able to predict the fate of the Universe in the distant future, are currently being investigated, with the likely possibility that we will be able to achieve them within the next few decades. And yet others, such as the behavior of the Universe in its earliest infancy, are still shrouded in mystery; though there are a variety of promising hypotheses and experiments planned to address this issue.

The Universe at a glance

When observing the night sky through a telescope, the field of vision presents a variety of luminous objects of different shapes and colors, separated by, apparently, the darkest of voids. After considerable study these images are now understood to represent a universe filled with galaxies each containing a billion suns (give or take a few million) tightly bound by their mutual gravitational attraction. This allows us to regard galaxies as solid objects of a given mass (in the same way that when you look at the gravitational pull of the Earth on the Moon you don't have to worry about the fact that they are made of atoms; the stars are the "atoms" which make up galaxies). Our Solar System is a member of one such galaxy, the Milky Way, as are all our neighboring stars.

Galaxies are bound to each other through gravitational attraction, forming galactic clusters. The Milky Way, together with the Andromeda Galaxy and about two dozen smaller ones, is a member of a cluster called (rather unimaginatively) the Local Group. Finally, clusters themselves are often bound into superclusters.

In this simplified picture the Universe consists of a dusting of galaxy groups that shine in visible light. There are also objects that can be seen only using detectors sensitive to other types of electromagnetic radiation: gas clouds that shine infrared radiation and super-massive black holes whose environments emit X-rays. And all these objects are submerged in a sea of weak electromagnetic radiation and low-energy neutrinos (discussed in the following chapter), both of which are remnants from the Universe's early days.

Curiously enough, all of this matter and radiation might be but a minority population in the universal census: there are strong indications that the Universe also contains an enormous amount of non-shining matter whose detailed properties are still unknown, but whose presence can be inferred through its gravitational effects. If this is confirmed this *dark matter* would represent a silent majority of colossal proportions.

Detailed observations show that the Milky Way is an average galaxy both in size and properties, and that the Local Group is also an average cluster. This suggests, as an extension of the Copernican view of the Cosmos, that most locations are equivalent with respect to the average properties of the Universe. In

Figure 8.1 Homogeneity and isotropy. The configuration on the left is homogeneous but not isotropic. The one in the center is isotropic, if one stands at the center, but not homogeneous. The last represents a homogeneous and isotropic ocean (by A. Dore).

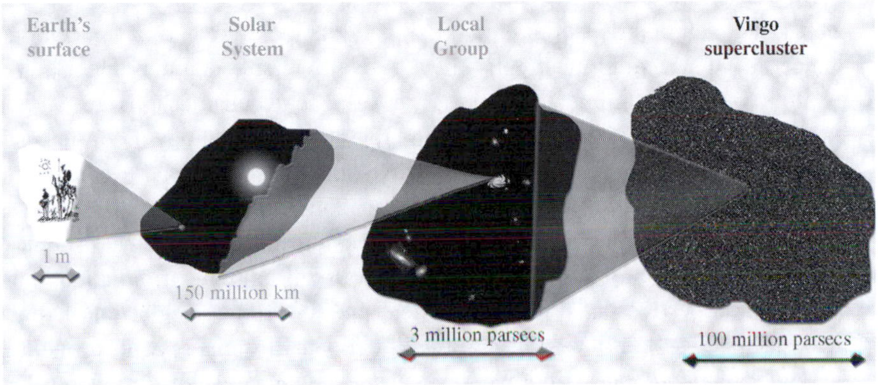

Figure 8.2 Illustration of the relative sizes of objects in the Cosmos.

other words, *on average* the Universe looks the same in every location and in every direction. In other words, that the Universe is *homogeneous* (from the Greek: *homo*-, the same, + *genos*, kind) and *isotropic* (from the Greek: *isos*-, equal, + *trope*, a turning) (see Figure 8.1).

This hypothesis, known as the *Cosmological Principle*, was originally formulated by Einstein, and is amenable to a measure of experimental verification. Though it is impossible to measure *all* possible qualities of the Universe at *every* location, we *can* select a number of accessible properties that can be precisely determined, and use the most sensitive instruments available to do so, probing as deep into space as we can. Using this admittedly limited set of data, it is found that, when averaging properties over distances of the order of 50 megaparsecs (1.6×10^{21} km), the Cosmological Principle does describe the observations (Figure 8.2).[1] Quantitatively we can then say that modern cosmology aims at the description of the average properties of the Universe, that is, at the properties and evolution of regions whose current size is about 100 megaparsecs.

The Cosmological Principle has allowed cosmologists to make specific predictions that have been subsequently verified by observations. These results have lent additional support to this hypothesis, so that the average homogeneity and isotropy of the Universe is now generally accepted. Had the Cosmological

Principle not been verified, the corresponding calculations would have been unmanageable. However, we will see below that this can be too much of a good thing: in a sense the Universe is too homogeneous and isotropic, to the point that the standard relativistic cosmology has difficulties accounting for its degree of uniformity.

The relativistic universe

Once the Cosmological Principle is adopted, the evolution of the Universe is determined by the initial conditions at some time in the distant past, and by the energy density and the pressure of matter and radiation. These properties of matter and radiation depend on the density and temperature and have been the subject of intense study. Currently they are well understood even under quite extreme conditions (temperatures as high as 10^{16} K and densities as large as 10^{17} that of water at sea level). These results are relevant in cosmology because these types of conditions prevailed in the young Universe. Justifying these statements is somewhat involved; I will have much to say about this in the sections below.

In his 1917 paper on cosmology Einstein considered a universe that is homogeneous and isotropic, containing a relatively small amount of radiation, and where matter behaves as a sparse gas whose pressure can be neglected (though the last two are certainly very strong simplifications, Einstein's results provide a very good qualitative description of the behavior of the Universe during most of its life). Of the various results he obtained one feature was completely unexpected: the solutions to the equations represented universes that were either expanding or contracting; there were *no* solutions representing a static universe. The perception that the heavens are unchangeable (a belief held dear since Aristotelian times and which had survived all the upheavals of the Copernican revolution and the Newtonian advances) is *inconsistent* with the simplest cosmological application of the General Theory of Relativity. It is only due to the very short span of human life that we see little change in the sky.

When one talks about an expanding universe, a universe in which all galaxies are moving farther and farther apart, the immediate picture is that in the distant past all the matter in the universe was originally concentrated in a very small region, and that this initial primordial egg suffered a stupendous explosion which set everything flying off in all directions. This description, however, is quite wrong.

The expansion of our relativistic universe does indeed indicate that in the distant past all of matter congregated. But in doing so matter did not congregate alone, but did so shrouded in space, for space-time and matter are indissolubly linked; the large density of matter in the distant past curled space around it, so that the *whole* of the Universe, matter, energy *and* space coalesced. The Einstein's equations do *not* describe the expansion of matter into space, but the linked spreading of space and matter. Space, as a dynamic component of the Universe, also evolves.

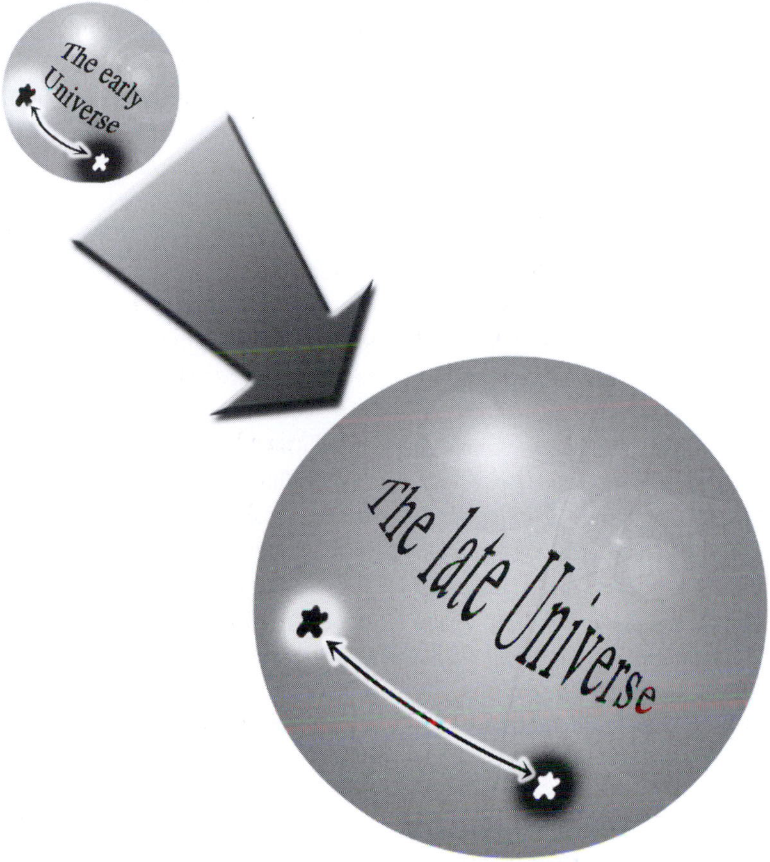

Figure 8.3 The expansion of the Universe as described by the General Theory of Relativity.

A consequence of this result is that the distance between any two galaxies (separated by 10^9 light-years or more in order to insure that the local effects are not important) will increase with time, not as a result of some repulsive force, but as a consequence of the expansion of the space that contains them. Like raisins in a cake they move apart because the dough that contains them rises. The standard parallel is that of two dots on a balloon, whose distance increases as the balloon inflates; the dots correspond to the galaxies and the balloon to the space that contains them (Figure 8.3).

Having obtained these non-static solutions (effectively the predictions based on the hypotheses of homogeneity, isotropy, and of a sparse universe), Einstein looked for observational confirmation. But the data available at the time provided no support for this result. Now we know that this is because at the time most observations measured the motion of stars within our galaxy,[2] while the effects predicted by Einstein are best seen when considering widely separated objects. In 1917, however, there was no reason to suppose the data was so biased.

Because of this Einstein felt compelled to modify his equations in such a way as to allow solutions describing a static universe, while still preserving the geometrical character of the theory. There is, in fact, a unique way of doing this (though this was proved much later), and consists in assuming that the curvature of space-time is determined not only by the mass–energy content of the universe, but, in addition, by a universal quantity we now call the *Cosmological Constant*. The presence of the Cosmological Constant is equivalent to the assumption that the universe is permeated with a uniform energy density and an associated pressure which can be adjusted so as to precisely balance (precariously) the expansionist tendencies of the universe.

The balancing of the universal expansion using the Cosmological Constant is a rather delicate proposition. If the latter is a bit too large then the universe soon re-contracts, if too small the expansion proceeds. In other words, Einstein's static universe (one that does not expand or contract) is *unstable*. Should this be an accurate description of our Universe, a slight perturbation would tilt the balance and send the Cosmos into a contracting spiral or into perpetual expansion; in such a universe anyone and everyone should tread very lightly, for they would tread on everyone's dreams.

Einstein was never too fond of the Cosmological Constant, and was happy to dispose of it once observations indicated that the Universe is, in fact, expanding. But, like the proverbial genie, once the Cosmological Constant was out of the bottle it was hard to get rid of, and so it lives on, even if it was disowned by its creator.

Building a universe

Einstein's 1917 paper generated a renewed interest in cosmology. By 1929 all solutions describing a homogeneous and isotropic universe had been completely characterized through the works of A. Friedmann, G. Lemaître, H. Robertson, A. Walker, R. Tolman, and others. The results are remarkably succinct: any universe satisfying the Cosmological Principle falls in one of only three categories, depending only on the total amount of mass–energy, including a possible contribution from the Cosmological Constant (Figure 8.4):

- If the total matter–energy density[3] lies below some critical value, space will be the three-dimensional equivalent of a saddle-shaped surface and is referred to as an *open* universe, for reasons that will become clear below. A distinguishing characteristic of this space is that the angles of a triangle add up to a number *below* 180°.
- If the total density of matter–energy equals the critical value then the space is the three-dimensional analogue of a plane, with the peculiarity that the angles of a triangle add up to precisely 180°. In this case the universe is *flat*.
- Finally, if the matter–energy density lies above the critical value, space will be the three-dimensional analogue of a sphere, and is characterized by the property that the angles of a triangle add up to a number *larger* than 180°. This is called a *closed* universe.

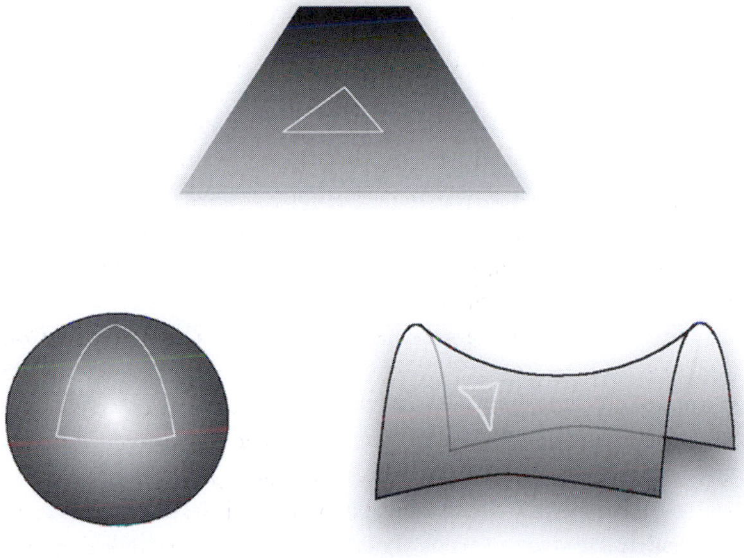

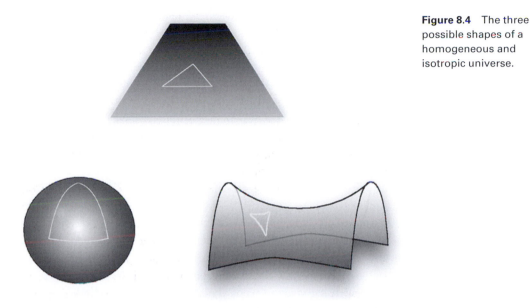

Figure 8.4 The three possible shapes of a homogeneous and isotropic universe.

We have already met all of these when considering the description of gravity as the curvature of space.

The geometry of these solutions depends only on the requirement that they describe homogeneous and isotropic universes, and it is quite independent of the Einstein equations. These equations describe the dynamics of the universe, how it evolves and its relationship with the matter and energy it contains. One of the most important results derived in modern cosmology is that under most circumstances the size of the universe *changes*. It is not only that stars are born and die, and during this evolution planetary systems are created and disappear, but the universe as a whole, including space, evolves. Nothing but change is permanent.

As the universe expands the matter and energy it contains will be dispersed into a larger and larger volume, and the mass–energy density will decrease. If in the distant future the universe should stop this expansion and start re-collapsing, the mass–energy density will then increase with time. One might therefore get the impression that early in its history the mass–energy density must have been larger than critical and the Universe was closed and sphere-like, then for a brief instant it was flat, and soon it became open (saddle-like), and will stay that way at least as long as the expansion continues. This, however, is *not* the case since the critical value of the density is not constant but depends on both the size of the universe and its rate of expansion or contraction. As the universe evolves the value of the critical density changes but in such a way that the geometry is preserved: if the mass–energy density is larger than critical it stays that way at all times and the universe will be forever closed, if it lies above its critical value the universe will be forever open, and if the mass–energy density attains its critical value at one time it will always be flat.

Box 8.1

Science and nature

The world is very complicated and it is clearly impossible for the human mind to understand it completely. Man has therefore devised an artifice that permits the complicated nature of the world to be blamed on something that is called accidental, and thus permits him to abstract a domain in which simple laws can be found. The complications are called the initial conditions, the domain of regularities is called the laws of nature. (E. Wigner, *Symmetries and Reflections*.)

The above considerations indicate that the mass–energy density plays a key role in determining the properties of our Universe. This is indeed the case, and because of this a large number of experiments have been devoted to measuring the actual value of this density in comparison to the critical one. This is a difficult enterprise since all types of matter and all types of radiation contribute, and they all must be taken into account. In addition, it is possible for there to be a non-zero Cosmological Constant that would then provide yet another contribution. The best current estimates of the critical density give a value of about 10^{-29} g cm^{-3} (which corresponds to having the mass of one Moon in each cubic parsec throughout the Universe); the mass–energy content of the Universe is about 70 percent of this value. The fact that the density has a value relatively close to critical has some interesting consequences that I describe later in this chapter.

The mass–energy density determines the shape of the universe, but the change in the size of the universe, its possible growth or contraction, is determined also by the way in which matter and energy react to compression and expansion, that is, to the *pressure* generated by the contents of the universe. The effects of pressure are rather complicated; on one hand a large pressure will tend to spread out the contents of the universe leading to a decrease in the energy density. But on the other hand a large pressure is also associated with a large amount of energy, this generates large gravitational effects which then *oppose* further expansion (the equations of General Relativity automatically take into account both of these effects).

In summary, General Relativity supplemented by the Cosmological Principle completely specifies the evolution and average properties of the Universe once the following ingredients are supplied:

- The total matter–energy density.
- The relationship between the energy density and the pressure generated by the material.
- The initial conditions.

The resulting predictions of this theory are amenable to experimental confirmation and are currently supported by all available data.

Before proceeding it is worth commenting on the role played by the initial conditions. The equations of General Relativity are of a type that determine the evolution of a system (be it a planet moving around a star or the Universe as a whole) in terms of the conditions of the system at one specific time. For example, if we know where the planet is and how fast it was moving last Monday, the equations determine how it will move in the future. The equations, however, do *not* determine the conditions that led the planet to be at the specific initial location moving with its specific initial velocity.

In the same way General Relativity, in conjunction with the amount and properties of matter, determines the evolution of the Universe given its configuration at some specific time. And, in fact, we now know which initial conditions in the very early Universe led (through the application of the equations of relativistic gravitation) to a universe with characteristics very close to the ones we observe. The validity of the theory is tested by comparing the predictions obtained in this way with the available data, but does not address the processes that gave rise to the initial configuration. Still, it is quite reasonable to ask what physical effects in the early Universe were responsible for generating the precise initial configuration that gave rise to the world around us. This is an interesting issue to which we have no definite answer (yet) and which is currently hotly debated.

A gaggle of universes

All universes consistent with the Cosmological Principle have been fully classified, but before going into a description of the results it is perhaps useful to note the significance of these investigations. There is (as far as we know) one Universe, which is presumably characterized by the properties of the matter it contains and, perhaps, by a Cosmological Constant. Its evolution is determined by whatever dynamics rules the Cosmos given a set of initial conditions. Should General Relativity be an accurate description of Nature, there will be a solution to Einstein's equations that closely resembles the real Cosmos. Unfortunately this has proved a difficult hypothesis to test, since data pertaining to the average properties of the Universe are hard to come by. More precisely, though astronomical observations are legion, they include effects that are due to the average properties of the Universe *and* others that are due to localized deviations from this average behavior. Separating these two effects has proven difficult.

Faced with a paucity of reliable data, cosmologists opted for studying *all* possibilities (Figure 8.5). In this way, once the measurements of the matter properties, the Cosmological Constant, etc. became sufficiently accurate, one could simply select the appropriate solution to the equations and compare the predictions of the theory with other observations. In the process of considering all these possibilities the role played by the various ingredients was significantly clarified. For example, a large Cosmological Constant produces a very

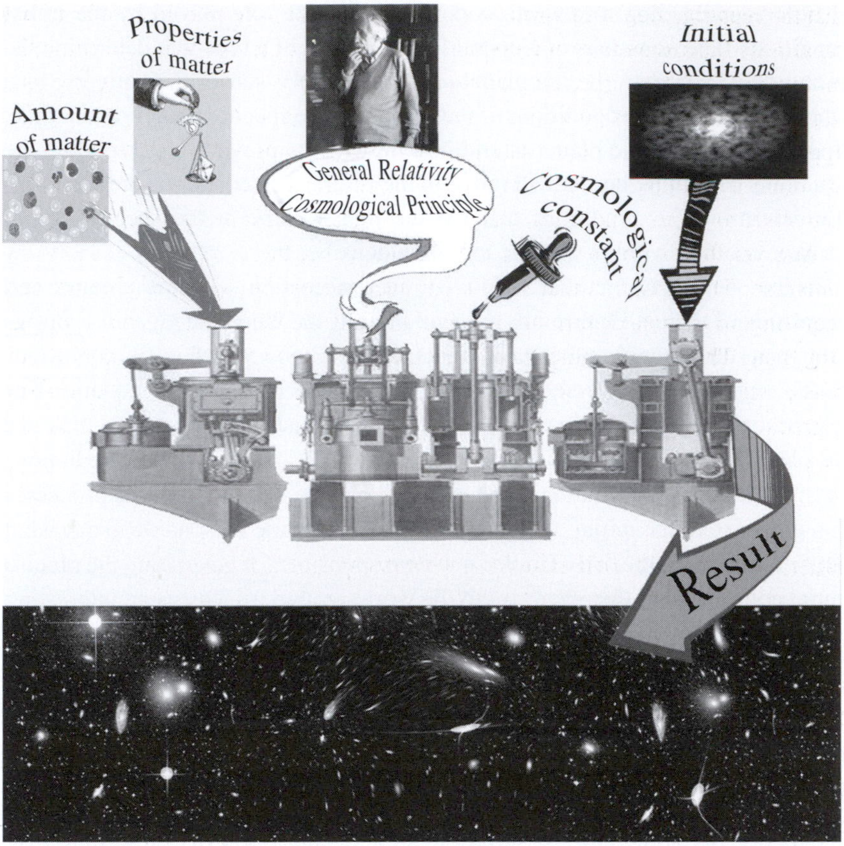

Figure 8.5 A depiction of the process through which the General Theory of Relativity, complemented by the Comological Principle, provides a description of the Universe.

rapid change in the size of the universe; if its effects dominate over those of matter then this change is exponential. If, on the other hand, the Cosmological Constant is small or zero the universe will expand or contract more sedately. For particular values of the mass–energy density and the Cosmological Constant the universe can be arranged to be static, or to linger without expansion for a considerable time.

The effects of matter and radiation can be guessed using the old Newtonian Theory of Gravity and Special Relativity. Newton's equations indicate that gravity is a universal *attractive* force, while Special Relativity leads to the equivalence of mass and energy. Combining these two results suggests that the general effect of mass and energy is towards *contraction*. This expectation is verified by explicit calculations: matter effects oppose expansion and foment contraction. For example, if the mass–energy density lies above its critical value and the Cosmological Constant is not too large, then an expanding universe will necessarily re-collapse.

There is by now a "benchmark" universe, that is, one where the assumed matter–energy density, the pressure, and the Cosmological Constant are such that the resulting behavior agrees closely with all the available data. It describes an almost flat expanding universe and is known as the Big Bang Theory (the motivation for this peculiar name is given below). It is noteworthy that the current data strongly favors a non-zero Cosmological Constant: the genie appears to have settled down out of the bottle. The goals of the Big Bang Theory are quite ambitious: to describe the evolution of the average properties of the Universe, to understand its past and to be able to predict its future.

And yet this is not, and does not pretend to be, the complete theory of the Universe. The reason is that such a complete theory would be self-contained, requiring no external information, while in order to obtain predictions from the Big Bang Theory one must supply initial conditions and the relation between mass, energy, and pressure. Strictly speaking the origin of these additional ingredients lies outside the scope of the Big Bang Theory, but one can still ask, and in this way cosmology advances.

Hubble's result: the Big Bang Theory

The uncertainty as to which (if any) of the many solutions describing homogeneous and isotropic universes describes *our* Universe was significantly reduced with the publication in 1929 of E. Hubble's results showing that our Universe is *not* static, but *expanding*. Hubble's measurements demonstrated that distant galaxies are receding from the Milky Way, and their speeds are proportional to the distance that separates us (Figure 8.6). This result agrees with the predictions of General Relativity for an expanding homogeneous and isotropic universe and provided a dramatic confirmation of the applicability of this theory to the Universe at large.

In order to understand the connection between Hubble's plots and the expansion of the Universe, it is important to recall that the motion of distant galaxies away from the Milky Way is not due to some repulsive force generated by our local environment, nor is it a consequence of some sort of cosmic conspiracy aimed at ostracizing us. What the Hubble plots do show is that space itself is increasing in size, and carrying with it the various galaxies it contains (Figure 8.3). Thus the average separation between *all* galaxies steadily increases. An observer in some galaxy in the Hydra cluster, 4 billion light-years away[4] from here would find all distant galaxies receding from him (it?) just as we see all distant galaxies receding from us. This average expansion of the Universe is commonly referred to as the *Hubble Flow*.

The results of Hubble can be expressed by the relation:

$$\nu = H_0 d$$

where ν denotes the speed of a galaxy and d its distance from Earth; this is refered to as *Hubble's Law*.[5] The number H_0 is called *Hubble's Constant* and is measured in units of *1/time*; the current best value for $1/H_0$ is 14 billion years.

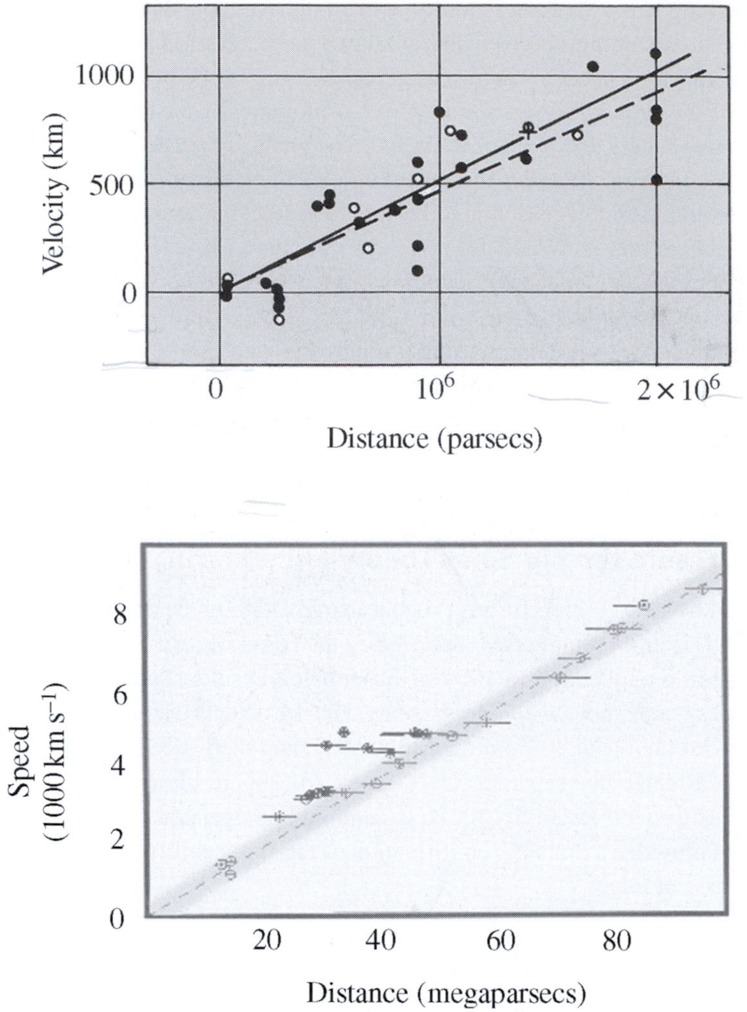

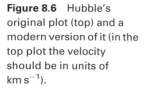

Figure 8.6 Hubble's original plot (top) and a modern version of it (in the top plot the velocity should be in units of km s^{-1}).

The Einstein equations show that an observer repeating Hubble's measurements at some other time would also find that the velocity v is proportional to the separation d, but the constant of proportionality would be *different*: earlier in the history of the Universe $1/H_0$ would be smaller, and later it would be larger: $1/H_0$ *increases* with the age of the Universe. Despite its name Hubble's Constant is not constant. Perhaps it should instead be called Hubble's Parameter, but by now the label is so widespread there is no hope of changing it.

The Big Bang Theory describes a universe that is growing steadily larger. According to the General Theory of Relativity, a homogeneous and isotropic universe with this property must have originated in a state of enormous density and temperature where all of space and its contents were reduced to a very small region. This initial configuration and the ensuing expansion are called the

Santa Clara County Library District
408-293-2326

Checked Out Items 8/6/2019 19:37
XXXXXXXXXX4423

Item Title	Due Date
1. Space-time, relativity, and cosmology 33305212321701	8/27/2019
2. Dream kitchens : recipes and ideas for modern kitchens. 33305206681896	8/27/2019
3. New kitchen idea book 33305209226541	8/27/2019
4. Small houses : big ideas for today's small homes 33305245056266	8/27/2019
5. Ideas for great kitchens 33305211592781	8/27/2019
6. Work surfaces and countertops 33305211577386	8/27/2019
7. Before _after kitchen makeovers 33305212488138	8/27/2019

No of Items: 7

24/7 Telecirc: 800-471-0991
www.sccl.org
Thank you for visiting our library.

Santa Clara County Library District
408-293-2326

Checked Out Items 8/6/2019 19:37
XXXXXXXXXX4423

Item Title	Due Date
1. Space-time, relativity, and cosmology 33305212321701	8/27/2019
2. Dream kitchens : recipes and ideas for modern kitchens 33305206681896	8/27/2019
3. New kitchen idea book 33305209226541	8/27/2019
4. Small houses : big ideas for today's small homes 33305245056266	8/27/2019
5. Ideas for great kitchens 33305211592781	8/27/2019
6. Work surfaces and countertops 33305211577386	8/27/2019
7. Before after kitchen makeovers 33305212488138	8/27/2019

No of Items: 7

24/7 Telecirc: 800-471-0991
www.sccl.org
Thank you for visiting our library.

Box 8.2

Hubble's Law

The relation $v = H_0 d$ is a consequence of the isotropy of the Universe. To see this imagine three observers I will call **L**, **C**, and **R** as the figure below and so placed that the distance from **L** to **C** is the same as that from **C** to **R**.

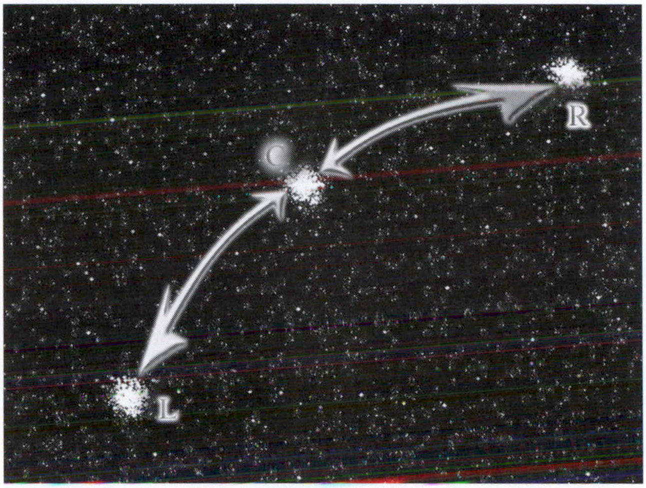

Now, suppose that the speed is proportional to the separation and that **L** finds **C** receding from her at a speed v. Then, since **R** is twice as far as **C**, **L** finds that the speed of **R** is $2v$ in the same direction as **C**. Combining these results it follows that **C** will find **L** receding from him at speed v and **R** receding from him at speed $2v - v = v$. This is in agreement with an isotropic universe, for the speed of recession is the same for both **L** and **R** as seen by **C**.

Suppose now that the speed depended on the *square* of the separation, then the speed of **R** with respect to **L** should be four times larger than that of **C** with respect to **L**. Then **C** finds **L** receding at speed v while **R** recedes at speed $4v - v = 3v$. In this case equivalent locations (**L** and **R**) recede from **C** at different speeds, which is *inconsistent* with the requirement of isotropy.

The Hubble flow determines a particularly convenient reference frame for studying the average properties of the Universe, namely, the frame in which the average speed of the galaxies is proportional to their separation. To a good approximation this is the same reference frame an average observer on an average galaxy would naturally use (unless the galaxy has large peculiar motions produced by an unusually active environment).

Big Bang,[6] a momentous event denoting the birth of our Universe, matter, energy, and space.

The Universe has been expanding since the Big Bang, and the quantity $1/H_0$ provides a rough estimate of the time separating us from that event, a lapse commonly referred to as the age of the Universe. Our Universe is then roughly 14 billion years old (actual calculations yield a slightly smaller value, 13.5 billion years). It is worth noting that Hubble's initial measurements gave a much smaller value, about 2 billion years, which was a serious problem, for many stars were known to be older than that; this was referred to as the "age problem." Initial proposed solutions to the age problem included a revival of the Cosmological Constant, whose presence would change the relation between H_0 and the age of the Universe. It took more than 40 years for the techniques of measuring cosmological distances and speed (see the appendix to this chapter) to reach the precision required for an accurate value of H_0; the current value of H_0 presents no age problem.

It is now generally accepted that we live in an expanding universe, and this realization naturally raises a desire to understand what has happened during the last 13.5 billion years (or at least the highlights), and what will happen to the Universe in the future. I now turn to the answers the Big Bang hypothesis gives to these questions.

The distant past: the Big Bang confirmed

The current state of expansion indicates that the Universe was smaller and denser in the past. If we imagine traveling back in time, we would see all galaxies coming closer together as space contracts and gets warped with increasingly intense gravitational effects. As we approach the Big Bang the average density and temperature of the Universe increase dramatically, eventually reaching billions and then trillions of degrees. Now, instead of going back all the way

Box 8.3

The Relativistic Heavy Ion Collider (RHIC)

Situated in the Brookhaven National Laboratory (New York) it studies the properties of matter at the highest temperatures yet achieved (10^{12} K). This is obtained through collisions of gold nuclei moving at 99.95 percent of the speed of light. The resulting fireball dissipates very rapidly by emitting a plethora of particles that carry information about the extreme environment that created them. The data from RHIC presents a diorama of the very early Universe: in a region smaller than 10^{-14} m^3 the early history of the Cosmos comes back to life.

to the Big Bang, let us for the moment stop at the time of 10^{-11} seconds after the Big Bang and follow the evolution of the Universe from that time on (later on I will talk about yet earlier times).

At this early epoch the density and temperature of the initial fireball were so high that all matter dissociated into a relatively small set of basic constituents. This happens because under these conditions collisions among bits of matter are very frequent and very violent, and as a result all but the sturdiest objects are broken up. As the temperature dropped so did the frequency and violence of the collisions and, eventually, composite structures were able to form (and survive). Thus, if we had been able to film the contents of the Universe as it cooled and then run the film backwards, we would first see atoms which are then broken apart into nuclei and electrons by the intense heat, then we would see the nuclei themselves decomposing into protons and neutrons, then the protons and neutrons decomposing into quarks.[7]

The storyline I present below is a simplified version of the *predictions* derived from General Relativity and the theory of matter at very high temperatures and densities. I will comment on the experimental evidence (or lack thereof) pertinent to the most important events in this history. All current observations are consistent with the predictions of the Big-Bang hypothesis which has now become the standard theory describing the Universe, its average properties, and its evolution.

The time before nuclei

The very early Universe (10^{-11} s after the Big Bang) presented an epoch of immense densities and pressures. In this chaotic environment no structure larger than 10^{-19} m in size could survive for any appreciable length of time and all matter was reduced to a super-hot "soup" of subatomic particles, quarks, electrons, neutrinos,[8] immersed in a sea of high-energy radiation. However, by the time the Universe reached the age of 100 microseconds, its temperature dropped to a (relatively) balmy 1.5×10^{12} K, and this allowed the formation of the basic constituents of our everyday matter: neutrons and protons.[9]

We do not have any direct experimental evidence dating from this period since the many transformations the Universe has since undergone have obliterated any relics from this era. What is being currently investigated instead is a recreation of this environment on a very small scale: two large atomic nuclei moving in opposite directions at speeds close to that of light are made to collide head on, creating a region whose density and temperature approach those of the early Universe. The observations derived from such experiments will allow us to understand the details of the processes that gave birth to the nuclear constituents.

Exit the neutrinos

Among the primordial soup of subatomic particles the Universe contained a very large number of neutrinos (a name that means "little neutral one"). These are very light and, under current conditions, very weakly interacting particles, that are produced by a variety of nuclear processes (more precisely by many processes involving the so-called "weak force").[10] Despite their weak interactions, the enormous densities of the very early Universe insured that even neutrinos scattered often, and were frequently absorbed and emitted in a variety of high-energy reactions.

But by the time the Universe was 1 s old, and its temperature had dropped to about 9 billion K, these effects ceased: the Universe had become too sparse for the neutrinos to experience significant interactions. The jargon phrase describing this change is that the Universe became "transparent" to the neutrinos, and these "decoupled." From this time on these neutrinos have been traveling unimpeded throughout space, being affected only by gravity.

Initially these early Universe neutrinos had very high energies and were moving at speeds close to c, in accordance with the high-temperature environment they inhabited. In addition, they experienced very strong gravitational effects, for the Universe was then very small and its curvature correspondingly large. Over the billions of years following their decoupling, these neutrinos witnessed a significant decrease in the gravitational effects, they have, so to speak, "traveled" from an environment where gravitational effects were very strong, to the present, where they are quite weak (on average). In this process the neutrinos' speed has been significantly reduced (to less than 10 percent of its initial value – depending on their mass); the corresponding energy loss is by a dramatic factor of 1/4 700 000 000. This is an effect related to the one observed when throwing a ball upward: when it starts the ball is close to the ground where the force of gravity is relatively large and its speed is also large, as it moves upward its speed decreases as it moves into regions where gravity is weaker.

The Big Bang Theory then predicts that the current Universe is bathed in a sea of slowly moving neutrinos, the earliest relics of the Universe's violent past. Unfortunately current technology is not sensitive enough to be able to detect these neutrinos, should future experiments overcome this difficulty, they will obtain images of the Universe when it was in its most tender infancy.

Enter the matter

At the respectable age of 200 seconds after the Big Bang, the temperature of the Universe had dropped to a mere 760 million K and the first atomic nuclei began to form. The first nucleus to form was that of deuterium, an isotope of hydrogen;[11] soon thereafter collisions among the deuterium nuclei created helium and lithium (and other unstable elements such as tritium). In this way the lightest elements appeared in the early Universe, and one might think that this process then continued along similar lines, populating all the slots in the periodic table, but this is not the case.

Heavier elements could have been created only through collisions where the components of participating helium nuclei are rearranged. These reactions, however, can occur only at very large temperatures (of the order of 350 million K), since the helium nucleus is very stable and requires much violence to break it apart. All the time helium was being "cooked," the universe kept expanding and cooling, and as a result, once the helium was ready, the temperature was too low to allow for the creation of heavier elements. By the end of this period, less than 10 minutes after the Big Bang, the manufacture of elements in the early Universe stopped. The nuclear constituency was very limited: 75 percent hydrogen, about 25 percent helium and trace amounts of deuterium, lithium and boron; heavier elements were not formed in the early Universe at all, but were the result of stellar evolution (Chapter 9). It is a remarkable success of the Big Bang Theory that these predictions agree with the observations.

Roughly 13.5 billion years ago most of the deuterium and helium present in our current Universe was created through the above processes. In the intervening eons the amount of helium has changed somewhat through nuclear reactions inside stars; deuterium, on the other hand, is *not* manufactured in stars (though it is destroyed there). As a result almost all the helium, and all the deuterium we observe today were created a few minutes after the Big Bang: these are the ultimate fossils.

A blast from the past

Over the following millennia matter in the Universe continued to be a mix of neutrinos (that are electrically neutral), light-element nuclei and electrons (both of which carry electrical charge), all submerged in a sea of radiation. With temperatures and densities still too high to allow atoms to form, all charged particles continually scattered, absorbing and emitting electromagnetic radiation. But once the Universe cooled below about 60 000 K, which happened about 350 000 years after the Big Bang, the conditions became less extreme and nuclei began to capture electrons and *keep* them, rapidly leading to the formation of neutral atoms (not that charge disappeared; it is just that positive and negative charges became bound tightly together).

A neutral environment has very little effect on electromagnetic radiation, for only charged particles are strongly coupled to it. As atoms formed, radiation and matter stopped interacting and parted ways, they *decoupled*, and the Universe became transparent to the radiation. From that time on this radiation has been traveling through the Universe essentially unimpeded (Figure 8.8), just as neutrinos had done a third of a million years earlier. The radiation that decoupled

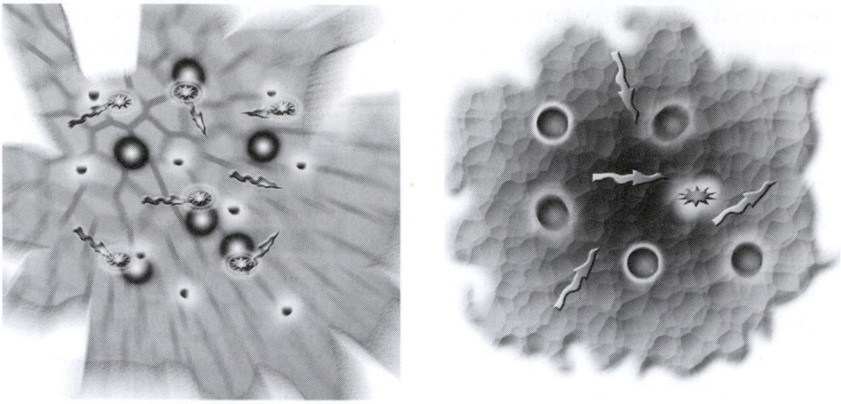

Figure 8.8 Left: at temperatures above 60 000 K atoms are dissociated into nuclei (large balls) and electrons (small balls) that interact strongly with radiation (arrows). Right: at temperatures below 60 000 K atoms can form, and these interact only weakly with the radiation.

Box 8.4

Cosmic background radiation

The discovery of the cosmic background radiation marks a curious coincidence of theoretical ideas and experimental results. In 1931 Lemaître first speculated on the possibility that the Universe is permeated by relic radiation from the early Universe; by 1948 G. Gamow predicted the existence of this radiation and argued that its main component should be in the form of microwaves. These calculations were revisited by a Princeton group of cosmologists in 1965 who confirmed the main results of Gamow.

At that time, in the Bell Laboratories at Homedel (30 miles from Princeton), the astronomers A. A. Penzias and R. W. Wilson (1978 Nobel laureates for this work), observed an unexplained noise in the antenna they were using to detect cosmic radio signals (generated by natural sources such as warm hydrogen clouds – this was not a search for extraterrestrial intelligence). This noise was independent of direction and impervious to seasonal changes. After eliminating all likely noise sources, Penzias and Wilson published their observations using the catchy title "A Measurement of Excess Antenna Temperature at 4080 Mc/s." Soon thereafter these observations came to the notice of the Princeton cosmology group who provided the explanation: Penzias and Wilson had observed one of the oldest fossils imaginable, a 13.5 billion year old memento from the early Universe.

in the early Universe was largely in the ultraviolet region, and inhabited an environment where gravitational effects were very strong. In evolving from these early conditions to the present environment this radiation lost a significant amount of energy. This did not result in its slowing down (for electromagnetic radiation moves at a fixed speed c), but in a dramatic increase of wavelength (by a factor of close to 50 000), so that currently the early Universe radiation has wavelengths in the microwave region. The Big Bang Theory *predicts* that our Universe is suffused by a homogeneous and isotropic sea of microwave radiation, called the *cosmic microwave background*.

This prediction has been amply confirmed. Earth-bound and satellite instruments have measured the characteristics of the background radiation and they all coincide with the predictions of the theory. In addition, extremely precise measurements indicate the presence of very weak inhomogeneities (of the order of 0.001 percent), that can be associated with regions in the early Universe that were not completely smooth (Figure 8.9). Those regions that had a slightly larger matter–energy density attracted material more strongly than sparser regions, and in doing so they grew into increasingly larger matter clumps. The very weak inhomogeneous structure in the cosmic background radiation represents the seeds of the structures (such as galaxy super-clusters) we see in the Universe today. The images of the cosmic background radiation inhomogeneities provide a picture of the Universe when it was but 350 000 years old, 1/40 000 of its current age.

The almost perfect smoothness observed in the cosmic background radiation is a feature that could not be achieved during the time it decoupled, but must be a consequence of earlier conditions. Any theory of the very early Universe must provide a coherent explanation for the conditions that led to both the very high degree of uniformity, *and* for all the features of the very weak inhomogeneities. This has proven a very difficult task to which I'll come later in the chapter.

The distant future

The Cosmological Principle provides a very limited number of options for the future of our Universe: it will either stop its expansion in a few billion years and re-contract (a closed universe); or it will expand ever more slowly and come to a stop at the infinite future (a flat universe); or it will continue expanding forever at some finite rate (an open universe). The amount of matter and the magnitude of the Cosmological Constant determine which of these possibilities corresponds to our Universe (Figure 8.10).

Recent observations of a class of exploding stars called Type Ia supernovae (see the appendix to this chapter) strongly favor the presence of a non-zero Cosmological Constant. This conclusion is based on the measurement of both the distance to and speed of a large number of such supernovae. The best estimates from these observations yield a *total* (including the contributions from all types of matter, radiation and the Cosmological Constant) mass–energy density very close to the critical value. If confirmed this would then imply that our Universe is, on average, quite "flat" (so that, on average, Euclidean geometry is quite accurate after all), and that it will continue expanding for all eternity.

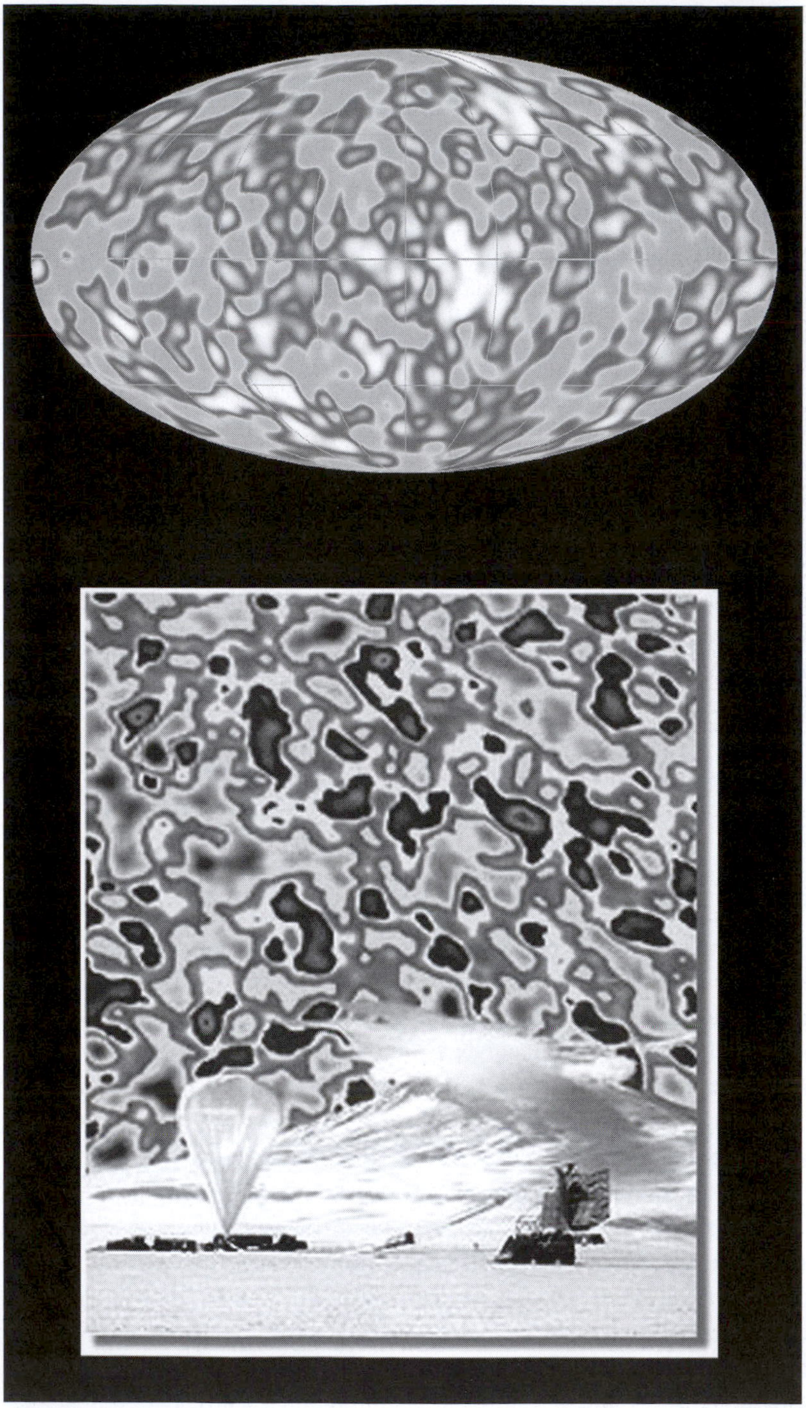

Figure 8.9
Inhomogeneities in the cosmic background radiation. Top: measurements by the COBE observatory, the inhomogeneities there depicted correspond today to structures of size close to 1700 megaparsecs. Bottom: the microwave sky; composite picture by the BOOMERANG experiment; the inhomogeneities in this case correspond to present structures a few dozen megaparsecs in size, roughly the size of galactic super-clusters. (Courtesy of NASA and the BOOMERANG Collaboration)

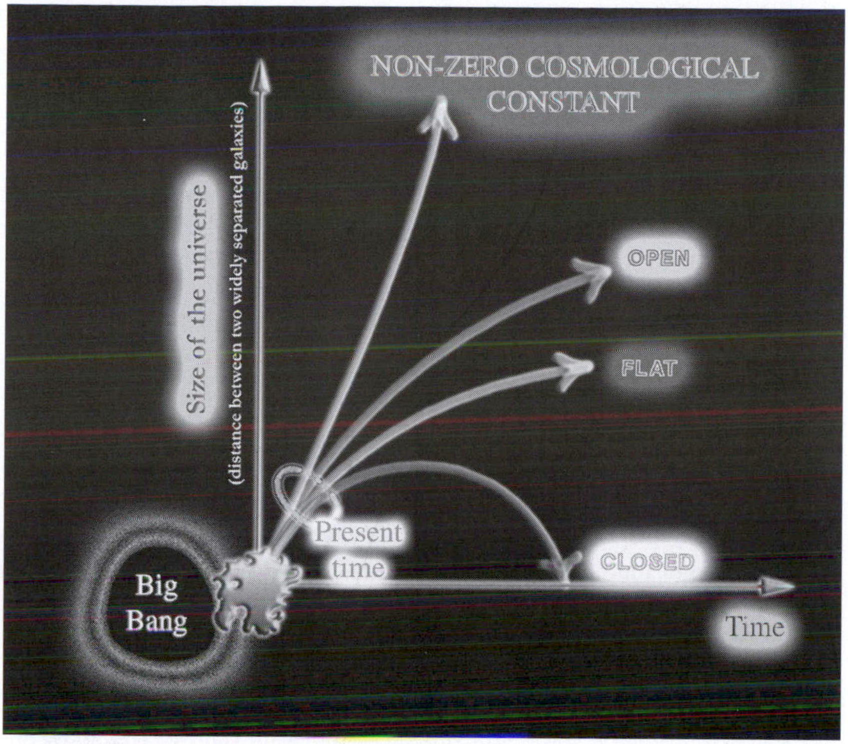

Figure 8.10 The possible fates of the Universe.

The fate of such a universe is one of empty darkness and forlorn death. Through the eons all galaxies will slowly lose energy by emitting radiation and, even if very weakly, gravitational waves. As a result, by 10^{30} years after the Big Bang, galactic clusters will have formed enormously massive and slowly contracting objects; by 10^{40} years after the Big Bang they will have formed super-massive black holes. But even black holes are unstable: through quantum effects they emit a very small amount of radiation which slowly depletes their mass, as the mass decreases the rate of radiation emission increases, so that in their final stages black holes blink out in a final burst of radiative splendor; 10^{100} years after the Big Bang, our Universe will witness the last of such final fireworks as the most massive black holes finally evaporate. After that space will only contain traces of radiation (the red-shifted relics from the cosmic background radiation together with the contributions from the now vanished black holes), and some stable subatomic particles, nothing more. And so it will stay 'til the end of time.

If, however, current data is flawed and the Universe is in fact closed, then there will be a time in the future where the current expansion will come to an end and the Universe will start to contract. For billions of years this will simply mean that future astronomers will see distant galaxies approaching the Milky Way as space itself becomes smaller. Eventually, however, the whole process described above will be played out in reverse: atoms will dissociate, the atomic nuclei will melt away, and then the protons and neutrons will break apart in the relentlessly increasing heat, until the whole Universe is reduced to a point, the "Big Crunch" (Figure 8.10). Our current understanding of physics does not allow an unambiguous determination of what would happen during the final stages of this process, nor of what would happen thereafter. Whether such a Universe would bounce back and start expanding, and so repeat a cycle of death and rebirth in a quantitative version of the Indian creation myth, is pure speculation.

Through a glass darkly

One of the most ingrained beliefs about the Universe is that it is infinite and eternal, an idea that appears as far back as Lucretius (55 BC). But these assumptions, natural as they may seem, have unexpected consequences. If the Universe is indeed infinite and it is populated by stars everywhere, then, when we look at the night sky in *any* direction, our line of sight will cross an infinite number of stars, and, if the Universe is also eternal, the light of all these stars would certainly have had time to reach us. As a result the night sky should be uniformly bright . . . which it is not.

This conclusion has come to be known as "Olbers' paradox" (in honor of H. Olbers, a nineteenth-century German scientist), but the very same question has occupied a large number of astronomers and philosophers who have provided a variety of explanations for the dark night sky. In the sixteenth century T. Digges, and later the seventeenth-century astronomer E. Halley, proposed that the explanation for the dark night sky lies in the enormous distance to the stars: the farther a star is, the dimmer its light (so they argued), to the point that stars lying sufficiently far will not yield any appreciable light on Earth. The error in this reasoning is that, while it is true that the contribution from any individual star dims with distance, it is also true that in an infinite universe uniformly filled with stars the number of stars at a given distance from us also increases with that distance, so we get dimmer light from more stars, and these two effects precisely cancel each other: the bright night sky remains.

J. Kepler was also aware of this puzzle and argued that it implied the impossibility of an infinite number of stars. These, like the trees in the forest, must be concentrated in some region outside of which there is nothing but an eternal void. This possibility would dispose of Olbers' puzzle, but at the price of raising another: by what agency did this conglomeration of stars come about? And for this Kepler had no answer.

The eighteenth-century Swiss astronomer Jean-Phillippe Loys de Chésaux proposed a different solution, envisioning that the Universe had an appropriate number of dark objects, such as dense dust clouds, which block the light of most stars. But, alas, this also failed to solve the problem: the opaque objects would necessarily heat up as a result of their absorbing all that light, and there would come a time (in an eternal universe there is plenty of that) where they would start shining themselves. So, again the puzzle remained.

In the nineteenth century Lord Kelvin noted that Olbers' argument also tacitly assumed that all stars shine eternally, and argued that the fact that stars have a finite lifetime solves the puzzle. However, just as stars die, others are born, and Kelvin's calculations failed to take this into account.

Before turning to the modern view of this old problem, note the three assumptions that lead to the paradox: first, that the Universe is infinitely large; second, that it is uniformly filled with stars and the star density does not change with time; and third, that it is infinitely old. This last condition is necessary for Olber's paradox to remain due to the finite speed of light c: if the Universe is now t seconds old, then we can see stars no farther than ct; for the Earth to receive light from stars infinitely far away, t must also be infinite. Within the context of the Big Bang Theory, the solution to this age-old problem lies in the fact that our Universe is expanding and not infinitely old. In an expanding universe the density of stars continually decreases, and its finite age implies that only the light of stars within a limited distance has had time to reach us. And so, the night sky should be dark as indeed it is, and this holds even when taking into account deaths and births of stars.

Alternatives to the Big Bang

All the data available at this time can be accommodated by the Big Bang Theory. This theory, however, is not the only one "in the market," and though competing proposals have by now proved deficient in one way or another, I include them to illustrate the importance of accurate measurements in being able to discriminate between hypotheses. I also note that some of the alternative ideas were proposed by some of the most prominent astrophysicists of the time, but even this illustrious parentage proved no enduring shield against the reality of the observations.

One possible alternative to the Big Bang was proposed by E. A. Milne in the 1930s, who was able to derive Hubble's relation between distance and speed by assuming that in the distant past all matter in the Universe was at one time located at one point in space, that an explosion ensued (sending matter into existing space), and that matter has been moving away from that point ever since. Such a scenario describes a very inhomogeneous universe (all matter was supposed to have coalesced at one point in space) and is incompatible with the measurements of the cosmic background radiation.

Similar to Milne's is the "plasma universe," proposed by H. Alfvén in the 1960s. This hypothesis assumes that the Universe began with a slowly contracting cloud of matter and antimatter; as the density increased these annihilated, producing a blast that reversed the initial collapse, and generated the observed expansion (note that within this and Milne's hypotheses matter expands into space, which remains unaltered, it does *not* describe the expansion of space and matter together). The problem with this scenario is that for the initial blast to have the required intensity, the original cloud must have reached enormous densities, so high, in fact, that it would have formed a black hole, and no expansion would have occurred.

A quite different hypothesis was proposed by H. Bondi, T. Gold, and F. Hoyle in the 1940s; it represented an eternal, homogeneous, and isotropic unchanging universe where not only all locations but all *times* are also equivalent. This hypothesis then envisages a never-changing universe and is therefore referred to as *steady-state cosmology*. In order to reconcile this hypothesis with Hubble's observations the proponents had to assume that in this universe matter is continually being created (out of nothing!), so that the pressure generated by this new material generates the Hubble Flow. This idea is troublesome because it violates energy conservation (recall that mass and energy are equivalent[12]), but the required rate of matter creation needed to explain the observed expansion is so small – about one hydrogen atom per cubic kilometer per year, that the hypothesis cannot be disproved on these grounds. Available technology does not allow the possibility of excluding all atoms from a region 1 km^3 in size, maintaining this quarantine for several years, and then probing in detail this volume in order to determine whether a few atoms have popped into existence.

The demise of steady-state cosmology came from other arguments. The first concerns the age of the galaxies: when we observe a distant galaxy, the images we receive today left the galaxy a long time ago (since light travels at a finite speed), so that we observe the galaxy as it was in the distant past (Figure 8.11). Steady-state cosmology assumes that the Universe has always looked the same, and so the images of distant galaxies should show galaxies in all stages of evolution. In contrast, within the Big Bang Theory the Universe had a beginning, so that it contained only young galaxies in the distant past, accordingly, the images of distant galaxies should show a predominantly young population. The observations unambiguously favor the Big Bang predictions.

The second argument against steady-state cosmology concerns the microwave background radiation. In order to explain the observations this hypothesis had to be modified to include the spontaneous creation of radiation, in addition to the spontaneous creation of matter. But it is incapable of explaining why this radiation should appear as microwaves (why not radio waves or gamma rays?); in contrast the Big Bang Theory does provide a coherent explanation, as described previously.

Puzzles and challenges

The success of the Big Bang Theory in explaining the presence of the cosmic background radiation, the abundances of light elements, and the Hubble Flow might give the impression that this provides a complete and consistent theory of

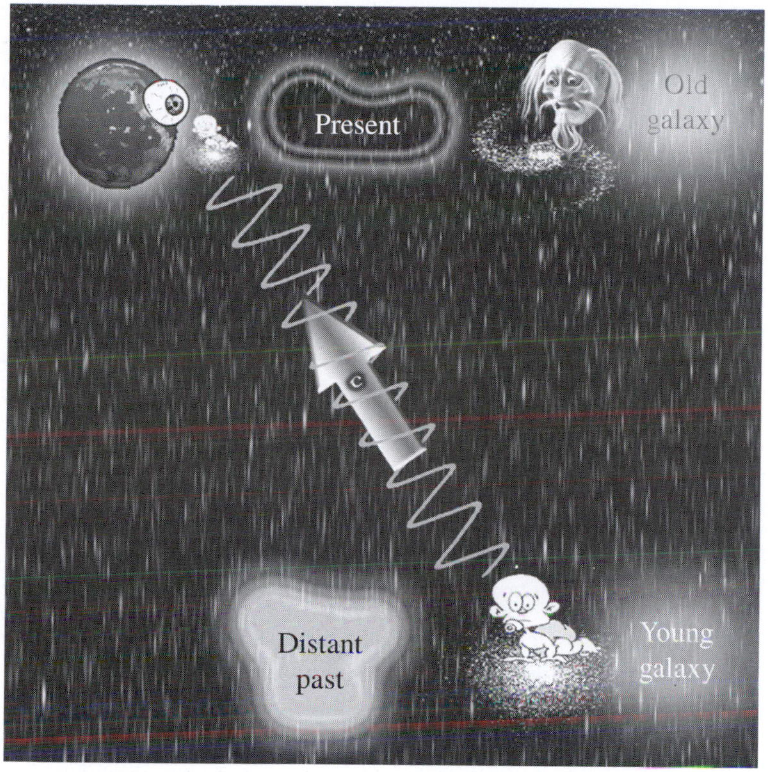

Figure 8.11 Due to the finite speed of light the images we receive from distant objects detail how these objects looked in the distant past.

the evolution of our Universe from its most tender infancy to our days. This, however, is not the case. What the theory provides is a description of the evolution of the Universe from the time it was 10^{-11} s old, and this description agrees with the observed average properties of the Universe provided we assume a certain density of matter and energy, and if we include a Cosmological Constant of the right magnitude (and sign), and if we also assume the appropriate (initial) conditions in the early Universe.

There are, however, puzzles connected with each of these items. The amount of matter required to match the observations is about 25 times larger than what can be accounted for by counting stars, galaxies, dust, and gas, so there is a missing mass problem. The value of the Cosmological Constant obtained from the observations is about 120 orders of magnitude (that is, a factor of 1 followed by 120 zeros!) larger than the calculated value, so there is also a Cosmological Constant problem. Finally, in order to explain the extreme uniformity in the microwave cosmic background radiation, the required initial conditions are ones of almost perfect homogeneity and isotropy – at least in those regions that evolved into our observable Universe. But in even as early as 10^{-11} seconds after the Big Bang the Universe was too big for all its parts to have had some sort of prior contact,[13] and yet only through such a contact can we understand the extreme uniformity we observe. So we also have a problem associated with homogeneity and isotropy: either we do not understand some fundamental properties of interactions in the early Universe or our observable Universe is but a fraction of the whole. I will review these issues in more detail below.

I hasten to mention that *none* of these issues represent a death blow for the Big Bang Theory, but represent open questions currently under investigation. Most of the hypotheses that aim at answering them can be described as extensions of the "standard" Big Bang Theory, such as those including modifications at very early times that could generate the desired degree of smoothness in the early Universe. The process of generating and testing these hypotheses will lead in the near future to a better understanding of the Universe at large and so, to use a tired phrase, the difficulties described below should be viewed as opportunities and not problems.

Where's the matter?

The amount of matter–energy in the Universe determines its future (the ditty reads: density is destiny). Recapitulating the results previously mentioned, this relationship allows for only three possibilities. A sparse, open universe where matter effects will slow the expansion, but would not be able to stop it; a flat universe where the density of matter–energy equals a precise critical value, and where the rate of expansion decreases with time, slowly coming to a stop in the infinite future; or a closed universe where the density lies above its critical value, and which will eventually re-collapse. Though these considerations are modified in the presence of a Cosmological Constant, it is clear that the amount of matter plays a decisive role in the evolution of the Universe. The obvious question is then: how much stuff is in the Universe? Or, more precisely, how does the average matter–energy density compare to its current critical value of $10^{-29}\,\mathrm{g\,cm^{-3}}$?

The cosmic census

Measuring the average mass density of the Universe is complicated and tedious. The methods devised can be separated into direct, indirect, and circumstantial. The direct approach simply attempts to count all objects we can observe and estimate their masses; indirect methods infer the amount of mass from its effects on certain objects, chosen so as to simplify the task; circumstantial methods use calculations to determine the effects of the density of matter on the characteristics of the matter itself, and then uses the observations to infer the density.

The evidence of our eyes

One obvious (and very tedious) way of gauging the amount of matter in the Universe is to count and classify stars, and to estimate the mass of each class using well-established theories of star formation and properties.[14] A long study of the population of the Coma cluster estimated it contains 3×10^{13} Sun-sized stars, which correspond to about 0.4 percent of the critical value, and an amount of interstellar gas, observed through its X-ray emissions, equivalent to about 2×10^{14} solar masses, and this corresponds to 3 percent of the critical value. Based on these observations one can estimate that the total amount of matter constituting usual cosmic objects corresponds to about 4 percent of the critical value.

It is unlikely that there are other large mass concentrations. For example, hot gas would show up through its infrared emissions, while cold gas or dust would show up in the way it absorbs light from any shining bright object behind them. A region with a large number of stars that are screened would also contain large amounts of elements heavier than lithium (for these are manufactured in stars), and these can be detected by the way they emit or absorb electromagnetic radiation. None of these effects are observed indicating that the above estimates are quite reliable.

The evidence from nucleosynthesis

As I mentioned above, most of the deuterium (an isotope of hydrogen) and helium are products of the nuclear reactions in the early Universe. This type of process, where heavier nuclei are created through reactions involving lighter ones go under the generic name of *nucleosynthesis*. In the specific case of the early Universe, it is referred to as Big Bang Nucleosynthesis.

If we could increase the amount of matter participating in nuclear reactions in the early Universe, the rate of expansion would decrease and the atomic nuclei would remain closely packed for a longer time, also there would more of them. As a result the reaction rate would increase and, in particular, the amount of helium and deuterium produced would be larger. The conclusion is that there is a close relationship between the current abundances of such light nuclei and the total amount of mass participating in the early Universe nuclear processes; measuring one determines the other. Combining this relationship with careful estimates of the amount of deuterium and helium currently present in the Universe yields the conclusion that the density of all matter that participates in nuclear reactions cannot be larger than a few percent of the critical value.

These arguments do not rely solely on observational evidence, but also use conclusions drawn from our knowledge of nuclear processes, which, fortunately, are quite well understood. The resulting estimate of the mass–energy density is independent of the one based on counting stars and estimating the mass of gas and dust clouds, and it is therefore quite remarkable that both methods give values so close to one another.

The evidence from the galaxies

A completely different way of measuring mass is through its effects on its surroundings. For example, planets will orbit a star due to their mutual gravitational interactions, and the characteristics of this motion are codified (to a very

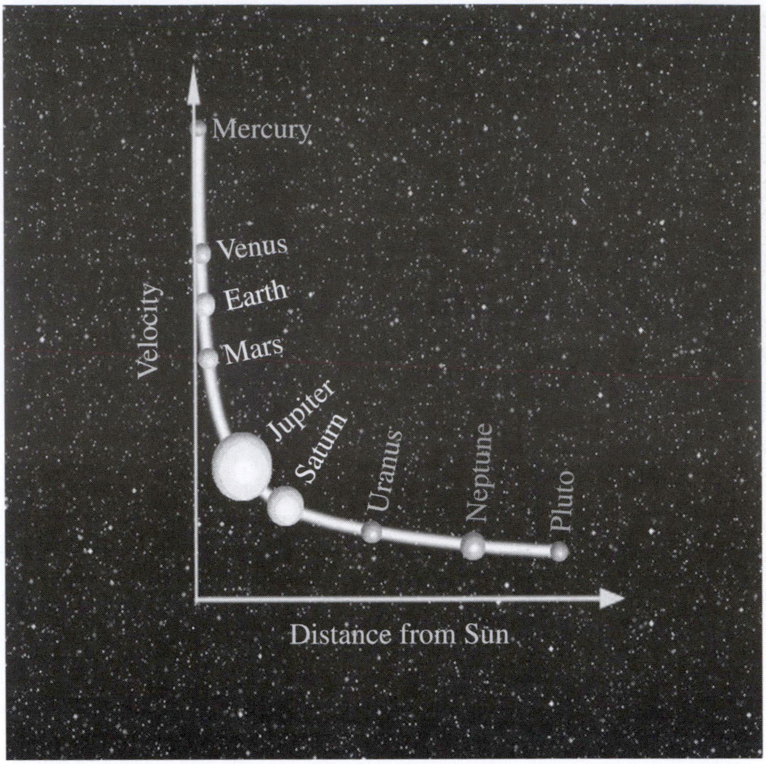

Figure 8.12 The relation between the orbiting speed and distance to the Sun.

good approximation) in Kepler's laws. In particular, a planet will move slower the farther it is from the Sun (Figure 8.12), and from the speed and orbit size we can determine the mass of the star.

We can now apply the same reasoning to determine the mass of a galaxy. In most galaxies the majority of stars are collected at the core, and so it is natural to assume that most of the galaxy's matter is also located there. Then this central core will act as an enormous "sun" being orbited by the outlying stars (that play the role of planets). We can then test the assumption that most of the mass is indeed located at the core by measuring the speed of outlying stars, which must drop with distance according to Kepler's Third Law. In addition, combining this speed measurement with that of the distance to the galaxy's center determines the core's mass.

These measurements were first carried out by the Swiss astronomer F. Swicky in 1933, and have been repeated a large number of times for a great number of galaxies. The results were extremely surprising: the speed of the outlying stars does *not* decrease with the star's distance to the core, but it is instead almost *independent* of the distance (Figure 8.13). These observations have been carried out for about 1000 galaxies, and in all but a few isolated cases such flat rotation curves have been obtained.

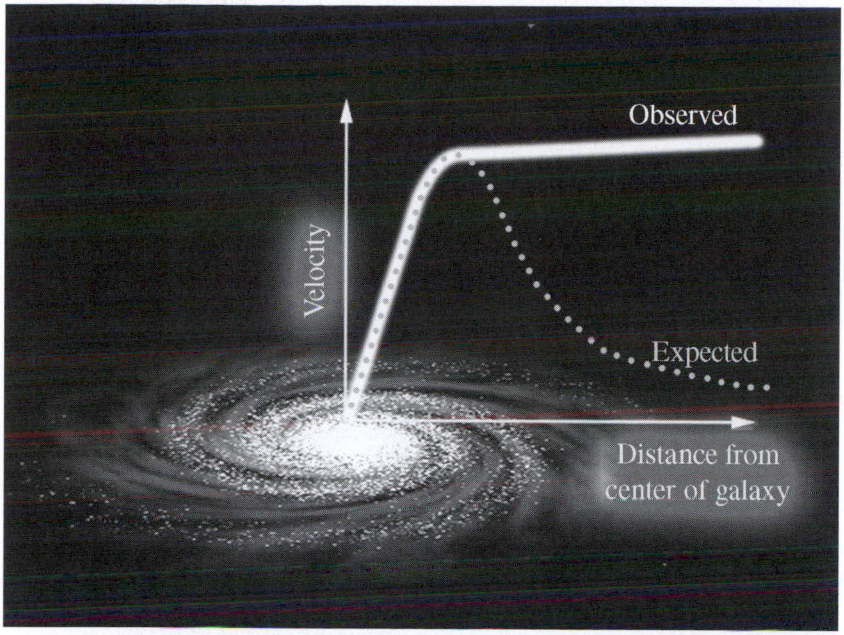

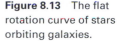

Figure 8.13 The flat rotation curve of stars orbiting galaxies.

Two explanations for this unexpected behavior have been proposed. The first and most dramatic is that the gravitational formulas used to make the initial prediction are inaccurate when considering objects the size of a galaxy. If this is the case, both Newton and Einstein's theories would have to be modified at distances of the order of 10^5 light-years (the typical size of a galaxy). This, however, presents an enormous challenge: since General Relativity is based on a single postulate (the Principle of Equivalence), it is very hard to change some of its predictions (such as the rotation curves) while leaving others intact (such as the behavior of the Solar System or the double pulsar). *Any* change would affect the Principle of Equivalence, and this then alters *all* the predictions. This hypothesis has not been pursued very actively partly because of these complications, but also because there is an alternative, less draconian, explanation.

The alternative explanation is based on the observation that flat rotation curves would automatically occur if the galaxy were surrounded by a halo of some (as yet unspecified) substance. This material should interact very weakly with electromagnetic radiation since, if present, it certainly does not shine in any significant way in any region of the electromagnetic spectrum. Because of these assumed properties, we best perceive its presence through its gravitational effects. In a spurt of creativity this hypothesized non-shining material was called *dark matter* (Figure 8.14).

If we assume the existence of dark matter, then the stars we believed to be in the outskirts of the galaxy are actually *inside* the main bulk, it just happens that we cannot *see* most of the material making up this bulk. Adopting a reasonable

Figure 8.14 Galaxy surrounded by the hypothesized halo of dark matter (depicted gray for clarity).

distribution of dark matter and using the equations of Newtonian gravity one then *predicts* a flat rotation curve. In this case most of the galactic mass would be in the form of dark matter, with estimates ranging from 10 to 40 times the mass of stars.

The flat rotation curves are not the only galactic puzzles that can be explained by the presence of large amounts of non-shining matter. For example, the average velocity of galaxies inside a cluster is consistent with gravitational forces which are significantly larger than those produced by the stars, gas, and dust seen in the cluster (first noted by Smith in 1936), but can be accounted for by assuming the presence of dark matter (in quantities consistent with those assumed to be present around galaxies). In addition, this increased gravitational attraction would explain the presence of gas inside galactic clusters: without this added attraction, the gas would have diffused into intergalactic space long ago.

The amount of dark matter necessary to account for the behavior and properties of galactic clusters is roughly 20 percent that of the critical density.

The dark matter solution

The dark matter hypothesis implies that "ordinary" matter, which is involved in nuclear reactions, and interacts with electromagnetic radiation, and which populates the periodic table of the elements, is quite rare. Most of the stuff in the

Cosmos is in the form of this substance that interacts very weakly and whose presence has been inferred through its gravitational effects.

This idea cannot be accepted simply because it explains the rotation curves of galaxies and the motion of galactic clusters; after all, that was what it was designed to do! The hypothesis must be combined with the General Theory of Relativity and used to determine the manner in which the presence of dark matter affects the evolution of the Universe and to verify that the predictions thus derived agree with independent observations. And, of course, even if very weakly, dark matter is supposed to interact with ordinary matter, and so we should be able to construct detectors sensitive to this most abundant of materials.

In parallel to these efforts alternative explanations are being considered. An example is the hypothesis that the Universe contains an enormous number of small rocks and stellar corpses (what remains from a star at the end of its life), all of which do not emit any significant amount of electromagnetic radiation but might form a galactic halo; these objects are known as MACHOs – for Massive Astrophysical Compact Halo Objects. If sufficiently abundant, MACHOs could account for the rotation curve of galaxies and the behavior of galactic clusters, but, since all these cosmic coprolites are made up of ordinary matter, their presence would be inconsistent with the abundances of light elements. In addition it is hard to imagine these objects being completely dark: planet-sized objects emit infrared radiation (even at masses as low as ten percent that of the Sun). Furthermore, a large number of stellar remnants would indicate a large number of stars in the past, but no indication of this has been observed.

Finally, there is direct observational evidence indicating that MACHOs are in fact quite rare. The observational technique is not based on detection through infrared or other such type of radiation since, by assumption, these objects are almost mute in such emissions; instead one uses gravitational effects. Specifically, when a MACHO crosses the line of vision between Earth and a star, its gravitational influence will affect the paths of the lightrays, and if the object is sufficiently massive it creates a gravitational lens; after the object moves on the effect disappears. Experiments then look for these itinerant gravitational lenses by continuously examining the images of several million stars. Though some of the lensing events have been observed, they are rare, and this indicates that MACHOs cannot account for dark matter (which is also consistent with the evidence from nucleosynthesis): there is no significant number of objects composed of ordinary matter with masses larger than 10^{-6} solar masses.

It is by now generally accepted that ordinary matter in any form cannot serve as dark matter. The alternatives being now considered include a variety of subatomic particles, some known, some hypothesized. Below I describe their properties and characteristics, and the way they would influence the evolution of the Universe. Which (if any) of these possibilities will provide an answer to the dark-matter puzzle is still a subject of debate.

Cold dark-matter hypothesis

Having ruled out ordinary matter, one might consider the possibility that dark matter is composed of certain as yet undiscovered subatomic particles, whose properties are constrained by the way dark matter is supposed to behave. One popular choice is to assume that a significant proportion of dark matter is composed of massive subatomic particles known as WIMPs (for Weakly Interacting Massive Particles),[15] which have not been yet observed. These particles are assumed to have had a rather inhomogeneous distribution at the time the Universe became transparent to electromagnetic radiation, but this unevenness did not affect the cosmic background radiation due to the (also assumed) weak interactions between WIMPs and ordinary matter and radiation.

On the other hand the gravitational effects of these inhomogeneities would have had its effects on the surrounding matter, serving as seeds for galactic clusters and super-clusters. The growth of structures seeded in this way is typically "bottom-up" (Figure 8.15): small inhomogeneities in the distribution of cold dark matter attract both ordinary matter and more dark matter, in this way the denser regions grow in size while their surroundings are depleted of matter. The resulting structures then attract each other, forming larger features, and the attractions between these form even larger ones, etc. The end result presents conglomerations of matter of all sizes whose distributions qualitatively reproduce the observations. Computer simulations also confirm the expectation that the weak interactions between dark matter and electromagnetic radiation lead to small inhomogeneities in the cosmic microwave background radiation (Figure 8.9).

Hot dark-matter hypothesis

An alternative possibility is that dark matter is composed mainly of very light particles traveling at speeds close to c. Such particles typically move through the distribution of matter and tend to homogenize the small-scale features; the result is a "bottom-down" process (Figure 8.16), the opposite as for cold dark matter: inhomogeneities in the hot dark-matter distribution quickly form large concentrations that eventually become unstable and fracture into smaller objects. The resulting structure is therefore tilted towards large-scale features. This peculiarity, combined with observations of the large-scale structure of the Universe, has been used to estimate the type of dark matter preponderant in our Universe.[16]

In contrast with the hypothesized WIMPs there are several known subatomic particles that could serve as hot dark matter candidates, with neutrinos being the favorite choice. But, since these are well-studied particles and their properties are known (in contrast with the cold dark matter candidates which have not been observed and may not even exist), the role they played in the early Universe is severely constrained. Despite their weak interactions an over-abundance of

Figure 8.15 Evolution of the Universe in the presence of significant amounts of cold dark matter from the early Universe (bottom) to the present day (top).

neutrinos would severely affect both the production of helium and the size of the early Universe inhomogeneities, and these constraints imply that neutrino hot dark matter cannot be sufficiently abundant to account for the observations. But hot dark matter could be composed of other types of as yet unobserved particles, and in this case the constraints are significantly laxer.

Figure 8.16 Evolution of the Universe in the presence of significant amounts of hot dark matter from the early Universe (bottom) to the present day (top).

The tepid-mix hypothesis

All the types of dark matter considered above provide a rational explanation for the flat rotation curves of galaxies and other related observations. But these results are far from sufficient for these ideas to gain acceptance, after all, explaining these observations is what these hypotheses were designed to do in the first place! In order to gain credibility, a dark matter hypothesis (when combined with the Big Bang Theory) must also provide predictions that are in agreement with the average properties of the Universe as observed today. Large-scale surveys reveal a rather characteristic structure of matter congregated into irregular filaments and cells separated by almost empty regions a few dozen megaparsecs in size.[17] This structure is difficult to generate assuming the presence of only cold or only hot dark matter, but computer simulations that include an appropriate *mix* of these substances in the early Universe do lead to predictions that agree with the observations. This is an encouraging sign, but is far from providing a definite proof for the whole scenario: that requires the discovery of the subatomic particles that supposedly confirm dark matter. Accordingly, several experiments[18] are currently underway aiming at confirming or disproving these ideas. At the moment we can only watch the stars and imagine them to be islands in this most abundant of materials.

The combined abundances of cold and hot dark matter used in successful simulations, together with the much smaller ordinary matter component, add up to about 30 percent of the critical density. If there are no other contributions to the matter–energy density of the Universe, this then implies that our Universe is *open* and will continue to expand indefinitely. Just as microscopic plankton determines the fate of the oceans, so the subatomic dark matter determines the fate of the Universe.

The Cosmological Constant

The Cosmological Constant has a rather convoluted history. As mentioned above, it was first introduced by Einstein in 1917 as a means for obtaining a static homogeneous and isotropic universe, but he was not happy with it. In a letter to P. Ehrenfest on February 4 1917 Einstein wrote about this idea: "I have again perpetrated something relating to the theory of gravitation that might endanger me of being committed to a madhouse." Six years before Hubble's momentous result, he wrote to H. Weyl, "... if there is no quasi-static world, then away with the cosmological term"; by 1931 he had completely dropped the Cosmological Constant from his equations (or, more precisely, had assumed it to be zero). From this period G. Gamow recalls[19] that, "... when I was discussing cosmological problems with Einstein, he remarked that the introduction of the cosmological term was the biggest blunder he ever made in his life."

Nonetheless this opinion was far from universal, especially since the data appeared to require a non-zero Cosmological Constant: Hubble's original data suggested a relatively large value for the Hubble constant H_0, which corresponds to a relatively young universe[20] (2 billion years old or so), younger, in fact, than some of the stars in it! But, as Fridmann and Lemaître first observed, assuming the presence of a sufficiently large Cosmological Constant can solve this "age problem." Later (in the 1960s) a more careful measurement of the Hubble constant yielded a value about seven times smaller than Hubble's; this corresponds to a much older Universe that appeared not to require a Cosmological Constant. And, again, it was dropped from the equations.

But this was not the end of the affair: recent measurements involving Type Ia supernovae strongly favor a non-zero Cosmological Constant, and its value is such that the total mass–energy density (comprising contributions from ordinary and dark matter, radiation, and the Cosmological Constant) is very close to the critical value. If confirmed, these measurements predict a flat, eternally expanding universe.

It might seem that this popping in and out of the Cosmological Constant indicates some fickleness in the work of cosmologists, whose results cannot be trusted on this account. This is not the case: despite Einstein's opinion, most researchers have kept an open mind concerning the presence or absence of a Cosmological Constant, and have carefully noted that all previous observations determined that its value lies below a certain limit, never that it is zero. Current, more refined, measurements indicate that, in fact, it is not zero, while still lying below the old boundary, so there is no real inconsistency.

The experimental techniques used in obtaining this latest value are carefully being examined and this has generated some debate. These discussions do not represent a schism among the cosmologists, but the natural process by which new scientific results are examined and accepted or rejected. In this case a final decision has not been yet reached: new independent experiments are needed (and are planned), new ideas are being floated (some supporting and some rejecting a non-zero Cosmological Constant) and examined, etc.

Though the evidence of a non-zero Cosmological Constant is still being argued, it is persuasive enough to have revived an old discussion concerning the origins of this effect. There are many hypotheses that predict a non-zero value for the Cosmological Constant, but the predicted value is inevitably enormous compared to the measurements. For example, the simplest, and most natural, arguments yield a value that is 10^{120} = 1 000 larger than the measured value![21]

One cannot simply ignore this issue since the currently accepted theories of the subatomic world predict that at some stage in the history of the Universe the Cosmological Constant was large. For example, during the time where protons and neutrons were formed the Cosmological Constant changed by 10^{44} times its current value. Such an enormous change must have occurred with some compensating mechanism, for if the Cosmological Constant was this large then, depending on its sign, the Universe would have either re-collapsed almost immediately or would have expanded so fast that there would have been no time to form any atoms, let alone galaxies. If the Cosmological Constant was small before neutrons and protons formed then the same undesirable effects would have occurred after they were formed. This inability to predict such a constant in Nature points to a lack of understanding at least some aspects of gravity and/or its interactions with matter.

Horizons and mass

One of the recurring themes in this description of the Universe, and, indeed, the most important additional assumption used when considering the average properties of the Universe, is the Cosmological Principle, which asserts that the Universe has the same average properties in all locations and in all directions. Of the myriad possible solutions to the equations of General Relativity only a handful obey the Cosmological Principle, and these are distinguished by the amount of mass and energy content; the mass–energy density determines the geometry and fate of the Universe.

All observations strongly support the Cosmological Principle. In particular, the almost perfect uniformity of the cosmic background radiation indicates that any deviations from uniformity are very small indeed. In addition, all current observations point to a mass–energy density very close to its critical value (within ten percent or so). These salient features can be accommodated within the Big Bang Theory *provided* appropriate initial conditions are imposed. It is universally assumed that some physical effects present in the early Universe must have been responsible for generating the appropriate initial configuration that led to our current Universe, and in this way such early effects also sealed the future evolution of the Universe. Logically speaking, the Big Bang Theory does not require the inclusion or justification of these initial effects for it to remain a valid and useful theory, and yet the origin of these early conditions *is* a valid question, for its answer contains an explanation of the global properties of the Cosmos. The present section contains a description of several current areas of research that aim at attaining a better understanding of that early Universe.

Perturbations and horizons; homogeneity and isotropy

The uniformity in the cosmic background radiation provided strong supporting evidence for the Cosmological Principle. But this initial satisfaction turned to puzzlement when it was realized that the observed uniformity covers the whole of the observable sky. This is surprising because it indicates that all points in the observable Universe must have been in close communication with each other some time in the past. A parallel is the following: if a few students in a school of thousands happen to wear plaid pants and checkered shirts, one might ascribe this to a coincidence of bad taste, but if *all* students show up one day wearing polka-dot pants and striped shirts, the playground during recess would look like a homogenous and isotropic sea of bad taste; this anomaly could be understood only if all students had previously agreed how to dress. Similarly, the close

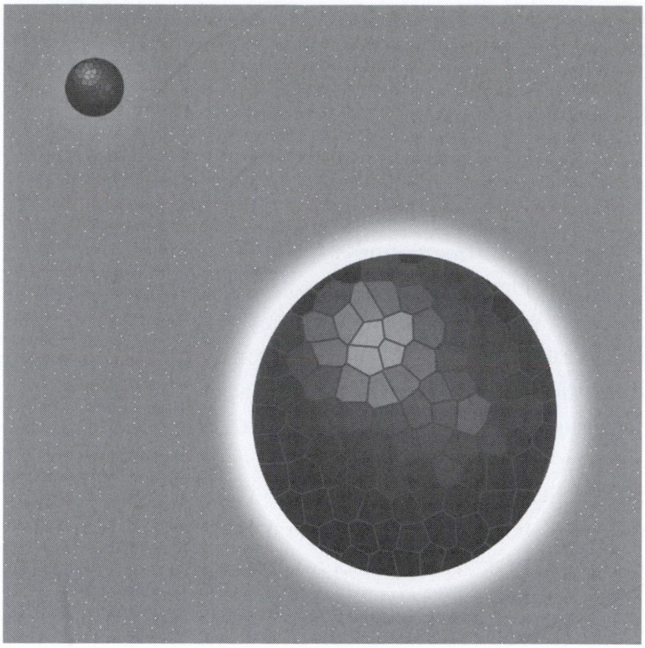

Figure 8.17 Illustration of the growth of fluctuations (represented by the facets in the spheres) in an expanding universe.

match in the properties of the microwave background radiation throughout the sky indicates that all regions of the currently observable Universe must have been in close contact some time in the past.

This, however, cannot have happened within the standard Big Bang Theory. Consider for example, two points opposite in the sky whose light has just reached us. Then, since we are in between them, there has been no time for them to have any sort of interaction, for nothing travels faster than light; in fact, within the Big Bang Theory regions of the sky separated by more than 2° could not have been in contact in the distant past. This is the *horizon problem*: the conditions leading to the observed smooth distribution of the cosmic background radiation are very unnatural: the Universe is too homogeneous and isotropic.

Admittedly the cosmic background radiation is not *perfectly* smooth, but even the 0.001 percent fluctuations are troublesome. The reason is that the expansion of the Universe magnifies any deviation from uniformity (Figure 8.17), so that the very weak inhomogeneities that the cosmic background radiation exhibits today must have evolved from rough patches that were positively puny (presenting deviations of less than 10^{-7} percent from perfect uniformity). So any description of the early Universe must predict not only this almost perfect smoothness, but also the presence of patches with very slight inhomogeneities that would grow into, among other things, our Milky Way.

It is worth noting that none of these arguments can be extended to regions farther than we can see; a constraint that, though obvious, is significant. Our (apparently) flat Universe is about 13.5 billion years old, and this implies that no light (or any other sort of information) from any object lying farther than 13.5 billion light-years away has reached our eyes. We are inescapably limited by this *horizon*; and, learning from the experiences of explorers in the olden times, we must allow the possibility that life beyond the horizon can take very different appearances. Of course, as time passes and our Universe grows older we will be able to see farther, and we will be able to determine whether distant regions are, in fact, so perfectly similar to the ones we see today. But that will have to wait. At the moment we can only assert that the region of the Universe accessible to us is homogeneous to the level of ten parts in a million, and leave open the possibility that outside our horizon much larger discrepancies might exist. The Universe might well be a quilt of regions 13 to 14 billion light-years in size, each quite smooth, and each quite distinct.

Mass and expansion: the flatness problem

More dramatic than the smoothness puzzle is the constraint due to the mass–energy density. I mentioned above that if this takes the critical value it will stay at the critical value for all times: a universe born flat stays flat (Figure 8.18). This, however, is a delicate condition: if the early Universe had a value that was slightly off the critical value, this difference is magnified by the expansion of the Universe; by now, 13 billion years later, it would be enormous. Such universes would have been quite different from ours: if the density happened to be larger than the critical value the Universe would have re-collapsed a long time ago and I would not be writing about this at all (if nothing else, due to lack of audience). If, on the other hand, the density had been somewhat smaller than critical, the expansion would have occurred at a much faster rate, and there would have been no time for forming the complex structures we observe today. In order to insure that the *current* energy–mass density is within ten percent of the critical value, then its value at the time the Universe became transparent to radiation must have matched the critical one to within a few trillionths of a percent! So it appears that either we are the result of a very lucky cosmic coincidence, or else the early Universe contained a mechanism that insured that the density was extremely close to the critical value. Discounting coincidences, the Big Bang Theory offers no such mechanism, a deficiency known as the *flatness problem*.

The inflation paradigm

Assuming the Big Bang Theory is an accurate description of the evolution of our Universe, the conditions in the very early Universe must have been exquisitely tuned: the mass–energy density must have had almost its critical value and the inhomogeneities must have been extremely small, and yet not completely absent. Such an initial configuration is so constrained that it is

Figure 8.18 Illustration of the manner in which mass affects the shape of space.

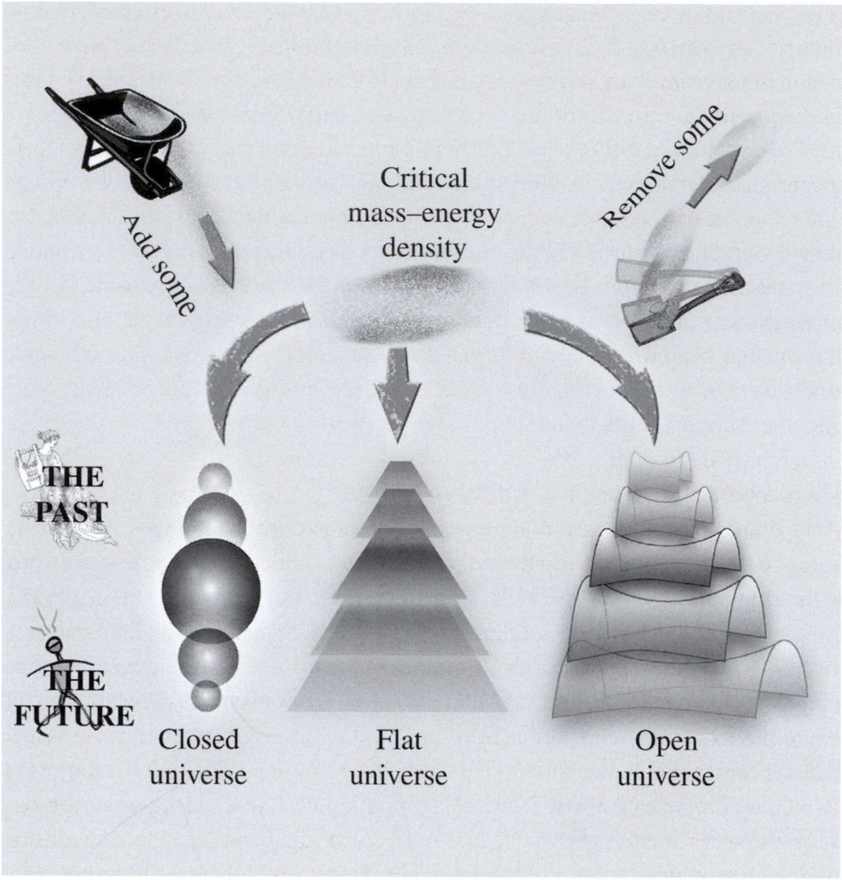

commonly believed that it was not accidental, but must be the inescapable result of whatever processes dominated the Universe during the first fractions of a second.

These early effects cannot be associated with any of the known interactions: the theories of subatomic physics combined in the Big Bang Theory do provide an accurate description of the evolution of the Universe starting at about 10^{-11} s after the Big Bang, when the temperature was about 10^{15} K, but they contain no mechanism for the Universe to evolve into an almost perfectly homogeneous system with critical density. In fact, these effects tend to *decrease* the degree of homogeneity and to drive the density *away* from its critical value.

Faced with such a puzzle one might take the draconian approach of declaring the Big Bang Theory irredeemably flawed and to start the search for a new hypothesis. One would then need to decide whether the alleged flaws are to be extended to the General Theory of Relativity, or whether they are simply due to the adoption of the Cosmological Principle. In the first case the task would be to devise a new theory of gravitation. In the second case it would be necessary to solve the equations of General Relativity without the assumptions of homogeneity and isotropy and produce a scenario that leads to the homogeneous cosmic background radiation, the almost critical value of the mass–energy density, and the observed structure of galaxy clusters and super-clusters; and do all this without fine-tuning the initial conditions. No such alternatives are currently available, and so, adherents to this possibility sit, and wait, and think.

A second more conservative view is that the Big Bang Theory is an accurate theory, but that requires extensions in order to be applicable at very early times (earlier than 10^{-11} s after the Big Bang). Within this approach the (as yet unknown) processes immediately after the Big Bang are assumed responsible for the smoothness and critical density. In addition these interactions, so important immediately after the Big Bang, should be of such a nature that by the time the Universe was 10^{-11} s old they became unimportant, allowing the Universe to evolve as in the standard Big Bang Theory. Within this second view the best studied possibility is known as the *inflationary universe*.

The basic idea of the inflation paradigm is that at some very early time (about 10^{-36} s after the Big Bang) the Universe underwent a period of very rapid expansion, increasing its size by a factor of *at least* 10^{27} in less than 10^{-11} s. To give an idea of this stupendous growth note that if we could blow up a microbe by such an enormous factor it would reach the size of a galaxy, and the nucleus of a single hydrogen atom the size of Pluto's orbit. This increase in the size of the Universe is known as inflation.[22]

The horizon puzzle is explained by the inflationary hypothesis by noting that due to the enormous increase in the size, the section of the Universe we can see today came from a region of the early Universe that was much smaller than previously thought (in fact, 10^{13} times smaller). Before inflation, this region was so small that, in fact, all points in it *were* in close contact through the exchange of radiation. And so, this speck of the Universe was almost perfectly homogeneous, and remained so throughout the inflationary period. After the inflationary era ended the standard Big Bang expansion proceeded, but then the initial conditions for *that* part of the Universe's evolution were ones of extreme homogeneity and isotropy.

The flatness puzzle is also solved by the inflation hypothesis through the simple observation that as the Universe expands, its curvature decreases. This is the same effect one would obtain if we looked at a part of the surface of a balloon that was being blown: as the size of the balloon increased the region we were concentrating on would look flatter and flatter. If we could manage to inflate the balloon by a factor of 10^{13}, any small part of its surface would be both enormously large and quite flat. In the same way the geometry of space after the inflationary epoch would approach that of the three-dimensional analog of a plane. Being almost perfectly flat, homogeneous, and isotropic the equations of General Relativity then imply that the mass–energy density should be almost identical to its critical value.

The inflation paradigm can then solve the problems described above without the need to invoke unlikely coincidences or miraculous adjustments. But there is a price to pay: one must specify the interactions responsible for the effect, and then verify all additional consequences they are supposed to produce. No single specific hypothesis has been yet able to fulfill all these requirements. One can then take the cynical view that the inflationary idea has simply exchanged one question for another, presenting no real gain. But this is not necessarily the case: the inflationary idea requires we ask *different* questions, and, as in the past, this has often proved a useful change. Only time will tell whether this will again prove so.

All inflationary hypotheses are based on the same basic observation: that a sufficiently large Cosmological Constant (of the right sign) will generate an exponential growth in the size of the Universe. All efforts have been then directed at providing mechanisms for first generating such a large Cosmological Constant very early in the history of the Universe, and then disposing of it a short time later;[23] while it was present the Universe inflated. This dance must have occurred significantly before the nucleosynthesis epoch, for otherwise inflation would ruin the accurate predictions for the abundance of light elements. Three specific hypotheses attempting to provide a dynamical explanation for inflation have been proposed, none is completely accepted, and one has serious problems (presented here for its historical significance). The evolution of these ideas presents one instance of science in the making.

Two of the inflationary hypotheses begin by assuming the presence of a large Cosmological Constant just after the Big Bang, and this leads to a short period of very rapid expansion. These hypotheses then propose mechanisms through which the energy contained in the Cosmological Constant is diverted into creating regions of "normal" space that behave according to the standard Big Bang Theory, once the average temperature becomes sufficiently low. In one case the regions of normal space appear as bubbles in the inflating background; these bubbles would form an ever growing aggregate that soon permeates the Universe. In another version of the hypothesis, domains of normal space of all sizes appear in a random fashion; these domains grow and, again, eventually fill up space. In either case, these transitions mark the end of the inflationary epoch, after which the standard Big-Bang expansion proceeds. In a third version of the hypothesis random quantum fluctuations in the early Universe generate the initial large Cosmological Constant needed to jump-start the inflationary engine, after the ensuing rapid expansion the hypothesis proposes a relaxation mechanism through which the Cosmological Constant dissipates as domains of normal space of all sizes appear, grow, and eventually fill the Universe.

In all three scenarios the region of space we can currently see had the minute size of 10^{-31} m before inflation (this is so small that many more of these specks would fit inside the nucleus of a hydrogen atom than there are stars in the sky); by the end of the inflationary era it had grown to about 1 cm. During inflation the

Cosmological Constant dominated over any other type of matter–energy; when inflation ended this energy was transferred to the dark and ordinary matter, producing an enormous rise in temperature. After this the Universe continued expanding as in the standard Big Bang Theory, with the modification that the conditions generated at the close of the inflationary epoch dispose of the flatness and homogeneity problems.

Despite their appeal for solving the horizon and flatness issues, all these ideas have unsavory properties. For example, the first hypothesis concludes that just after the inflationary epoch the Universe looked like a conglomeration of bubbles, so that it contained inhomogeneous structures of the size of a typical bubble. These features are predicted to be small enough to lie within our observable Universe, but the data give no inkling of their presence. The other two scenarios avoid this problem but at the cost of requiring that the various interactions responsible for inflation be very precisely matched, so that in that case we exchange fine-tuning the density and fluctuations for fine-tuning the size of the forces present in the early Universe.

There is also the issue of the observed fluctuations in the cosmic background radiation: under inflation all fluctuations are smoothed out by a factor of 10^{27} or more, so that even large inhomogeneities (of the order of 100 percent) would be reduced to the level of 10^{-11} percent after inflation, too small compared to the observed 0.001 percent effect. Whether a new clever modification of the inflation idea will solve all these problems in an elegant and convincing way, or whether we will have to leave the inflationary idea as a brave but ultimately unsuccessful attempt at describing the early Universe is at present an open question.

At a different level the inflation paradigm presents philosophical problems. It is the contention of this idea that the Universe is, in fact, quite *in*homogeneous and *an*isotropic, and that the only reason we are not aware of this is that all the evidence lies beyond our horizon. This then provides a hypothesis some of whose predictions cannot be fully tested *even in principle.* Future evidence supporting some inflationary-universe hypothesis might become so overwhelming that it will be accepted despite this deficiency, or, perhaps, a completely new alternative will enlarge our understanding of Nature and will make this an irrelevant conundrum.

Quo vadis?

The General Theory of Relativity has produced a generous amount of verified predictions. Its application to the Universe at large has provided a deep understanding of the evolution of our Universe, and, simultaneously, it has created a series of questions that probe the most remote and significant of events, the Big Bang. Detached as this avenue of research is from our everyday life, it is hard to underestimate its significance, not because it aims at providing a cosmic solution to our social and economic problems, but because it pretends to provide at least a partial answer to that most ancient of questions, where do we come from, and where are we going? We might not have the tools, patience and ingenuity to achieve this goal, and even if we do, we might not like the answers we obtain; but if we succeed at least we would *know*!

Appendix: measuring the Universe

Introduction

The General and Special Theories of Relativity discussed in the previous chapters are the tools currently used in the investigation and description of the Universe. Most of the objects in the Cosmos are rather mundane: stars, planets, rocks, and gas clouds. Yet in many respects the Universe is far from being placid and peaceful. There are stars which explode with the energy of a billion suns; black holes with millions of times the mass of a star which routinely devour whole planetary systems, generating in one day as much energy as our galaxy puts out in two years; there are enormous dust clouds where shock waves trigger the birth of new stars; and intense bursts of gamma rays whose origin is still uncertain. The Universe itself is expanding, and all the evidence points to its having originated in a state of unimaginable temperature and density.

Most of the information we obtain about these phenomena comes to us in the form of electromagnetic radiation, though recent experiments have begun to use neutrinos, and even gravitational waves, to open new "eyes" to the Universe. When comparing the theoretical predictions of the current hypotheses with observations, we are limited to whatever information we can ferret out of these messengers that reach our telescopes after traveling millions, and sometimes billions, of light-years.

Pitting the theoretical predictions of relativistic cosmology and astrophysics against the data can be done only after we measure the distance to the object of study, its velocity (with respect to us) and composition. But this is much easier said than done, for the Universe is so vast that obtaining this information can be a very complicated proposition indeed. How do we know, for example, that a certain galaxy is a billion light-years away? A ruler that long would weigh more than 600 trillion tons and how would we build it? If we imagine sending light beams and timing their return trip (using the known speed of light to extract the distance), how do we arrange for them to be reflected back to Earth? And, even if we did, who will wait around to record their return? How can we determine the composition of a star without scooping some material from it? These difficulties have been met by developing a variety of methods based on our knowledge of the properties of waves, quantum physics, mechanics, and electromagnetism. Some of these methods are described below.

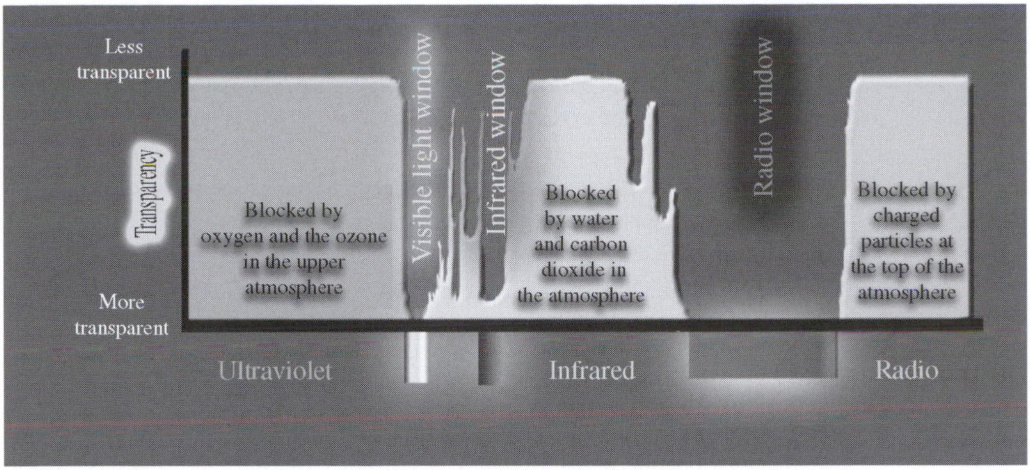

Figure 8.19 Absorption of electromagnetic waves by the atmosphere (observations in the ultraviolet, X, and gamma rays must be done outside our atmosphere).

In pursuing the arguments presented below the reader will perhaps draw a parallel between those interested in this careful measuring of distances, speeds and densities with the Businessman, the sole inhabitant of the fourth planet visited by the Little Prince (in A. de Saint Exupéry's novel of that name). This Businessman spends his life counting and recounting stars, deluding himself that in doing so he will own them. It is true that many astronomers and astrophysicists painstakingly count the stars, and the distances to them. But their aim is not to put the tabulation in the bank as the Businessman would do, but to elicit from these dry figures the story of the Universe, its evolution from the Big Bang to our days and unto its final state, be it one of perpetual expansion or of calamitous re-collapse. The goal is to better understand the Universe, and in doing so, to better appreciate its wonders. Perhaps even the Little Prince would be fascinated by such a story.

Light revisited

Since the earliest antiquity we have used visible light to observe the Universe, a preference motivated by our physiological constraints. But with the advent of a technological society we have widened our horizons using all types of electromagnetic radiation, from radio waves to gamma rays, to probe the recesses of space. Electromagnetic radiation has two important advantages: it travels freely through most of interstellar space (unaffected, for example, by the endemic magnetic fields), and it is easy to detect (though admittedly there are complications since our atmosphere is not transparent to all frequencies – Figure 8.19, and to observe certain wavelengths telescopes have to be placed in orbit). In this section I will describe some basic properties of the light we receive and the manner in which it can be used to extract information about its sources.

The inverse-square law

A source of light will look dimmer the farther away it is; the farther away a star is the fainter it will look. To understand this property in a quantitative way imagine constructing two spheres around a given star, one ten times larger than the other (if the radius of the inner sphere is R, the radius of the outer sphere is $10\,R$). Now let me subdivide each sphere into little parcels, 1 square foot in area, and assume than on the inner sphere I could fit one million such squares. Since the area of a sphere increases as the square of the radius, the second sphere will accommodate 100 times the number of squares, that is, 100 million of them (all 1 square foot in area).[24] Now, since all the light from the star goes through both spheres, the amount of light going through one little square in the inner sphere must be spread out among 100 similar squares on the outer sphere. This implies that the brightness of the star drops by a factor of 100 when we go from the distance R to the distance $10\,R$ (Figure 8.20). If we go to a distance of $20\,R$ the brightness would drop by a factor of 400, which is the square of 20; for $30\,R$ there would be a decrease by a factor of $900 = (30)^2$, etc. Thus we conclude that:

Light intensity drops as $1/(\textit{distance})^2$.

This relation can be used, for example, to determine the brightness at the distance of one light-year when knowing the brightness observed on Earth and the distance to the star (in units of light-years):

$$\textit{brightness 1 light-year away} = (\textit{brightness on Earth}) \times (\textit{distance to star in light-years})$$

I will use this relation repeatedly in the following.

The Doppler Effect

We have seen that light always travels at a fixed speed of about 300 000 km s^{-1}, and that no amount of persuasion will induce a beam of light to move faster; in particular, the speed of light is quite impervious so the motion of its source. Yet there is a different property of light that is quite sensitive to the speed of the source: its color.

Imagine standing by the train tracks listening to the train's horn. As the train approaches the pitch of the blast has a higher pitch, while it becomes lower as the train recedes from you. This implies that the frequency of the sound waves changes depending on the velocity of the source with respect to you: as the train approaches the pitch is higher indicating a higher frequency and smaller wavelength, as the train moves away the pitch is lower, corresponding to a smaller frequency and a longer wavelength.

This behavior, called the *Doppler Effect*, is common to *all* waves, including light waves. If a light bulb at rest gives off pure yellow light, the light reaching your eye will have a smaller wavelength (shifted towards the blue) when it moves towards you, and a longer wavelength (shifted towards the red) when moving away

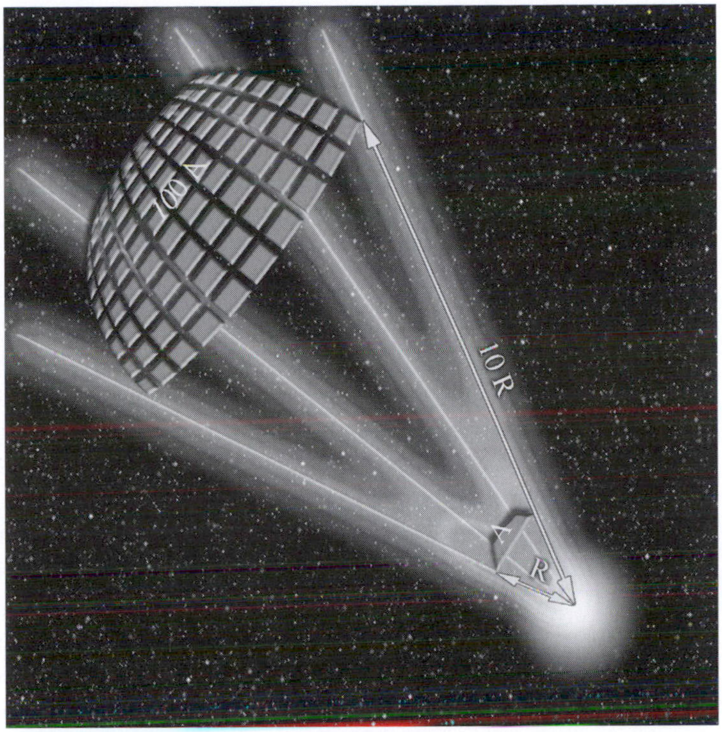

Figure 8.20 Illustration of the inverse-square law: the light through 1 square foot is dispersed into 100 square feet when the distance is 10 times larger, so the light intensity *per square foot* is 100 times smaller. The intensity drops as 1/distance^2.

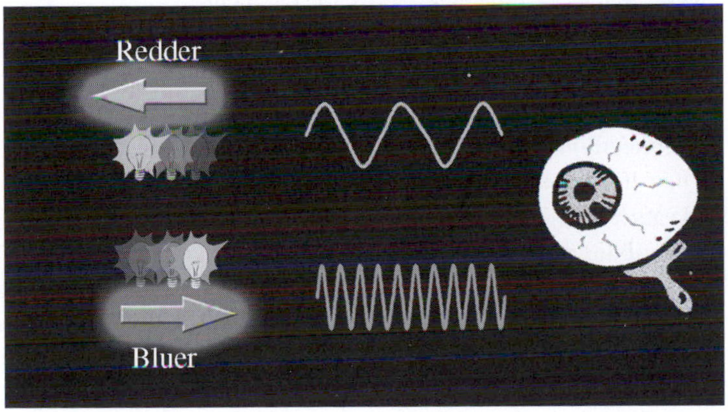

Figure 8.21 Diagram illustrating the Doppler Effect. If the source is moving towards the observer the wavelength decreases (blue-shift); if the motion is away from the observer the wavelength increases (red-shift).

(Figure 8.21). If you have a source of light of a known (and pure) color, you can determine its velocity[25] with respect to you from the color you observe. The Doppler shift will affect light of all colors, changing all wavelengths by the same factor; that is, the *percental* change in the wavelength is the same for all colors. If the source is moving sufficiently fast towards you the yellow light will be received as, for example, X-rays; in this case, however, the source must move at 99.99999999 percent of the speed of light. For most situations the frequency shift is small; even for a source speed of 300 km s^{-1} it is less than 0.1 percent.

Interesting as this might be, it seems useless for the purposes of measuring the speed of a distant galaxy. For how can we know the wavelength of the light it emits *when at rest* with respect to us, except by the impracticable expedient of going there and measuring it? In fact there *is* a way, which curiously enough relies on the properties of atoms.

Emission and absorption lines

When heated every pure element gives off light, but when this light is decomposed using a prism one finds only a few lines, not a smooth spectrum of colors. This set of colors is unique to each element and provides a unique fingerprint: if we heat a mixture of gases and decompose the light it gives off, we can determine the chemical composition of the mixture by observing the colors it emits. The decomposed light given off by any object is termed its *emission spectrum* (Figure 8.22).

Similarly, when we shine white light through a cold cloud of gas of a given element, the gas blocks, or absorbs, some colors. When the "filtered" light is decomposed using a prism, the spectrum is not smooth but shows a series of black lines, and these correspond precisely to the colors the gas gives off when heated. The decomposed white light with some of the colors blocked by the cold gas is called an *absorption spectrum*.

After the discovery of emission and absorption spectra scientists came to rely heavily on their being an inimitable signature of their parent element. For example, in 1868 the English astronomer J. N. Lockyer observed a yellow emission line in the solar spectrum that did not correspond to a known element. He (correctly) hypothesized that this line belonged to a new element he called helium (from the Greek *helios*: Sun)[26] and which was isolated on Earth only 25 years later by Sir W. Ramsay.

Identifying new elements in this way requires great care since some elements produce a large number of lines, some of which are quite hard to detect in the laboratory or even to predict (using quantum mechanics). As an example of the possible pitfalls we have the story of "coronium": in 1869 Harkness and Young discovered a weak green emission line in the solar light (more precisely in the light from the corona, the brilliant "crown" one observes during an eclipse) that did not seem to correspond to any known element. This led to the hypothesis that a new element, dubbed coronium, was responsible for it. It took 70 years for that mysterious line to be identified with one generated by iron under the condition that 13 of its electrons are stripped from it, a state that is relatively easy to achieve in the enormous temperatures present in the outer solar atmosphere, but very hard to see in the laboratory.

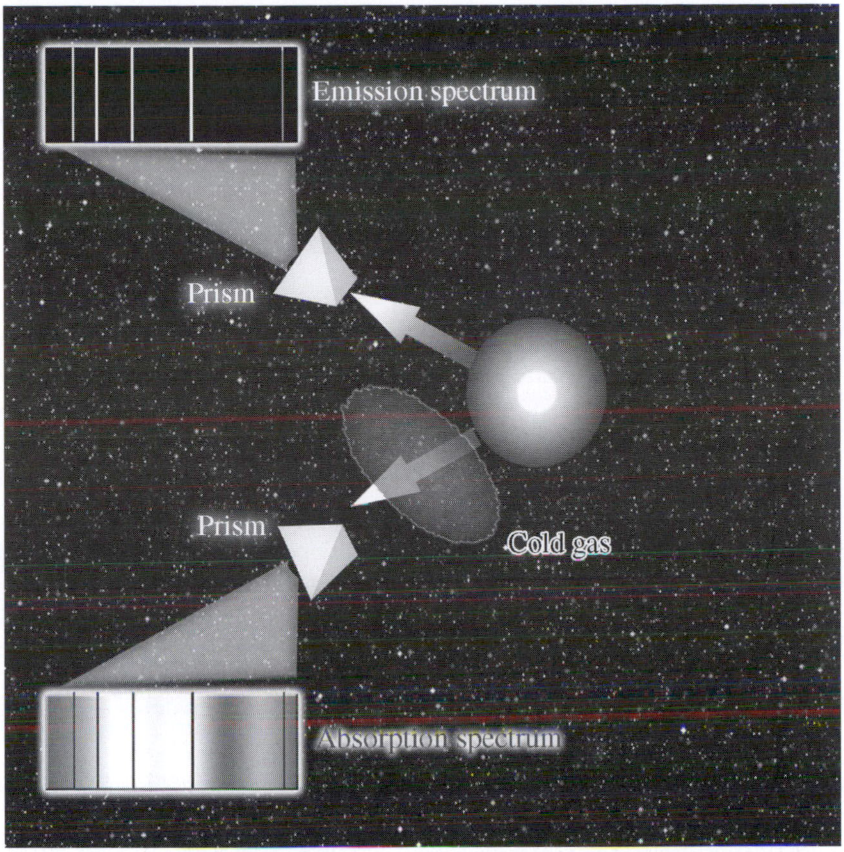

Figure 8.22 The emission and absorption spectra of a pure element. For a realistic situation where many elements are present, the emission and absorption spectra can be very complicated, but they can always be decomposed into a superposition of the spectra for individual elements.

A happy marriage

When observing stellar light from various distant stars (decomposed using a prism) it was found that, just as the Sun, they presented a variety of lines, but for many stars these lines corresponded to no known element. This may imply that each such star carries a new set of elements; but the simplest hypothesis is that the mismatch between the stellar lines and those observed in the laboratory is due to the Doppler Effect, which will shift the lines towards the red or blue according to the motion of the star (which is the source in this case) with respect to Earth. Testing this second possibility is straightforward: the Doppler Effect depends on a single unknown quantity, the speed of the star; therefore an astronomer only needs to determine whether there is a value for the speed such that the lines generated by a combination of (known) elements are shifted to the positions observed in the spectrum from the star. In all cases the matching is perfect: in one fell swoop the astronomer determines the composition and the speed of the star.[27]

It is noteworthy that the philosopher A. Compte asserted that there are certain facts that, though in principle accessible to investigation, will in fact never be known. He cited as an example the chemical composition of stars, which are so distant one cannot imagine obtaining samples of the stuff they are made of for analysis. The ingenious use of scientific facts proved him wrong, showing that it is often our lack of imagination that limits our knowledge.

Cosmic distance ladder

Understanding the Universe, even in its average properties, is a very difficult enterprise, plagued by a variety of technical and theoretical issues. The best theory available for this endeavor is the General Theory of Relativity, which does provide concrete predictions for future states of the Universe and inferences about its configuration in the distant past. These predictions, however, depend on data derived from observations, in particular, its current size.

The basic idea behind most distance measurements can be illustrated by a simple example. Imagine that a student is in her room and would like to find the distance to the building across the street without leaving her desk (like an astronomer wanting to measure the distance to a star without leaving Earth). This she achieves by a series of steps that we call the student's "distance ladder":

- *The first rung*: she gets a ruler and laboriously measures the distance from the desk to the window.
- *The second rung*: using a very accurate chronometer she finds out how long it takes for sound to travel from her desk to the window; using then the fact that *distance* = *velocity* × *time* and the data from the previous step she determines the speed of sound.
- *The third rung*: she stands at her (open) window and claps, measuring the time it takes for the echo from the building across the street to come back to her. Then, knowing the speed of sound from the previous step, she finally determines the distance to the building.

The same idea is used when measuring the distance to objects deep in space: one finds a reliable method to determine the distances to near-by stars (the equivalent of using the ruler). Then one uses this data to measure certain useful quantities (the equivalent of determining the speed of sound). These measurements then provide the calibration necessary to measure the distance to objects outside the range of the first method. This procedure is iterated by measuring different quantities which are useful for larger and larger distances. In the discussion below I will use the *light-year* as a unit of distance; 1 light-year is the distance covered by light in one year, equal to about 9.5 trillion km.

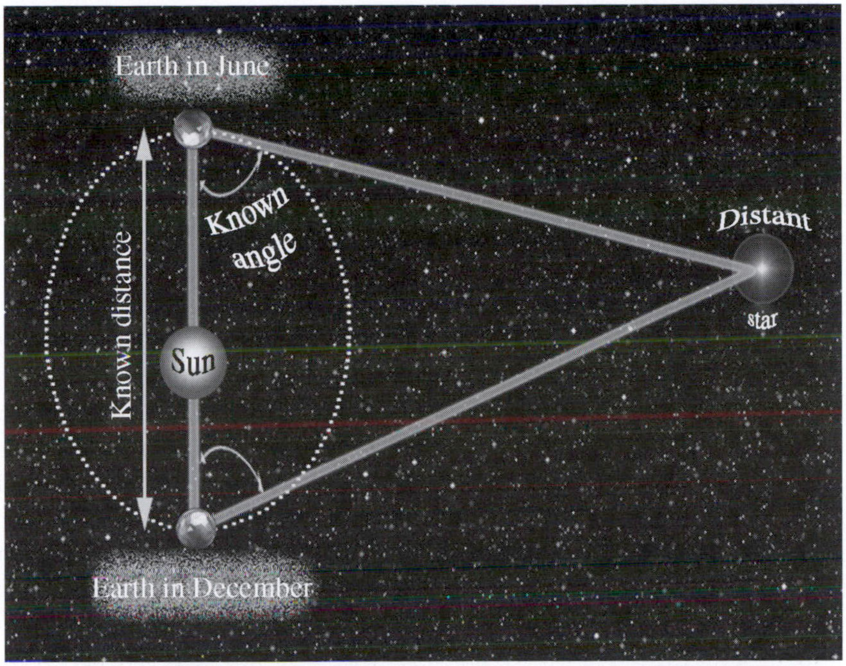

Figure 8.23 Measuring distances using parallax (knowing the size of Earth's orbit and measuring the angles of the light from the star at two points in the orbit, the distance to the star can be derived).

Step 1: distances up to 100 light-years

For near-by stars their distance is measured by parallax: the star is observed in, say, December and then in June, and the direction of the star with respect to the Sun is measured in both cases. Combining these measurements and the known diameter of the Earth's orbit with elementary geometry one can determine the distance to the star (Figure 8.23).

For very distant stars the measured angles will be very close to 90° and the measurements become more and more difficult. For example, Earth-bound observations suffer from atmospheric effects generated by air layers with different temperatures that make the image of the star jitter randomly (and are also responsible for the twinkling of the stars), leading to an uncertainty in the measurement of angles. This type of distortion prevents measuring angles with a precision below 0.033 arc-seconds,[28] which implies that the method fails for stars farther than 100 light-years. Using parallax we can find distances to our nearest neighboring stars.

Step 2: distances up to 300 000 light years

During the decade 1905–15 E. Hertzsprung and H. N. Russel observed a group of nearby stars whose distances they knew (using parallax). Using this distance, the observed brightness, and the $1/(\text{distance})^2$ law, they derived the brightness of each star at a distance of 1 light-year. Then they proceeded to plot this brightness as a function of the star's color (that they got by direct observation). What they

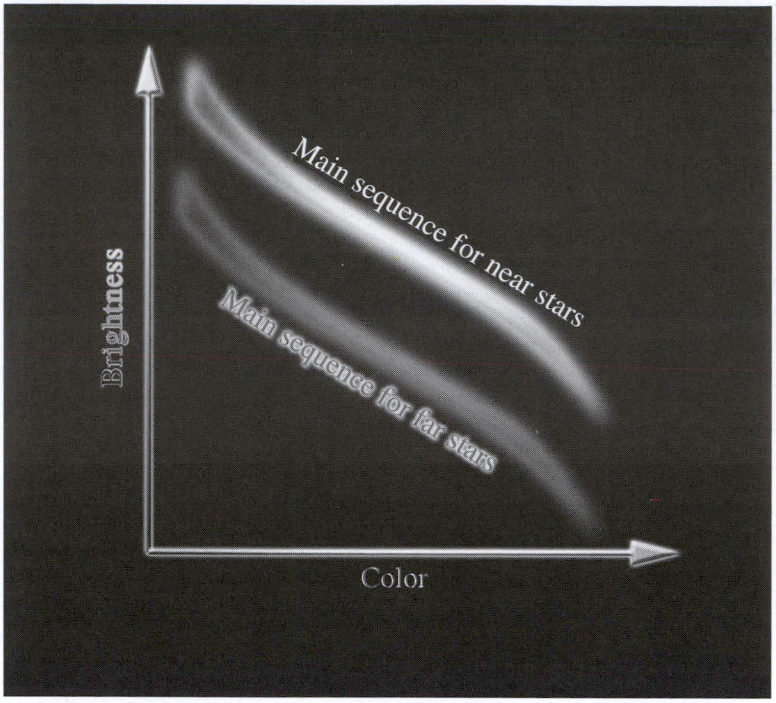

Figure 8.24 The main sequence for near and distant stars; the distance to the latter can be calculated using the difference in brightness.

found is that most stars (90 percent of them) lie on a narrow band in this type of plot which they called the *main sequence*. This result had important implications in the field of stellar structure and evolution, but these points have little bearing for the matter at hand: measuring distances.

Suppose we are interested in a group of distant (farther than 100 light-years) stars that we know are clustered (for which there are good astronomical indicators). Though parallax cannot be used to measure their distance (they are too far away), a method based on the Hertzsprung–Russel plot can remedy this deficiency. The procedure is the following: for each star we plot its *observed* luminosity (here on Earth) and its color. The result is a graph similar to Hertzsprung and Russel's original one, but with one important difference: due to the distance separating us, the main sequence for such stars will be much dimmer (Figure 8.24). This decrease in brightness is due to the $1/(\text{distance})^2$ and from it we can derive the distance to such stars.

This method can be used to determine distances up to 300 000 light-years (the size of a galaxy). For larger distances some of the main sequence stars are too dim, and it becomes hard to sufficiently populate the main sequence to obtain a reliable distance estimate. This method can then be used to measure distances within our galaxy and to nearby satellite galaxies such as the Large and Small Magellanic Clouds, and the dwarf galaxy in Sagittarius.

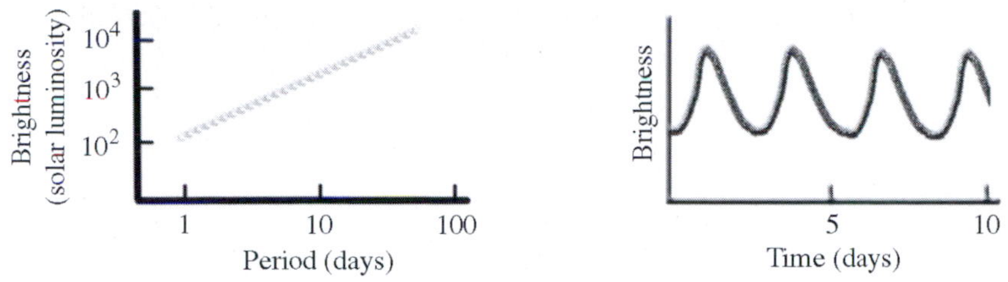

Figure 8.25 Relationship between the brightness and period of the Cepheid variable stars (left) and their brightness as a function of time (right).

Step 3: distances up to 13 000 000 light-years

In 1912 Henrietta Swann Leavitt made a series of careful observations of a set of 25 variable stars, called Cepheid stars[29] located in the Large Magellanic Cloud;[30] her results are the key for measuring intergalactic distances.

Cepheid variables are so called because they brighten and dim periodically and quite regularly (Figure 8.25). Leavitt used the known distance to her chosen stars (obtained using the Hertzsprung–Russel method) and their observed brightness to derive the brightness of each star at some standard distance (e.g. at 1 light-year). Then she noted that this near-by brightness is simply related to its period (the time for it to undergo a bright-dim-bright cycle). Using Leavitt's plot (Figure 8.25) one can infer the brightness of a Cepheid variable at a distance of 1 light-year once the period is measured.

If one requires the distance to a given galaxy one first locates the Cepheid variables in this galaxy and determines the period of each of these stars, and from this (using Figure 8.25) their brightness at the distance of 1 light-year. Comparing this calculated brightness to the one observed on Earth allows us to obtain the distance to these stars using the $1/(\text{distance})^2$ rule.

Using ground-based observatories this method works up to 13 million light-years; for larger distances these stars become too dim to be observed from Earth's surface. Much more recently the Hubble telescope has used these same type of indicators to measure much larger distances (the Hubble telescope is outside the Earth's atmosphere and can detect much fainter stars), for example the distance to the galaxy M100 was estimated to be 56 million light-years using this method.

The Cepheid-variable method allows us to measure the distance to our nearest galactic neighbors, the so-called "local group" consisting of a small cluster of galaxies containing the Milky Way and the Andromeda galaxy (M31), which are bound gravitationally.

Step 4: distances up to 1 000 000 000 light-years

For distances above 6×10^7 light-years all of the above methods fail, for the stars used as standards, even the Cepheids, are too dim. Yet a thorough verification of the predictions of General Relativity and a determination of the ultimate fate of the Universe requires we measure distances up to a billion light years, the average separation between galactic super-clusters. In order to measure such vast expanses one needs an enormously bright source whose properties are well understood. Both of these requirements are fulfilled by certain types of catastrophic stellar explosions that are called (with characteristic flare for the dramatic) “Type Ia supernovas.”

In general a supernova is a stage in the evolution of certain stars during which the stellar core collapses unto itself and in the process becomes enormously bright for a brief period of time, often out-shining a galaxy for a few days. These supernova events can be observed over enormous distances and some of them are sufficiently regular to be used for distance measurements.

Type Ia supernovas occur in binary systems (a system of two stars) where the two stars are at different stages of evolution. One of them was originally relatively small, similar in size to our Sun, but has reached the final stage in its evolution: it has shed its outer layers and its core has collapsed to the size of a planet. This stellar remnant is called a white dwarf and is stable against further collapse *only* if its mass lies below 1.4 solar masses. The second member of the system is at an earlier stage of evolution during which it has grown to many times the size of our Sun, becoming a red giant.

Under these conditions the enormous gravitational pull of the white dwarf will siphon material from the outer layers of the red giant, and the white dwarf's mass grows steadily. Eventually, it will surpass the 1.4 solar masses threshold and will become unstable: its internal configuration cannot balance the enormous self-gravitational pull produced by the increased mass, and the star catastrophically implodes, reducing its size to a few kilometers in a few seconds (at which point it is abruptly stopped by a pressure generated by the highly compressed nuclear components). The energy released produces a brilliant display of fireworks we call a Type Ia supernova event.

Since all Type Ia supernovas have the same “detonation” mechanism and similar white dwarf parentage, their evolution is relatively easy to distinguish through the way in which their brightness changes as a function of time (its “lightcurve”) during the flare-up and the succeeding days (Figure 8.26). The observation of nearby Type Ia supernovas, whose distance is measured using one of the previous methods allows us to determine the maximum brightness these objects reach when observed at some standard distance (of, say, 1 light-year). This requires a repetition of the previous procedure: we know the distance through the application of one of the previous methods, and this, compared to the observed brightness will give the desired value with the use of the $1/(\text{distance})^2$ rule.

Once a supernova event is identified as being Type Ia by its lightcurve, a comparison of the observed maximum brightness with the known value at the distance of 1 light-year yields the distance to the object (using the $1/(\text{distance})^2$ rule once again). The enormous brightness of these events implies

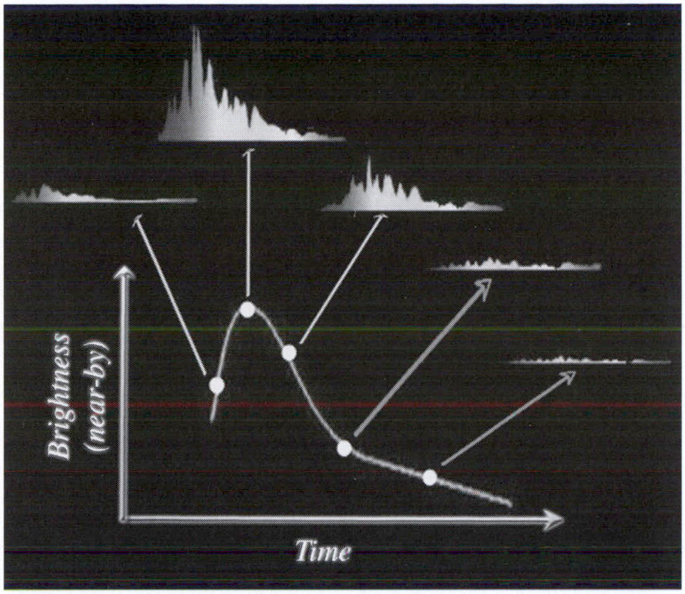

Figure 8.26 Lightcurve (brightness as a function of time) for a Type Ia supernova. Also included are the spectra (the light intensity for each color) at different times. (Courtesy of the Supernova Cosmology Project)

they can be observed across vast distances, making this an applicable method up to 10^9 light-years.

The expanding Universe at last

We have finally reached the stage where measurements can be used to compare observations with the predictions of the General Theory, and the results unequivocally support the theory. These results indicate that the Cosmological Principle is satisfied (over distances of the order of a billion light-years), and the Universe expands as predicted by the Big Bang Theory. In addition, the rate of expansion indicates the presence of a non-zero Cosmological Constant, and that the total mass–energy density (including the Cosmological Constant effects) of the Universe is equal to its critical value ($10^{-29}\,\mathrm{g\,cm^{-3}}$): the Universe is isotropic, homogeneous, expanding, and flat.

Dramatic as they are, these results must be taken with a measure of caution. The reason is that all images of very far supernovae also correspond to events that occurred in the distant past, and characteristics of supernovae could have been different during those early times. Such a modification could radically alter the above conclusions. To balance this skepticism it is important to note that the *lightcurves* from these early supernovae do match those obtained from near-by supernovae. So, if the supernovae characteristics were modified in the early Universe, this modification would only affect the maximum brightness of the event, leaving all its other distinguishing properties untouched. Though this is not inconceivable, it has proven difficult to generate a theory that meets these constraints.

Notes

1. The average properties of smaller regions (for example those of the size of a galactic cluster – 3 megaparsecs) do exhibit significant differences.
2. In fact, galaxies were "discovered" only in 1924!
3. For a homogeneous and isotropic universe the amount of matter–energy in a given volume is the same no matter where the volume is located. The density of matter–energy is the amount of this quantity in a volume of convenient size (like 1 cubic megaparsec, for example).
4. One might rightly wonder how is it possible to measure the distances to such distant objects (sometimes reaching hundreds of megaparsecs), and how can we then determine the speed of such objects. The methods developed to achieve these goals are highly non-trivial and are summarized in the appendix to this chapter.
5. The label "Hubble's Law" was given to honor E. Hubble; it is *not* an additional constraint imposed on the properties of the Universe but a *consequence* of the General Theory and the Cosmological Principle.
6. Alas! Another understatement, in this case, one of cosmic proportions.
7. Our current understanding reaches back to about 10^{-11} s after the Big Bang (temperatures of about 10^{15} K). There are many hypotheses about the way the Universe looked at earlier times, but none has been accepted yet; this is an area of active research.
8. See Chapter 9 for a brief description of the properties of these particles.
9. All atoms are made of a tiny very dense nucleus surrounded by a cloud of electrons; the nucleus is made of a number of neutrons and protons (neutral and positively charged, respectively). Neutrons and protons are about 2 000 times heavier than an electron.
10. See Chapter 9 for a more detailed description of neutrinos and their properties.
11. This means that deuterium has the same chemical properties as hydrogen but its nucleus is made of a proton and a neutron bound together (the nucleus of hydrogen is a simple proton).
12. This however is no conclusive counter-argument: even if one *dislikes* disposing of the principle of the conservation of energy, future unambiguous experimental results might force a revision of these ideas; our personal proclivities carry no weight in Nature, only experimental evidence counts.
13. No effect travels faster than light speed, so the size of the regions that were in contact 10^{-11} seconds after the Big Bang is roughly the distance that light can cover in this time.
14. When studying stars we get much information from the electromagnetic radiation we receive from them. The current theory of star formation and evolution is able to account for all the observations, so that the mass estimates thus obtained are quite reliable. There *are* deviants, but their numbers are too small to be significant when estimating the *average* mass density.
15. The contrast between the WIMP and MACHO acronyms is, of course, not a coincidence.
16. Such considerations assume the existence of dark matter, and would become irrelevant should this hypothesis be disproved.
17. This irregular structure does not contradict the Cosmological Principle, on a scale of 100 megaparsecs or so the Universe is quite homogeneous.
18. For example, the Cryogenic Dark Matter Search (CDMS) experiment at Lawrence Berkeley National Laboratory (USA); the Dark Matter (DAMA) experiment based in

Rome and Beijing; the NaIAD experiment at the Boulby mine (UK); with others being planned.

19. G. Gamow, *My World Line* (New York: Viking Press, 1970), p. 44.
20. The quantity $1/H_0$ is a measure of the age of the Universe.
21. An inaccuracy that defies even the budget deficit predictions of the most reckless of governments.
22. There are many variants of the ideas presented here, in some cases, for example, the expansion factor is 10^{50}; if a single hydrogen atom were to be enlarged by this amount, it would become a billion times the size of our observable Universe.
23. In this case the label "constant" is, of course, misused.
24. This increase as the square of the distance is a consequence of our space having three dimensions.
25. More precisely this is the velocity along the line of sight.
26. The same line has been previously observed by the French astronomer P. Jenssen, but was misidentified as a sodium line.
27. More accurately, the composition of the outer layers of the star.
28. One degree contains 60 arc-minutes, and each arc-minute contains 60 arc-seconds.
29. The name is derived from the constellation in which they were first observed.
30. This is a small galaxy (of only 10^8 stars) bound to the Milky Way.

9
The lives of a star

Introduction

All stars have a definite "intrinsic" luminosity (the luminosity measured close-by) and a definite color (determined by the wavelength of the predominant electromagnetic radiation coming from the star), and because of this each has its particular location in the Hertzsprung–Russel diagram described in the previous chapter. Should a star change characteristics over time, this evolution can be plotted as a curve in the Hertzsprung–Russel diagram (Figure 9.1). Since most stars are found in the main sequence, it is reasonable to hypothesize that stars spend most of their lives in that region of the Hertzprung–Russel plot, evolving into it when they are born and out of it when they are about to die. Models of stellar evolution confirm this.

Of the four known forces[1] it is gravitation that determines the evolution and final fate of very large objects. This might seem puzzling since gravity is the weakest of these forces,[2] so one might guess, for example, that electric effects will completely dominate gravitational ones. Curiously enough this is not the case: it *is* true that the electrostatic repulsion of a pair of stars with a large negative charge can overcome the gravitational attraction if the charge in the stars is large enough (for two Sun-like stars this would be the case if we removed only one electron out of every 10^{18}), but such hypothetical stars cannot exist since they are *unstable*. The reason is that the charges inside each star would also repel one another and this repulsion will be larger than the gravitational attraction that keeps the star together: either the charges are expelled from the star or else the star itself disintegrates. Since electric forces are so much larger than gravitational ones, even relatively small excesses of positive or negative charge are incompatible with stellar stability; as a result all stars are electrically neutral (to an accuracy of better than one part in 10^{18}).

Though stars cannot be electrically charged, they can (and usually do) support very large magnetic fields. This is because magnetic forces are never purely attractive or repulsive since individual magnetic poles are not found in Nature, and, in addition, because the forces generated by such fields are much smaller than their electric counterparts.[3] Stellar magnetic fields reach intensities several

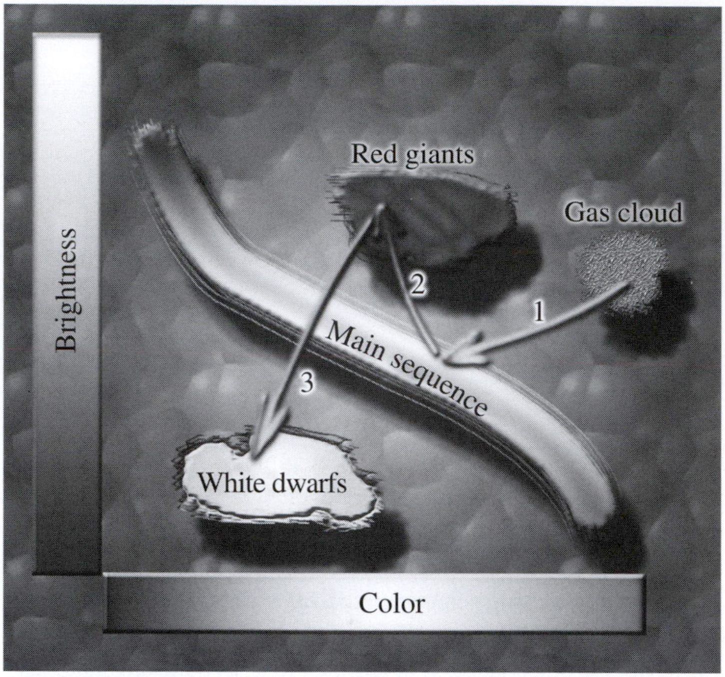

Figure 9.1 Diagram illustrating the evolution of a Sun-like star. Born from a gas cloud it moves towards the main sequence (1) where it spends most of its life. After all hydrogen is consumed in its core, the star burns helium and becomes a red giant (2). Finally, when the helium is consumed nuclear reactions subside and the star becomes a white dwarf (3) where it will spend its remaining (billions of) years.

thousands of times larger than Earth's, and neutron stars have fields reaching a trillion times stronger than Earth's, and even then the star is not destabilized by them.

Gravity generates a force that permeates the whole of the star, and leads to a compression of the stellar material. When subject to such treatment this material is far from passive and reacts by generating an opposing pressure that tends to balance the contraction generated by gravity. Stellar stability or evolution results from the greater or lesser efficiency of this opposition. Were it not for the (fortunate) presence of such balancing effects, all matter would have long ago collapsed into one or more pinpoints of enormous density under the universal and relentless gravitational attraction.

Pressure

At any stable stage in its life, the basic properties of a star are largely determined by the type of pressure that is most effective in balancing gravity. The interactions among atoms and among subatomic particles are ultimately responsible for generating all types of pressure, and so are indirectly responsible for the main characteristics of stars. It is then worthwhile to take a brief detour in order to provide a very cursory view of the subatomic world, before describing the way in which the inhabitants of the subatomic zoo conspire to oppose gravity.

A quantum pit stop

Our material world is populated with a great variety of substances which are now understood to be composed of a hundred or so chemical elements. Each element comes in packages we call atoms, and these combine into compounds according to a series of well understood affinities. Atoms, however, are not the indivisible geometrical objects envisioned by Democritus and the early Greek atomists, but are complicated entities composed of a heavy and very small nucleus surrounded by a cloud of electrons. Atomic nuclei are composed (mostly) of neutrons and protons, and these are composed of quarks. If we had a microscope capable of resolving objects of sizes 10^{-10} m we would be able to see individual atoms. If we increased the resolving power by 100 000 we would be able to see objects of sizes 10^{-15} m, and we would be able to resolve individual protons and neutrons (Figure 9.2); quarks would show up if we managed another increase in the magnification by a factor of 100. The most powerful microscopes available can observe distances down to 10^{-19} m, without any indication that quarks or electrons are made up of some even smaller objects (though even more powerful microscopes may certainly uncover such a structure).

The proton, electron, neutrinos, and lightest quarks are stable: if left alone they apparently will remain unchanged indefinitely.[4] Heavier quarks (they come in six "flavors") and neutrons are unstable: if left alone they will spontaneously decay producing two or more lighter particles in the process; for the neutron this takes about 15 min, for the quarks less than 10^{-8} s.

Quarks are held together inside protons and neutrons by what is unimaginatively called the *strong force*, which is attractive. This same force affects neutrons and protons and binds them into nuclei. Inside a stable atomic nucleus the neutrons and protons are distributed so as to maximize the average distance between protons, so that the electric repulsion is as small as possible, making it easier for the strong force to keep the system together. However, if the number of protons is very large the nucleus eventually becomes unstable and will spontaneously disintegrate into smaller daughter nuclei and a few free neutrons. For example, the nucleus of uranium 238 has 92 protons and 146 neutrons and is relatively stable (if we have a collection of uranium 238 atoms it takes about 4.5 billion years for half of them to decay); its cousin[5] uranium 235 also has 92 protons but only 143 neutrons, and having a smaller neutron "buffer" is less stable (half a population will decay in 704 million years, six times faster than uranium 238).

The quantum world has several other voting members, among which neutrinos play an important role in many stellar processes. The neutrino history is rather curious. It has been known since 1914 that certain radioactive nuclei decay, not by breaking into smaller nuclei, but through the emission of electrons (though at the time the electrons were not identified as such and were called "beta rays"). One surprising characteristic of such processes is that the total

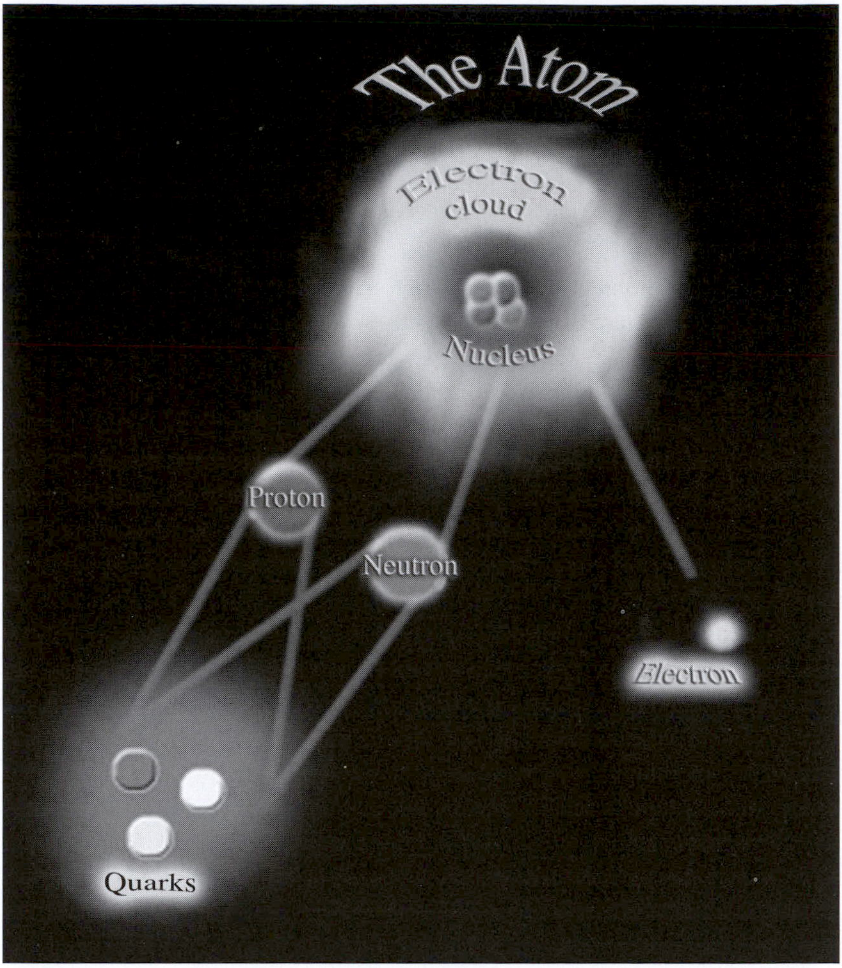

Figure 9.2 The layers of atomic structure.

energy of the emitted electron and final nucleus varies within a certain range, and is consistently smaller than the initial energy of the decaying parent nucleus (equal to $m_{\text{parent}}c^2$). This led W. Pauli in 1930 to hypothesize that the energy deficit is carried away by an unknown particle which happens to be very hard to detect, and which is now called a *neutrino* ("little neutral one"). Though the neutrino hypothesis came to be quite generally accepted, it had to wait 26 years before incontrovertible evidence of the existence of neutrinos was obtained. Neutrinos are far from rare, being copiously produced in stellar processes (through the ongoing nuclear reactions) and in the early Universe. These particles interact only very weakly with all other types of matter (on average a neutrino will undergo a single interaction while traveling through 10^{13} km of lead). So despite this continual neutrino bombardment from the Sun and other sources, the overwhelming majority of these "projectiles" simply go through the

Earth and its inhabitants without causing any damage. They are doing so even as you read this.

The above citizens of the subatomic world, electrons, neutrons, protons, quarks, and neutrinos are members of a family of quantum particles called *fermions*. These particles can be characterized by their charge, their position and an additional property called *spin*, which is akin to their being perpetually rotating about an axis.[6] A remarkable peculiarity of fermions is that no two of them can have identical properties. For example, two neutrons of the same spin cannot, by their very nature, come close together for then they would have the same properties (position and spin). When a group of fermions of the same type and with the same spin are compelled to approach closely they will generate a very stiff opposition to further contraction, *in addition* to the electric repulsion present when these particles are charged. The impossibility of two identical fermions being in close proximity is called the *Pauli Exclusion Principle* and is responsible for the electron degenerate pressure described below; it is also the reason we are – usually – supported by the chairs we sit on.

Gas pressure

Almost everyone is familiar with air pressure: it keeps car tires and balloons inflated, it affects the weather, etc. To understand how this common type of pressure is generated imagine a poor traveler who misses the train in the middle of the monsoon season, and, having no alternative, he opens his umbrella and starts walking to his destination. As he walks, myriad water drops fall on his umbrella and bounce off, and because of this he feels the umbrella being pressed down; the heavier the downpour, or the higher the speed of the drops, the stronger the effect (Figure 9.3). A similar effect keeps balloons and tires inflated: their insides are filled with billions and billions of air molecules, and these bounce off the walls, pushing them outward in the process. Of course there are also air molecules doing the same on the outside, so if, for example, one wants to inflate a tire, the trick is to have more molecules inside than outside bouncing off each bit of surface every second, or to have the inside molecules move faster. In the first case the gas inside is denser (Figure 9.4), in the second it is hotter. A star that is stabilized by gas pressure working against gravity is called a "normal" star (only because this is the most abundant type).

Degenerate[7] fermion pressure

Imagine that we take an absolutely unbreakable air-tight box filled with air. The top of the box is fitted with a piston so that we can decrease the box's volume at will (Figure 9.4). As we push down on the piston the gas pressure increases (for the reasons mentioned above); continuing in this endeavor we soon find that the air molecules break into their constituent atoms due to the now frequent collisions

Figure 9.3 Under a downpour the bouncing raindrops increase the pressure on the umbrella in the same way as gas molecules produce a pressure on a surface. (Courtesy of Malcolm Saunders)

with the walls and among themselves. At still smaller volumes we will find that the atoms break apart into nuclei and electrons; the temperature inside the box is also much higher, having reached about 10 000 °C. Continuing the contraction we eventually reach a density of about 15 000 times that of water at room temperature, at which point the electrons find themselves in close proximity. Electrons, being fermions, are controlled by the Pauli Exclusion Principle and will accordingly generate a very strong pressure, the *degenerate electron* pressure, which is much stiffer than the still-present gas pressure or the electric repulsion.

The onset of the degenerate electron pressure is rather sudden and will offer a rapidly increasing resistance to further reduction of our box's volume. But let us assume that we have the means of overcoming it, and proceed with the contraction. When the temperature reaches a few million degrees the atomic nuclei collide so violently that their neutrons and protons are rearranged, so that new nuclei are produced. When the temperature reaches several million degrees the atomic nuclei will break apart into their constituent protons and neutrons, and the neutrons will proceed to decay. But not all neutrons disappear, for at these enormous temperatures and densities there are additional nuclear reactions that

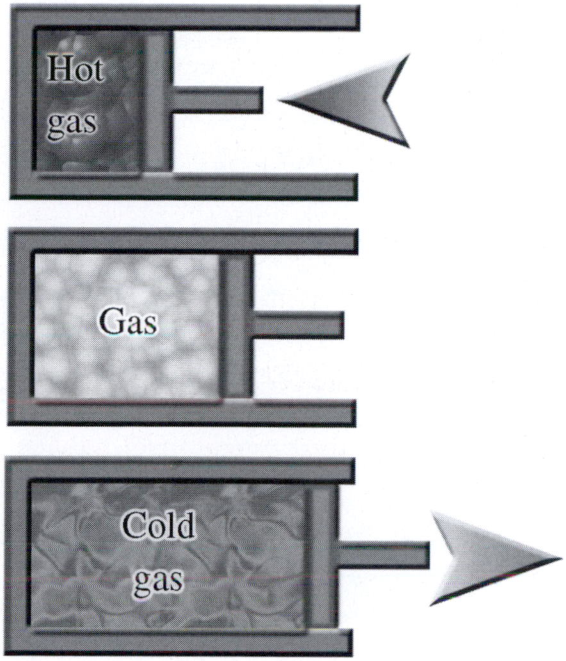

Figure 9.4 A device for studying gas pressure. As the piston is driven in, the gas inside becomes denser and the pressure and temperature increase; the opposite effects occur when the piston is moved out.

produce neutrons: when an electron and a proton collide, they can produce a neutron and a neutrino (Figure 9.5). Though both neutron decay and production occur copiously, under the conditions described above the latter is favored, so eventually we end up with a box full of neutrons and neutrinos,[8] with a sprinkling of electrons and protons – since most of these were "used up" in creating neutrons. Once the density inside the box reaches 10^{17} times that of water at room temperature the neutrons and protons are forced to be very close together and the corresponding degenerate pressures abruptly set in. As before, this will present an added stiff resistance to further contraction (much harder to overcome than the one generated by the electrons).

If we manage to overcome even the neutron degenerate pressure, the material inside the box eventually reaches a temperature of 10^8 °C, at which point the protons and neutrons themselves will break apart into their constituent quarks.[9] While there are speculations that the core of certain very massive stars will contain this "quark matter" (at least at some stage in their evolution), this is at best a rare occurrence.

A star which is stabilized by the electron degenerate pressure is called a *white dwarf* (that this peculiar name is appropriate will become clear soon). If the star is stabilized by the neutron and proton degenerate pressures (where the neutron component, being the most abundant, dominates) it is called a *neutron star*. I will sketch below the conditions that determine which configuration corresponds to a given star, but before going into that we need to understand what makes stars tick.

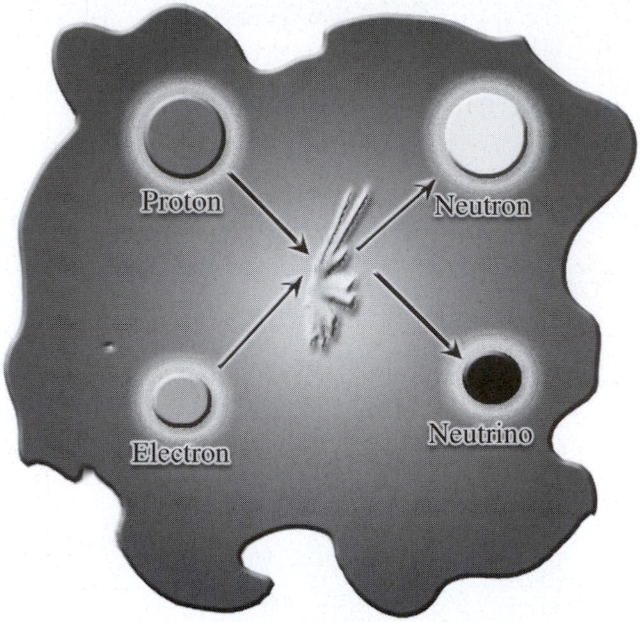

Figure 9.5
Transformation of electrons and protons into neutrinos through nuclear processes.

Stellar power

If there is one feature that makes stars noticeable it is the fact that they shine, that they do so under their own power, and that they are capable of sustaining their light and heat output for an extremely long time (as we know from the geological evidence which shows that our Sun has been shining upon our planet for at least a few billion years). For the Sun, "normal" methods of energy generation, such as coal-burning, cannot account for such a stupendous energy output for more than 15 million years; the source of stellar energy must lie elsewhere.

The clue to the mystery was provided by various nuclear processes that appear to generate enormous quantities of energy with no apparent source. This miracle was soon explained by noting that the conversion of a tiny amount of matter into energy could account for the observations, a hypothesis that was soon verified. This led Arthur Eddington, in 1920, to hypothesize that nuclear reactions were also responsible for making the stars shine (Figure 9.6); by 1938 the detailed nuclear reactions that should occur in the stellar interior were elucidated by H. Bethe (Figure 9.7). These reactions occur only in the densest and hottest section of the star, the central core; the energy generated then seeps outward increasing both the temperature and the gas pressure of the stellar material. For a sufficiently large energy output the gas pressure will balance the gravitational tendency toward collapse and the star will stabilize.

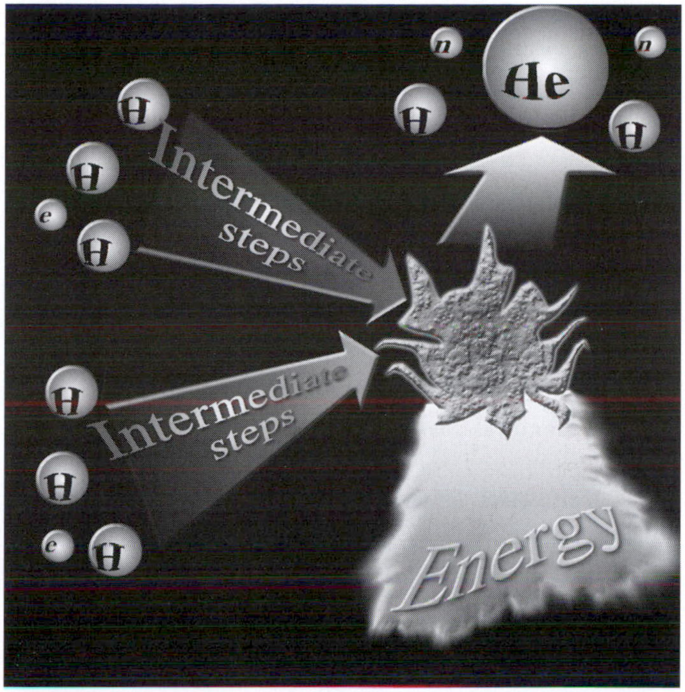

Figure 9.6 Illustration of the nuclear reactions that create helium (He), energy, and neutrinos (n) from hydrogen (H) fusion; "e" denotes electrons (during the "intermediate steps" some unstable nuclei are created).

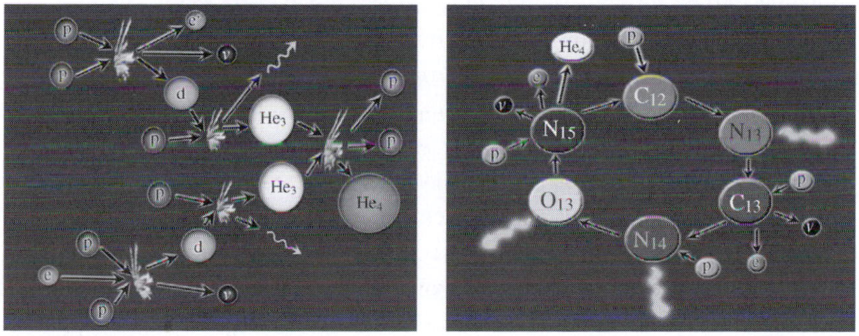

Figure 9.7 The two main nuclear reaction chains inside stars. The pp-cycle (left) dominates in stars up to about one solar mass; the CNO-cycle (right) dominates in heavier stars. Wiggly arrows denote emitted radiation, e denotes electrons, e^+ anti-electrons, ν neutrinos, p protons, d deuterium nuclei – a hydrogen isotope, He_4 denotes a helium nucleus and He_3 one of its isotopes, C_{12} a carbon nucleus and C_{13} one of its isotopes, N_{14} a nitrogen nucleus and N_{13} and N_{15} two of its isotopes, O_{13} represents an unstable oxygen isotope.

In the pp-cycle two protons collide to form e^+, ν and d; alternatively two protons and an electron collide to form ν and d. The deuterium collides with a p to form a He_3 nucleus, and two of these combine to form a He_4 and two protons.

In the CNO-cycle a proton combines with a C_{12} nucleus to form N_{13}, which then emits radiation and is transformed into C_{13}. This nucleus then collides with a proton to form e, ν and N_{14}, which upon collision with a proton is transformed into O_{13} while emitting radiation; the O_{13} then decays spontaneously into N_{15}, again emitting radiation. Finally the N_{15} collides with a proton and produces e, ν, He_4 and C_{12}, and the cycle repeats itself.

Satisfying as this story might be one might wonder how it is possible to decide whether nuclear processes are indeed the origin of a stellar brightness, for this apparently requires us to probe the very core of the stars, something not even science fiction writers like to envisage. The evidence supporting the nuclear hypothesis of stellar power is indirect: it is possible to construct computer models of stars that include all the details of the nuclear furnace at the core (as provided by the hypothesis), together with the physical properties of the outlying stellar material. The calculations then predict various observable properties of the stars that can be compared with observations. For our Sun, the models predict both the luminosity, and the various modes of oscillation of the solar atmosphere;[10] these predictions are amply confirmed with observations. In addition the recent observation of the predicted neutrino flux all but unambiguously confirms that nuclear power is responsible for the generation of solar energy.

The main power source for all stars is then furnished by nuclear reactions. Possibly the most familiar of these reactions are the ones used in nuclear power plants; these, however, are *not* the ones occurring in stellar interiors.[11] The relevant reactions present inside stars go under the name of *nuclear fusion*. At the center of the stars, temperatures reach a few million degrees and pressures top several trillion times atmospheric pressure on Earth. Under such extreme conditions, atoms break apart resulting in a "soup" of atomic nuclei and electrons, all of which experience frequent and violent collisions. Nuclear collisions will be forceful enough to overcome the electric repulsion (all nuclei have a positive charge), allowing the participants to come very close together. At such short distances (10^{-15} m) the strong force comes into play and will rearrange the components of the colliding nuclei. When this happens the result is *new* nuclei whose combined mass is smaller than that of their parents, and the mass deficit appears as energy. For example, slamming hydrogen nuclei can produce the nucleus of a new element, helium (Figure 9.6); the jargon is that this reaction "burns" hydrogen and that resulting "ashes" are mainly helium.

Hydrogen is the most abundant element in stars and the most significant fusion reactions involve this element. The details are somewhat involved and the dominant reaction depends on the stellar mass (Figure 9.7); but in all cases the end product is the creation of helium atoms, electromagnetic radiation, and neutrinos. Most of the radiation heats up the core, generating the pressure that balances gravity and also fosters further nuclear reactions. The left-over energy makes a tedious pilgrimage, repeatedly scattering off the stellar material as it makes its way towards the surface; eventually some of this radiation is emitted into space. Neutrinos also leave the star but they do so quite easily for they interact very weakly.

The processes that convert hydrogen into helium constitute the main reactions in the mature stages of stellar evolution. But, as the star ages and starts speeding up towards its death, other nuclear processes become important. At core temperatures of about 600 million °C helium nuclei will fuse to form carbon. Then, as temperatures increase further, nuclei will "eat up" a helium nucleus and form heavier ones. In this way oxygen, neon, magnesium, and silicon are produced. Finally, when the central temperature reaches three billion °C, other more involved nuclear reactions produce iron from silicon through a long series of intermediate steps. Iron is of particular interest due to its stability: all previous nuclear reactions produced energy, but those involving iron require that energy be *supplied* in order to occur. As a consequence of this, once iron is created in the core of a star it will simply stay there, impervious to its extreme surroundings during all but the most violent of stellar death throes.

Stars lose mass as they pour out electromagnetic radiation (cf. the equivalence of mass and energy) in addition, most stars exhibit violent flares that result in their spewing out particles which also results in a decrease of the stellar mass. For our Sun the loss is of "only" 1.35×10^{14} (135 trillion) tons per year (which, large as it sounds, is only about 7×10^{-12} – 7 trillionths – of a percent of the total solar mass).

The lives of a star

With these preliminaries I can now describe qualitatively the highlights in the life of a star. This description will be but a very rough approximation to the intricate processes that characterize the various stages of stellar evolution, and which are responsible for the stupendous phenomena that mark the transitions between these stages. The sections below should be read keeping this in mind.

Birth

The life of stars begins with a swirling cloud of dust, gas (composed mainly of hydrogen, since this is the most abundant element), and miscellaneous debris, located in a region not isolated from other cosmic objects. A significant perturbation (such as a shockwave from a near-by exploding star or a tiny inhomogeneity inherited from the early Universe) will compress the cloud in a non-uniform manner, resulting in regions of higher and lower density. The regions where density increases will tend to attract more material due to the increased gravitational pull they produce; as a result these regions will grow at the expense of their surroundings: the cloud will both contract and clump into separate blobs, with the largest one at the center.[12]

As the cloud contracts it will speed up its rotation (much as an ice-skater turns faster when he draws his hands towards his body), possibly ejecting some of the outlying blobs due to centrifugal effects, while the remaining blobs will gradually become more and more compact. This process takes about one billion years to complete and produces a primitive planetary system: a protostar (which is very big but too cold to produce nuclear reactions) circled by protoplanets.

As time passes both protoplanets and protostar will become more and more compact (again due to gravity's pull), and this contraction drives up their internal temperature.

A rising star

If the protostar mass is large enough (above about ten percent of the mass of the Sun) its central temperature will eventually reach 15×10^6 K and nuclear reactions at the core turn on. The energy released then heats up the stellar material, which increases the pressure, and the contraction stops. The star has reached a stable plateau in its evolution; it has come of age and accordingly takes its place in the main sequence of the Hertzsprung–Russel diagram (Figure 9.1). The main nuclear reaction occurring at this stage of the star's life consumes hydrogen and produces helium: the star "burns" hydrogen into helium.

By now most of the debris from the original cloud has been expelled by the radiation emanating from the star, or has been swept by the already fully formed planets; the stellar system has reached maturity. This stage lasts a relatively long time depending on the mass of the star, if it is as light as our Sun it proceeds for about 10 billion years; but heavier stars use up hydrogen much faster: for a star of 25 solar masses this stage is reduced to 10 million years, and for the heaviest ones to "only" one million years.

A giant appears

After the supply of hydrogen in the core is depleted, the corresponding nuclear reactions stop (there are other fusion reactions, but they can occur only at temperatures higher than the ones present at the center of the star at this stage). Without this energy supply the pressure drops and the gravitational collapse proceeds. As the core contracts its material is compressed more and more, increasing the central temperature until, at about 10^8 K, nuclear reactions involving helium "turn on": helium atoms slam together, and, through a complicated reaction, produce carbon and copious amounts of electromagnetic radiation. The energy output raises the core temperature, while the radiation pressure pushes out the external layers, resulting in a star many times its previous size consisting of a very hot core surrounded by cooler outer layers that shine red. The star has become a *red giant*.

Our Sun will undergo this metamorphosis in about 4.5 billion years. The bloated result will grow to the point that it will engulf the orbits of Mercury, Venus and, possibly, the Earth (more massive stars can grow to spectacular sizes – Figure 9.8). The red-giant phase lasts a relatively short time (about one percent of the hydrogen-burning phase), which the star spends off the main sequence (Figure 9.1).

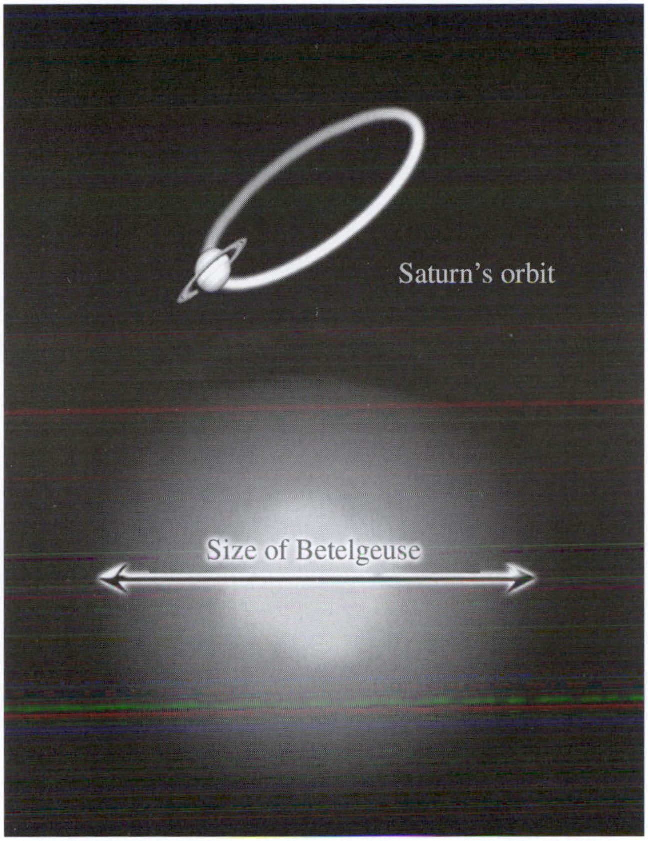

Figure 9.8 The red super-giant Betelgeuse in the constellation Orion (Saturn's orbit included for comparison).

And so it goes

When the supply of helium is used up the story is repeated: gravitational contraction takes over and the star collapses further. Eventually other nuclear reactions become viable, energy output increases for a time until the various nuclei are depleted, and then contraction takes over again. In this way the star produces oxygen, silicon, and, finally, iron. Each new nuclear reaction burns in a smaller portion of the core than the previous one, so that the stellar interior resembles an onion whose somewhat fuzzy layers detail the series of processes the star has lived through (Figure 9.9).

When the core of the star turns into iron all nuclear reactions stop permanently. The reason is that iron is a very stable nucleus and all nuclear processes involving iron use up more energy than they produce. Without an energy source the core can no longer sustain itself against the overwhelming gravitational attraction, and the star collapses. One must imagine a ball of iron containing about half the star mass, which simply falls into itself, reaching the size of a planet in a few minutes. For our Sun the core will contain about one

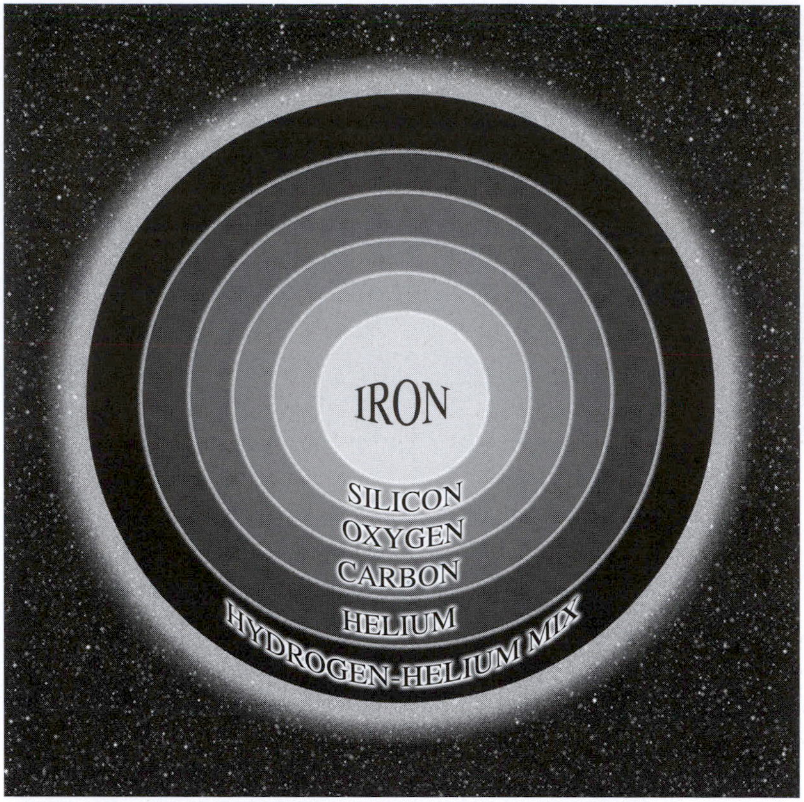

Figure 9.9 Illustration of the stellar layers created by nuclear processes characterizing the various stages of its evolution.

thousand-trillion-trillion tons (10^{30} kg) of iron,[13] and will contract from a radius of about 100 000 km to about 5000 km in one hour, at an average speed of 30 km s^{-1}, 100 times the speed of sound in air.

At the subatomic level the electrons are stripped off the atoms and are then squashed closely together. As the density reaches 15 000 times that of water on Earth, the electron degenerate pressure will rise abruptly and present a stout opposition to gravity. This effect is "powered" by the electrons' dislike for close contact and so it is not depleted over time (as the nuclear reactions were). The question is whether this electronic opposition is stiff enough to break the runaway gravitational contraction.

Light stars

For stars whose cores are lighter than 1.4 solar masses the electron degenerate pressure *will* balance gravity; in doing so the core will suddenly decrease its breakneck speed from many km s^{-1} to almost zero. This will involve a certain amount of overshoot and bouncing back and forth before stability is achieved, and in the process all the outer layers of the star are ejected. The final result is a

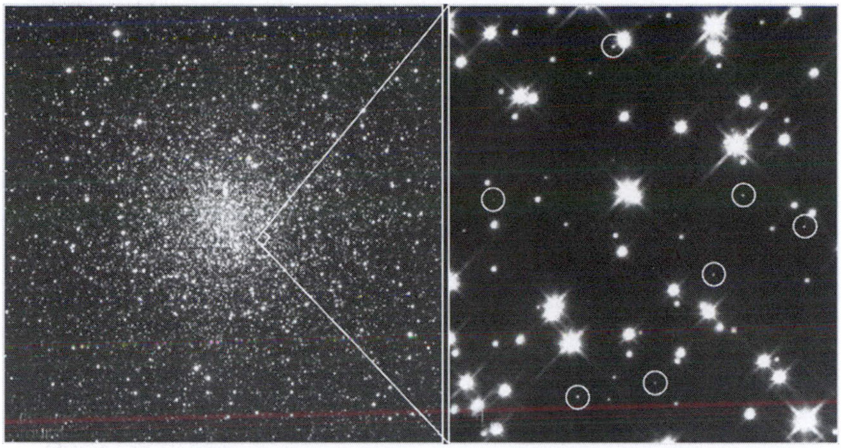

Figure 9.10 Globular cluster M4 (left) with a close-up of several white dwarfs near its center. (Courtesy of STSCI NASA)

beautiful expanding cloud of stellar material, at the center of which there remains a small bright star, a *white dwarf*, the reminder of the core.

By the time contraction is stopped and the star stabilizes, its core will have contracted to the size of a small planet (like ours), and its material will be so dense a teaspoon of it would weigh 1 ton on Earth. Though their racy days of nuclear reactions are forever gone, white dwarfs remain bright for a very long time, shining by slowly depleting their left-over heat from their earlier history. Thus they very gradually cool and eventually darken into dead cinders, coprolites of their past glory (Figure 9.10). This is the end of a star whose core mass is smaller than 1.4 times the mass of the Sun (Figure 9.11).

It is interesting to note that the theory *predicts* that white dwarfs will always be lighter than 1.4 solar masses; beyond this value the gravitational pull will be too strong for the electrons to resist. Observations have confirmed the existence of this boundary on the mass of white dwarfs, a prediction, known as the Chandrasekhar[14] limit, after its discoverer. The fate of stars with cores masses beyond the Chandrasekhar limit is much more dramatic, as I now describe.

Medium-size stars

The final stages in the life of a massive star are much more spectacular. Should a star end up with an iron core whose mass is larger than 1.4 times that of the Sun, but below 3–4 solar masses (the calculations are still somewhat uncertain), it will collapse from its initial size, several times larger than our Sun, to the size of a city, producing one of the most spectacular events in the heavens: a Type II supernova explosion.[15]

Imagine an iron sphere weighing 5×10^{27} tons (about 2.5 solar masses), which contracts from a radius of a million km down to 1000 in a few minutes,

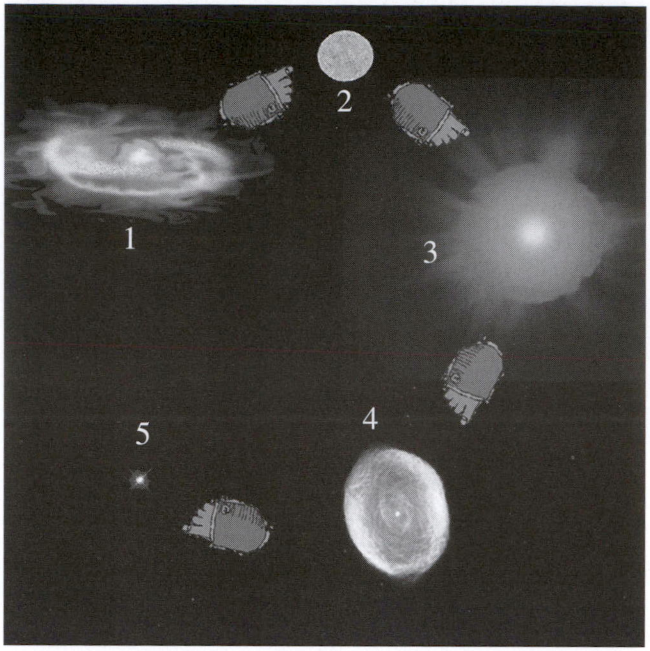

Figure 9.11 Milestones in the evolution of a star of mass below 1.4 times the solar mass: protostar (1), mature star (2), red giant (3), white dwarf with planetary nebula (4), white dwarf (5).

and then to 10 km in a few seconds. During these last few seconds the electrons are crushed together, but the degenerate pressure they generate is unable to withstand the gravitational attraction, and the catastrophic collapse proceeds. The iron nuclei are broken into their proton and neutron constituents, and the protons rapidly react with the very abundant electrons to produce more neutrons and neutrinos (Figure 9.5). The neutrinos leave the star in a compact burst or "flash," so intense that it will fry anything in the star's vicinity, notwithstanding the neutrinos' very weak interactions. As they leave the neutrinos carry with them a significant portion of the total energy emitted during this dramatic process.

When the star reaches a radius of a few dozen kilometers and the appalling density of 10^{17} that of water on Earth,[16] the neutrons, which are by now the main constituents of the core, are crushed together. At this point the neutron degenerate pressure very abruptly rises and stops the contraction, and this sudden break produces a tremendous shock wave that blows away the outer layers of the star, obliterating in its wake any near-by object that survived the neutrino flash. The heat released triggers a series of nuclear reactions that create all elements heavier than iron, and these, together with the lighter elements previously present in the star, are strewn about by the blast. A fraction of the stupendous amount of energy released during the contraction is transformed into visible light, and as a result this one star will suddenly shine with the brilliance of 100 000 Suns (Figure 9.12). The shock wave and the ejected material will plow through the surrounding medium, seeding the region with heavy elements and

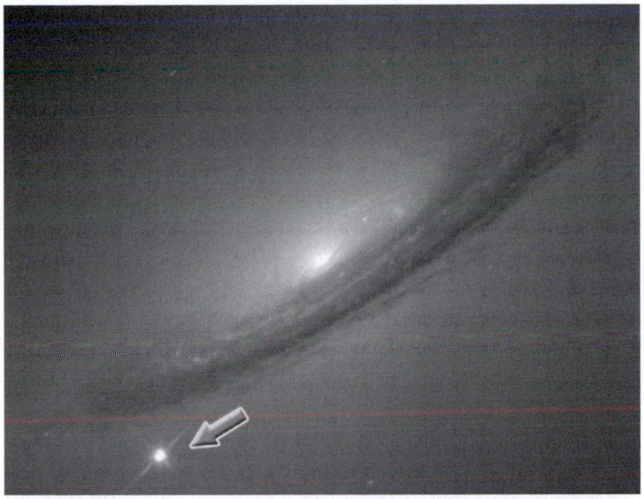

Figure 9.12 Supernova (lower left corner) in the galaxy NGC 2546 (note that brightness of the supernova is comparable to that of the rest of the galaxy). (Courtesy of P. Challis. The High-Z Supernova Search Team, Mt Stromlo Siding Springs observatoriés)

triggering the collapse of neighboring gas clouds. In death supernovae give birth to new stellar generations.

After the neutron degenerate pressure comes into play, and after the exterior shells are ejected, the stellar core stabilizes permanently. But even now the star still has enough energy to perform a rather fancy dance before its final demise. Just as the original gas cloud increased its rate of rotation as it contracted, so the core will speed up as it reduces in size by a factor of 100 000, often reaching a rate of 30 times per second (remember that this an object with more than twice the mass of the Sun!), so that the star's surface moves at about 2000 km s^{-1}. In addition the collapse also concentrates the magnetic field by a factor of 10^{10}. The large magnetic field, the rapid rotation rate, and the small but significant number of electrons still present combine to produce a very intense and collimated beam of X-rays that rotates with the star. Supernovae remnants become X-ray lighthouses.

Whenever the X-ray beam shines on Earth we detect an X-ray pulse, which is repeated regularly since it is driven by the star's rotation, and this type of neutron star is called a *pulsar*. The first observation of such a regular signal raised the tantalizing speculation that this was not a natural phenomenon, but the powerful signals of an alien transmitter, saying, in effect, "Please pick up the phone . . . it's us!" But after the same type of signal was discovered in many locations in space, and after it failed to change over long periods of time, a less fanciful explanation was sought . . . and found.

As time goes on, the remnants of the star's energy slowly radiate away. The rotation will slow until, eventually, the star stops generating its X-ray beam. All its energy is now spent, and it has reached its final stable configuration as a dense and dark cinder, the gloomy reminder of an explosion that matched a galaxy in brilliance (Figure 9.13).

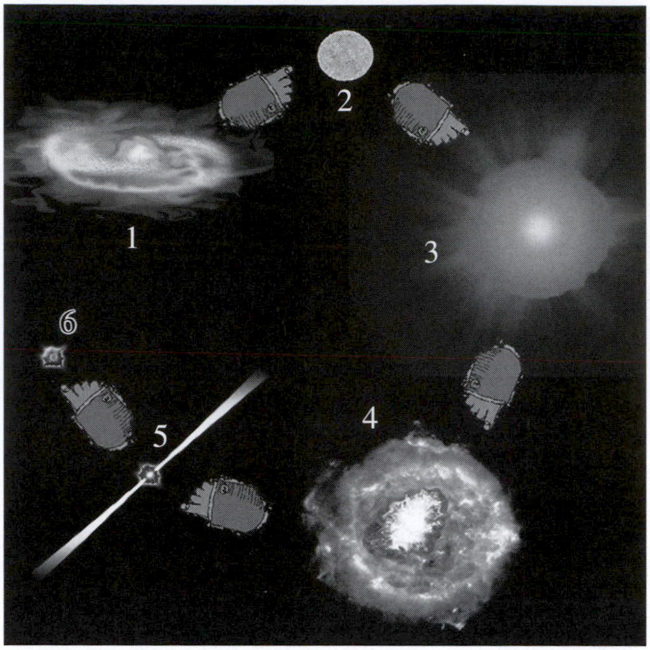

Figure 9.13 Milestones in the life of a star of mass between 1.4 and 3–4 times that of the Sun: protostar (1), mature star (2), red giant (3), supernova (4), pulsar – with X-ray lighthouse beams (5), non-rotating neutron star (6).

Box 9.1

The Crab nebula

This is a remnant of a supernova explosion (Figure 9.14). It was observed on July 4 1054 AD by Chinese astronomers, being about as bright as the full Moon, and was visible in daylight for a short period; after a month or so the star dimmed and disappeared. This event was probably also recorded by Anasazi Indian artists (in present-day New Mexico and Arizona) in a pictograph of a hand-shaped object, presumably the supernova, located near the Moon (Chaco Canyon National Park).

The Big Bang Theory explains correctly the abundances of helium and lithium, and predicts negligible production of any of the other elements in the early Universe; these are all produced inside stars. Every bit of carbon in a flower's DNA, every bit of oxygen we take in every breath, every bit of silicon in a sandy beach, was created in a star; life itself (at least here on Earth) is made of stellar ashes.

The most famous supernova was observed by Chinese astronomers more than 1000 years ago, its remnants are what we now call the Crab nebula (Figure 9.14). The beautiful kaleidoscope of gases we see are the remnants of the parent-star's

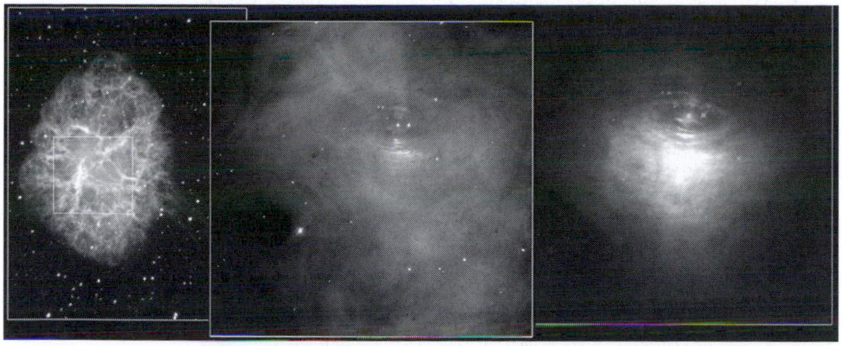

Figure 9.14 The Crab nebula in visible light (left; Palomar and Hubble observatories) and in a composite of X-ray and visible light which shows the pulsar at the center of a funnel-like cloud (right; Chandra observatory). (Courtesy of STSCI NASA)

outer layers, still spreading outward, and in the middle of the cloud a strong X-ray pulsar has been detected: after ten centuries the core of the original star is still active, rotating at the rate of 30 times per second. We have also met other supernovae: one, observed by Tycho Brahe in 1572, and another by Galileo in 1604 (known as Kepler's supernova) which shook the geocentric model. In 1987 a star in our galaxy "went" supernova, from which we observed the portion of the neutrino flash that passed through the Earth, the ejecta are still clearly visible, though no pulsar has been (yet) detected. Type II supernova events are not uncommon, averaging about once every 50 years in a galaxy similar to the Milky Way, so these events and their remnants have been and are assiduously studied using all available instruments. The evolution of a middle-size star is illustrated in Figure 9.13.

It must be said that the full details of the above supernova story have not been worked out mathematically. The complications reside in the very large number of processes which are relevant in the extreme environment surrounding the collapse of these stars. Pencil and paper calculations are out of the question, and the computer codes must include macroscopic effects (such as gravitation, turbulence and the interactions of matter with very large magnetic fields) and microscopic processes (such as those responsible for the creation of neutrinos, nuclear reactions, and the interaction of radiation with various atoms), in order to reproduce all the properties of these stupendous events. An additional obstacle is the paucity of data: though supernovae occur regularly, *near-by* events (within our galaxy) are seen only twice per century on average. It is therefore difficult to compare theoretical predictions with observations. Though the above description does fit all observations, future data might require some alterations.

The heavyweights

For stars whose core is heavier than about 3–4 solar masses the outcome of the collapse is even more dramatic. As for lighter stars the core will collapse to a

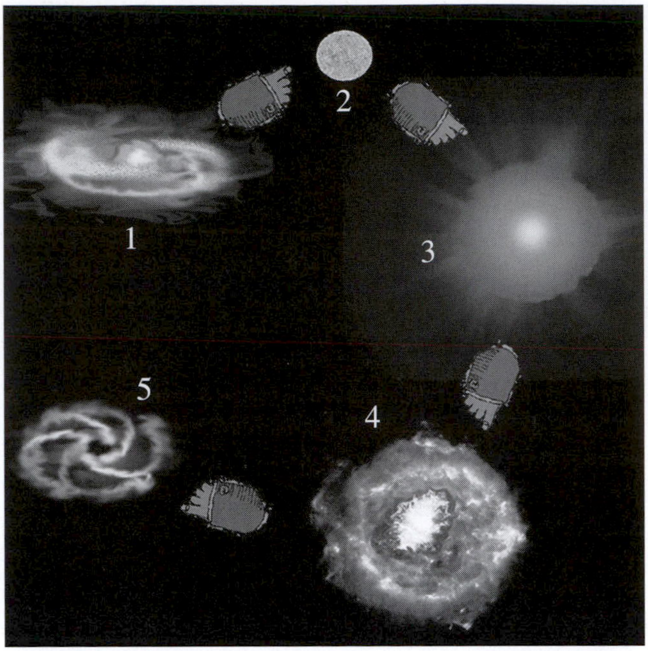

Figure 9.15 Milestones in the life of a star of mass heavier than 3–4 solar masses: protostar (1), mature star (2), red giant (3), supernova (4), black hole – with material swirling into it (5).

planet size within several minutes, and, again, the compressed electrons are unable to break the contraction. Over the next few seconds the nuclei are broken into individual protons and neutrons and the former react with the electrons to produce neutrons and neutrinos, and a very intense neutrino burst ensues. When the star has contracted to the size of a city the neutrons and protons put up one last futile opposition to gravity, but they are overwhelmed and the collapse continues.

All this time the mass of the core is being concentrated in a smaller and smaller region, the density increases and with it the intensity of the gravitational field. Scarcely 30 millionths of a second after reaching a radius of a few kilometers, a horizon is formed,[17] and a black hole is born. As before, this process is accompanied with a stupendous release of energy which blasts away the outer layers of the collapsing star, seeding heavy elements and triggering stellar formation; all this with galactic brilliance. The life of heavy stars is summarized in Figure 9.15.

The fate of core material once its surface crosses the horizon is unknown. The densities will soon increase beyond any value that has been probed to date and the collapse might proceed indefinitely, or it might be stopped by new physical effects currently unknown to us. But whether the contraction proceeds indefinitely or not, all these processes will happen inside the horizon; the horizon itself will remain. The point is that *current* knowledge predicts the creation of black holes through the process described above, and this prediction is independent of whatever novel processes occur inside the horizon.

Box 9.2

Grey holes

The assertion that nothing can escape a black-hole horizon is modified by quantum effects, as first noted by S. Hawking. Within quantum theory it is possible for there to appear in a vacuum a particle–antiparticle pair that survives for a very short time. If the particle is charged its antiparticle has the opposite charge, and this effect can be observed by introducing an electric field that will separate the pair once it is produced (which also provides the energy needed to "create" the particles). If this unhappy pair happens to be born just on the rim of a black-hole horizon it is possible for one member to move outwards while the other falls in through the horizon. In this case the black hole will have effectively emitted one particle, and since the quantum effects responsible for this process affect all particles, black holes will be completely democratic in this type of emission, called after its discoverer, Hawking radiation. The energy for this to occur is provided by the black hole itself which becomes correspondingly lighter.

This type of emission is very weak and is usually overwhelmed in even the mildest of environments. However, it becomes more significant as the black-hole mass decreases; for a mass of 2×10^{22} kg (corresponding to a horizon radius of barely 0.06 mm) the emission will match the cosmic microwave background radiation, for a mass of 2×10^{19} kg (in which case the horizon radius is minuscule, the size of a large molecule) it will match the brilliance of our Sun. As the black hole emits Hawking radiation it loses energy and so its mass decreases, this in its turn increases the amount of radiation and makes for a larger mass decrease. The resulting runaway process ends in the final evaporation of the black hole in a final burst of energy. The whole process of black-hole evaporation is very slow, despite its runaway dramatic final moments. A 1 solar mass black hole will evaporate in about 10^{71} s, very much longer than the current age of the Universe (about 5×10^{17} s).

There is apparently no upper limit to the size of a black hole. Strong evidence exists for the existence of black holes of millions and even billions of solar masses in the center of various galaxies, including our own Milky Way. These objects could have been created either through the collapse of an enormous amount of matter, through an unparalleled appetite, or, more likely, through the coalescence of many smaller black holes.

As for all black holes, the evidence for the existence of these behemoths is circumstantial: observations show that certain galaxies exhibit enormous outbursts of energy from very compact regions. So stupendous is this energy output that no nuclear process can be responsible (for it would require more mass than that known to reside in the area – a quantity that we can estimate by watching the way in which surrounding stars move about). The only known processes capable of producing this amount of energy occur as matter disappears into the horizon. This hypothesis matches the observations provided the black-hole mass lies in the million or even billion solar-mass range, and if they devour several Sun-size stars year by year. Super-massive black holes are believed to reside in the relatively dense environment of the galactic centers, where the ready availability of stars in these regions allows them to satisfy their voracious appetites.

Notes

1. These are the weak and strong interactions (which are of very short range and affect subatomic particles), electromagnetism, and gravity.
2. For example the electric repulsion between the nuclei of two hydrogen atoms is 10^{36} times larger than their gravitational attraction; for two top quarks – the heaviest subatomic particle known – this number drops to 1.5×10^{31}, which is still enormous.
3. If v is the typical value of the velocity of a charged particle, the magnetic force is suppressed by a factor of (v/c) compared to the electric one.
4. Electrons will suffer no change for at least 10^{26} years, protons for at least 10^{31} years.
5. The technical name is *isotope* (not cousin).
6. This parallel is somewhat misleading: the spin can only take a limited set of values, while a "real" rotation can take any value; the picture of an electron as a spinning sphere is but a poor attempt at describing a certain quantum property that has no analogy in our everyday experience.
7. This choice of adjective is conventional and has a technical origin, it does not reflect on the moral quality of the objects being discussed.
8. Neutrinos are hard to keep contained since they interact very weakly, and as a result they can easily go through most materials.
9. In case you are wondering, the energy responsible for all these transformations is supplied by whatever machinery we use to drive the contraction.
10. The solar atmosphere behaves as a fluid being pounded by the relentless energy output of the core; as a result it will vibrate with characteristic frequencies.
11. Most power plants use nuclear *fission* to generate energy. In these processes a heavy atomic nucleus splits; the daughter nuclei have a smaller mass than the parent with the deficit being converted into energy.
12. In some cases there will be more than one blob near the center of the cloud, these give rise to systems with two (or perhaps – though more rarely, three) stars orbiting each other, surrounded by a retinue of planets.
13. This is equivalent to over one million trillion times the yearly world iron production (at the level of the year 2000).
14. S. Chandrasekhar (1910–95) physicist and astrophysicist awarded the Noble Prize in 1983 "for his theoretical studies of the physical processes of importance to the structure and evolution of the stars."
15. Type I supernovae are discussed in the previous chapter.
16. A teaspoon of stellar material will now weigh one trillion tons on the Earth's surface.
17. The horizon radius is proportional to the mass of the core; for 6 solar masses it is about 18 km from the center of the core.

Bibliography

The list of publications listed below is intended for readers interested in pursuing some of the topics covered in the text. The publications are separated into three levels, introductory, medium-level, and advanced. Publications in the first category require no previous preparation in science or mathematics. Medium level publications assume familiarity with calculus. Advanced references provide expert description of the topics and require familiarity with classical and relativistic physics.

Basic references

Adler, Mortimer J. *Aristotle for Everybody: Difficult Thought Made Easy* (New York: Macmillan, 1978)

Aleksandrov, D., A. N. Kolmogorov and M. A. Lavrentév, eds. (trans. S. H. Gould and T. Bartha). *Mathematics, Its Content, Methods, and Meaning* (Cambridge, Mass.: MIT Press, 1964)

Angrist, S. W. *How Our World Came to Be* (New York: Thomas Y. Crowell, 1969)

Aristotle (ed. Jonathan Barnes) *The Complete Works of Aristotle: the Revised Oxford Translation* (Princeton, NJ: Princeton University Press, 1984) Seris: Bollingen series 71:2

Bacon, F. *The Advancement of Learning* (New York: Modern Science Library, 2001) Modern Library Science Series (ed. S. J. Gould)

Bartlett, Robert. *Medieval Panorama* (Los Angeles, CA: J. Paul Getty Museum, 2001)

Bennett, Jeff. *On the Cosmic Horizon: Ten Great Mysteries for Third Millennium Astronomy* (San Francisco, CA: Addison Wesley Longman, 2001)

Bennett, Jeffrey, Megan Donahue, Nicholas Schneider, and Mark Voit. *Essential Cosmic Perspective* (2nd College Edn) (San Francisco, CA: Addison Wesley, 2002)

Bergamini, David and the editors of *Life*. *The Universe* (New York: Time, inc., 1966) Series: Life nature library.

Bergamini, David and the editors of *Life* (consulting eds.: René Dubos, Henry Margenau, and C. P. Snow). *Mathematics* (New York: Time, inc., 1963) Series: Life science library.

Brecher, Kenneth and Michael Feirtag, ed. *Astronomy of the Ancients* (Cambridge, MA: MIT Press, 1980)

Bulfinch, Thomas. *Bulfinch's Mythology* (New York: Avenel Books, distributed by Crown Publishers, 1978)

Butterfield, Herbert. *The Origins of Modern Science, 1300–1800* (revised edition) (New York: Free Press, 1957)

Institute of the History of Natural Sciences, Chinese Academy of Sciences, compilers. *Ancient China's Technology and Science* (Beijing: Foreign Languages Press, 1983) China knowledge series

Cimino, Barbara. *Galileo* (Mexico: Editora cultural y educative, 1967) Series: Los grandes de todos los tiempos, vol. VIII

Cohen, M. R. and E. Nagel. *An Introduction to Logic and the Scientific Method.* War Dept. education manual EM 621, published for the US Armed Forces Institute by Harcourt, Brace and Co., 1934

Coles, Peter. *Cosmology: A Very Short Introduction* (Oxford, New York: Oxford University Press, 2001)

Dantzig, Tobias. *Number, the Language of Science; a Critical Survey Written for the Cultured Non-mathematician.* (4th edn, rev. and augm.) (New York: Macmillan, 1954)

de Gortari, Eli. *La Ciencia en la historia de México.* (Mexico: Editorial Grijalbo, 1980) Series: Tratados y manuales grijalbo

Descartes, R. (trans. M. Clarke). *Discourse on Method and Related Writings* (London: Penguin Books, 2000) Series: Penguin classics

Durant, Will and Ariel. *The Story of Civilization* (New York: Simon and Schuster, 1935)

Dyson, Freeman. *Disturbing the Universe* (New York: Harper and Row, 1979)

Einstein, Albert and Leopold Infeld. *The Evolution of Physics; the Growth of Ideas from Early Concepts to Relativity and Quanta* (New York: Simon and Schuster, 1961)

Feynman, Richard P. *The Meaning of it All: Thoughts of a Citizen Scientist* (Reading, MA: Perseus Books, 1998)

Freemantle, Anne, ed. *The Age of Belief: The Medieval Philosophers.* (New York: New American Library, 1954) Series: Mentor Books

Friday Locke, Raymond. *The Book of the Navajo* (5th edn) (Los Angeles: Mankind Pub. Co., 1992)

Galilei, Galileo (trans. Henry Crew and Alfonso de Salvio; with an introduction by Antonio Favaro). *Dialogues Concerning Two New Sciences* (New York: Macmillan, 1933)

Gamow, George (rev. and updated by Russell Stannard). *The New World of Mr. Tompkins* (New York: Cambridge University Press, 1999)

Garcia Font, Juan. *Historia de la ciencia* (3ra edicion) (Barcelona: Ediciones Danæ, 1968) Series: Biblioteca de la cultura

García Martínez, Bernardo *et al. Historia General de México* (México: El Colegio de México, 1981)

Gardner, Martin. *Relativity Simply Explained* (New York: Dover, 1962)

Goldstone, Lawrence and Nancy. *The Friar and the Cipher: Roger Bacon and the Unsolved Mystery of the Most Unusual Manuscript in the World* (New York: Doubleday, 2005)

Graves, Robert. *The Greek Myths* (London, New York: Penguin Books, 1984)

Hacyan, S. *El descubrimiento del universo* (México: Fono de Cultura Económica, 1986) Series: La ciencia desde Mexico, Vol. 6

Hamilton, Virginia (ill. Barry Moser). *In the Beginning: Creation Stories from Around the World* (San Diego: Harcourt Brace Jovanovich, 1998)

Harrison, Edward. *Cosmology: The Science of the Universe* (2nd edn) (Cambridge, New York: Cambridge University Press, 2000)

Huff, Darrell (illustrated by Irving Geis). *How to Lie with Statistics* (New York: Norton, 1993)

James, E. O. *The Ancient Gods: The History and Diffusion of Religion in the Ancient Near East and the Eastern Mediterranean* (New York: Capricorn Books, 1964)

James, Jamie. *The Music of the Spheres: Music, Science, and the Natural Order of the Universe* (New York: Grove Press, 1993)

Jeans, J. (trans. M. H. Barroso) *Historia de la física* (Mexico-Buenos Aires: Fondo de Cultura Económica, 1948) Series: Breviarios de fondo de cultura economica vol. 84

Kak, Subhash. Babylonian and Indian Astronomy: Early Connections and Greek and Indian Cosmology: Review of Early History. In *The Golden Chain*, ed. G. C. Pande (New Delhi: CSC, 2005) (arXiv:physics/0301078)

Kragh, Helge. *Cosmology and Controversy: The Historical Development of Two Theories of the Universe* (Princeton, N.J.: Princeton University Press, 1996)

Krickeberg, Walter (ill. Rudolf Heinisch; trans. Sita Garst and Jasmín Reuter) *Las antiguas culturas mexicanas* (México: Fondo de Cultura Económica, 1961)

Krupp, E. C. *Echoes of the Ancient Skies: The Astronomy of Lost Civilizations* (Old Saybrook, CT: Konecky & Konecky, 1983)

Kuhn, T. S. *The Copernican Revolution: Planetary Astronomy in the Development of Western Thought* (Cambridge, MA: Harvard University Press, 1957)

Lanczos, Cornelius. *Albert Einstein and the Cosmic World Order*. Six lectures delivered at the University of Michigan in the Spring of 1962 (New York: Interscience Publishers, 1965)

Landau, Lev and Yury Rumer *¿Qué es la teoría de la relatividad?* (Santiago de Chile: Editorial Universitaria, 1969) Series: Colección El Mundo de la ciencia, vol. 1

Lapp, Ralph E. and the editors of *Life*. *Matter* (New York: Time, inc., 1963) Series: Life science library

León-Portilla, M. *La filosofía náhuatl (estudiada en sus fuentes)* (México: UNAM, Dirección general de publicaciones, 1956) Series: Instituto de Investigaciones Históricas, UNAM; serie Cultura Náhuatl, Monografías, vol. 10

Lindberg, David C. *Roger Bacon and the Origins of Perspectiva in the Middle Ages: A Critical Edition and English Translation of Bacon's Perspectiva, with introduction and notes* (Oxford: Clarendon Press; New York: Oxford University Press, 1996)

Lindberg, David C. *Studies in the History of Medieval Optics* (London: Variorum Reprints, 1983)

Malin, Stuart. *The Greenwich Guide to Stars, Galaxies and Nebulae* (Cambridge, New York: Cambridge University Press, 1989)

Mandeville, Sir John (trans. C. W. R. D. Moseley). *The Travels of Sir John Mandeville* (Harmondsworth, Middlesex, England; New York: Penguin Books, 1983) Series: Penguin classics

Maspero, G. (ed. A. H. Sayce; trans. M. L. McClure). *The Dawn of Civilization: Egypt and Chaldæa*, 5th edn (London: Society for Promoting Christian Knowledge, 1910)

Newton, Isaac (trans. Andrew Motte). *The Principia*. (Amherst, NY: Prometheus Books, 1995) Series: Great minds series

Newton, Roger G. *The Truth of Science: Physical Theories and Reality* (Cambridge, MA: Harvard University Press, 1997)

Pais, Abraham *"Subtle is the Lord . . .": The Science and the Life of Albert Einstein* (Oxford, New York: Oxford University Press, 1982)

Park, Robert *Voodoo Science: The Road from Foolishness to Fraud* (New York: Oxford University Press, 2000)

Physical Science Study Committee. *Physics*, 2nd edn (Boston: Heath, 1965)

Popper, Karl R. *Objective Knowledge: An Evolutionary Approach* (Oxford: Clarendon Press; New York: Oxford University Press, 1979)

Robertson, H. P. *et al. The Universe*. (New York: Simon and Schuster, 1956) Series: Scientific American Books

Russell, Bertrand. *A History of Western Philosophy, and its Connection with Political and Social Circumstances from the Earliest Times to the Present Day* (New York: Simon and Schuster, 1945)

Jeon, Sang-woon. *Science and Technology in Korea; Traditional Instruments and Techniques*. (Cambridge, Mass.: M.I.T. Press, 1974) Series: M.I.T. East Asian science series; vol. 4

Schmidt, Peter, Mercedes de la Garza y Enrique Nalda (ed.). *Los Mayas* (México: Americo Arte Editores; Landucci Editores; CONACULTA, INAH, 1998)

Schwinger, Julian. *Einstein's Legacy: The Unity of Space and Time*. (New York: Scientific American Library, distributed by W. H. Freeman and Co., 1986) Series: Scientific American Library series; no. 16

Smith, Mark. Ptolemy's theory of visual perception. *Transactions of the American Philosophical Society*, **86** (2), (1996)

Smith Williams, Henry (ass. Edward H. Williams). *A History of Science* (New York, London: Harper and Brothers, 1904–10)

Snow, C. P. *The Two Cultures: and a Second Look* (New York: Cambridge University Press, 1986)

Snow, C. P. (intro. William Cooper). *The Physicists* (Boston: Little, Brown, 1981)

Sobel, Dava. *Galileo's Daughter: A Historical Memoir of Science, Faith, and Love* (New York: Walker and Co., 1999)

Straumann, Norbert. The history of the cosmological constant problem. In *18th IAP Colloquium On The Nature Of Dark Energy: Observational And Theoretical Results On The Accelerating Universe*, ed. P. Brax *et al.* (Paris: Frontier Group, 2002) (arXiv:gr-qc/0208027)

Taton, Rene (trans. A. J. Pomerans). *History of Science* (New York: Basic Books, 1964–6)

Thuan, Trinh Xuan. *The Birth of the Universe: The Big Bang and Beyond* (New York: H. N. Abrams, 1993)

Wells, H. G. (written originally with the advice and editorial help of Mr. Ernest Barker, Sir H. H. Johnston, Sir E. Ray Lankester, and Professor Gilbert Murray, ill. J. F. Horrabin) *The Outline of History, Being a Plain History of Life and Mankind* (New York: P. F. Collier and Son Co., 1925)

White, S. *Why Science?* (Commack, NY, Nova Science, 1998)

Wightman, W. P. D. *The Growth of Scientific Ideas* (New Haven: Yale University Press, 1953)

Wilson, Mitchell and the editors of *Life*. *Energy* (New York: Time, inc., 1963) Series: Life science library

Yanez, A. (ed.). *Mitos indígenas* (2nd edn) (México: Ediciones de la UNAM, 1956) Series: Biblioteca del estudiante universitario, vol. 31

Medium-level references

Bergström, Lars. Non-baryonic dark matter: observational evidence and detection methods. *Rept. Prog. Phys.,* **63** (2000), 793–841 (arXiv:hep-ph/0002126)

Born, Max (prepared with the collaboration of Günther Leibfried and Walter Biem). *Einstein's Theory of Relativity*, rev. edn (New York: Dover Publications, 1962)

Feynman, Richard. *The Character of Physical Law* (Cambridge: MIT. Press, 1967)

Ohanian, Hans C. *Special Relativity: A Modern Introduction* (Lakeville, MN: Physics Curriculum and Instruction, 2001)

Ryden, Barbara. *Introduction to Cosmology* (San Francisco: Addison-Wesley, 2003)

Tsujikawa, Shinji. Introductory review of cosmic inflation. In *2nd Tah Poe School On Cosmology: Modern Cosmology; Naresuan University, Phitsanulok, Thailand; 17–25 April 2003* (unpublished) (arXiv:hep-ph/0304257)

Weinberg, Steven. *The First Three Minutes: A Modern View of the Origin of the Universe* (New York: Basic Books, 1977)

Advanced-level references

Albrecht, Andreas. Cosmic inflation. In *NATO Advanced Study Institute: Structure Formation In The Universe 26 Jul–6 Aug 1999, Cambridge, England* (ed. Robert G. Crittenden and Neil G. Turok Dordrecht) (Boston: Kluwer Academic Publishers, 2001) (arXiv:astro-ph/0007247). NATO science series. Series C, Mathematical and physical sciences; v. 565

Bernstein, Jeremy and Gerald Feinberg (ed.) *Cosmological Constants: Papers in Modern Cosmology* (New York: Columbia University Press, 1986)

Bergmann, Peter Gabriel (with a foreword by Albert Einstein). *Introduction to the Theory of Relativity* (New York: Dover Publications, 1976)

Böhm-Vitense, Erika. *Introduction to Stellar Astrophysics* vols. 1 and 2 (Cambridge, New York: Cambridge University Press, 1989)

Brandenberger, Robert H. Inflation and the theory of cosmological perturbations. In *Summer School in High Energy Physics and Cosmology: ICTP, Trieste, Italy, 2 June–4 July 1997* (ed. E. Gava *et al.*) (Singapore; River Edge, NJ: World Scientific, 1998) (arXiv:astro-ph/9711106) Series: ICTP series in theoretical physics; vol. 14

Chandrasekhar, S. *An Introduction to the Study of Stellar Structure* (New York: Dover Publications, 1967)

Clayton, Donald D. *Principles of Stellar Evolution and Nucleosynthesis* (Chicago: University of Chicago Press, 1983)

Dolgov, A. D. Dark matter in the universe. In *The Universe of Gamow: Original Ideas in Astrophysics and Cosmology (Gamow Memorial International Conference)* (Odessa: Astroprint, 1999) (arXiv:hep-ph/9910532)

Einstein A., H. A. Lorentz, H. Weyl, and H. Minkowski (notes by A. Sommerfeld). *The Principle of Relativity: A Collection of Original Papers on the Special and General Theory of Relativity* (New York: Dover, 1952)

Einstein, Albert. *The Meaning of Relativity* (Princeton: Princeton University Press, 1955) Series: The Stafford Little lectures, 1921

EROS collaboration. Limits on Galactic Dark Matter with 5 Years of EROS SMC Data. *Astron. Astrophys,* **400** (2003), 951–956 (arXiv:astro-ph/0212176)

Feynman, Richard P., Robert B. Leighton, and Matthew Sands. *The Feynman Lectures on Physics*, Vol. II (Reading, MA: Addison-Wesley Pub. Co., 1963–5)

Kolb, Edward W., and Michael S. Turner. *The Early Universe, and The Early Universe – Reprints.* (Redwood City, CA: Addison-Wesley, 1990) Series: Frontiers in physics; vols. 69 and 70

Misner, Charles W., Kip S. Thorne, and John Archibald Wheeler. *Gravitation* (San Francisco: W. H. Freeman, 1973)

Padmanabhan, T. Cosmological constant – the weight of the vacuum. *Phys. Rept.,* **380** (2003), 235–320 (arXiv:hep-th/0212290)

Panagia, N. Supernovae. In *Proceedings of the International Summer School on Experimental Physics of Gravitational Waves; Urbino, Italy, September 6–18, 1999* (ed. M. Barone *et al.*) (Singapore, River Edge, NJ: World Scientific, 2000) (arXiv:astro-ph/0003083)

Pauli, W. (trans. G. Field). *Theory of Relativity*. (New York: Dover, 1981)

Peebles, P. J. E. *Principles of Physical Cosmology*. (Princeton, NJ: Princeton University Press, 1993) Series: Princeton series in physics

Peebles, P. J. E. *The Large-scale Structure of the Universe*. (Princeton, NJ: Princeton University Press, 1980) Series: Princeton series in physics

Peebles, P. J. E. and Bharat Ratra. The cosmological constant and dark energy. *Rev. Mod. Phys,* **75** (2003), 559–606 (arXiv:astro-ph/0207347)

Purcell, Edward M. *Electricity and Magnetism*. (New York: McGraw-Hill, 1985) Series: Berkeley physics course; vol. 2

Schwarzschild, Martin. *Structure and Evolution of the Stars*. (New York, Dover Publications, 1965) Series: Dover books on astronomy and astrophysics

Shapiro, Stuart L. and Saul A. Teukolsky. *Black Holes, White Dwarfs, and Neutron Stars: The Physics of Compact Objects* (New York: Wiley, 1983)

Tolman, Richard C. *Relativity, Thermodynamics, and Cosmology* (New York: Dover Publications, 1987. Reprint – originally published: Oxford: Oxford University Press, 1934)

Vilenkin, Alexander. Cosmological constant problems and their solutions. In *The Dark Universe: Matter, Energy and Gravity; Proceedings of the Space Telescope Science Institute Symposium; Held in Baltimore, Maryland, April 2–5, 2001* (ed. Mario Livio) (Cambridge, New York: Cambridge University Press, 2003) (arXiv:hep-th/0106083)

Weinberg, Steven. *Gravitation and Cosmology: Principles and Applications of the General Theory of Relativity* (New York: Wiley, 1972)

Zeldovich, Ya. B. and I. D. Novikov (trans. Eli Arlock; ed. Kip S. Thorne and W. David Arnett). *Relativistic Astrophysics* (Chicago: University of Chicago Press, 1971–83)

Index

T

U

V